Strategies for
Green Organic Synthesis

Strategies for
Green Organic Synthesis

V. K. Ahluwalia

Visiting Professor
Dr. B. R. Ambedkar Center for Biomedical Research,
University of Delhi, India

CRC Press
Taylor & Francis Group
Boca Raton London New York

CRC Press is an imprint of the
Taylor & Francis Group, an **informa** business

Strategies for Green Organic Synthesis

First published 2012 by Ane Books for CRC Press

Published 2024 by CRC Press
2385 NW Executive Center Drive, Suite 320, Boca Raton FL 33431

and by CRC Press
4 Park Square, Milton Park, Abingdon, Oxon, OX14 4RN

CRC Press is an imprint of Taylor & Francis Group, LLC

British Library Cataloguing in Publication Data
A catalogue record for this book is available from the British Library

ISBN: 978-1-4398-7050-1 (hbk)

Visit the Taylor & Francis Web site at
http://www.taylorandfrancis.com

and the CRC Press Web site at
http://www.crcpress.com

For distribution in rest of the world other than the Indian sub-continent

Preface

The book entitled 'Strategies for Green Organic Synthesis' is an attempt to discuss the strategies for organic synthesis. A brief introduction about green synthesis is given in the *First chapter*. The *Second chapter* deals with carbon-carbon bond formation. In this, various reactions used for carbon-carbon single bond, carbon-carbon double bond and carbon-carbon triple bond formation have been discussed. *Third chapter* deals with various reaction for the formation of carbon-oxygen single and carbon-oxygen double bonds. *Fourth chapter* deals with the reactions for the synthesis of carbon-halogen, carbon-sulphur, carbon-nitrogen (single, double and triple bonds), nitrogen-nitrogen double bond, nitrogen-oxygen bond formations. The *Fifth chapter* deals with interconversion of functional groups and *Sixth chapter* deals with protecting groups. Enantioselective Synthesis are described in *Seventh chapter*. All the above strategies are necessary for Retrosynthesis, which forms the subject matter of the last chapter.

In all, the reactions used for various bond formations, interconversion of functional groups, protecting groups, enantioselective synthesis and retrosynthesis, *green conditions* have been described. These green conditions include used of microwave irradiations, sonication, use of green solvents (like water, ionic liquids, polyethylene glycol, supercritical carbon dioxide and fluorous solvents) and green catalysts like phase transfer catalysts, crown ethers and enzymes. Polymer supported reagents have also been used. Besides, some of the reactions have been performed by photoirradiation.

The book will be extremely useful for chemists all over the globe, who are involved in developing procedures for green synthesis, including green synthesis not only the new synthesis but also for existing synthesis. Besides, the book will be useful for all postgraduate students in organic chemistry.

Any suggestions from students, teachers and Research chemists are most welcome in order to incorporate these in the next edition of the book. All suggestions will be gratefully acknowledged.

Finally, the author takes the Opportunity of expressing his thanks to Prof. Sukh Dev, FNA, for help in the organisation of the book and to Mr. Sunil Saxena of Ane Books Pvt. Ltd. for the help in the publication of this book.

V.K. Ahluwalia

Contents

Chapter 5: Interconversion of Functional Groups   221

Chapter 6: Protecting Groups   289

Chapter 7: Enantioselective Synthesis   301

Introduction

1.1 WHAT IS GREEN CHEMISTRY

Green chemistry is defined as environmentally benign chemical synthesis. It focuses on a process (whether carried out in industry or chemical laboratory) that reduces the use and generation of hazardous substances or byproducts. Strict laws have been passed by various governments particularly in advanced countries to strictly follow the procedures for various syntheses so as to reduce or eliminate the products (or byproducts) that are responsible for the pollution of the environment. The chemists all over the globe are motivated not only for the environmentally benign synthesis of new products, but also to develop green synthesis for existing chemicals. This has been possible by the replacement of the organic solvents, which are hazardous by water or eliminate the use of solvent altogether.

1.2 PRINCIPLES OF GREEN CHEMISTRY

As already stated, green chemistry deals with chemical syntheses which are benign. This is one of the important ways to prevent or reduce the pollution.

Following are the twelve principles of green chemistry* developed by Paul T. Anastas and John C. Warner.

(*i*) It is better to prevent waste than to treat or clean up waste after it is formed.

(*ii*) Synthetic methods should be designed to maximize the incorporation of all materials used in the process into the final product.

(*iii*) Wherever practicable, synthetic methodologies should be designed to use and generate substances that possess little or no toxicity to human health and the environment.

(*iv*) Chemical products should be designed to preserve efficacy of function, while reducing toxicity.

(*v*) The use of auxiliary substances (solvents, separation agents, *etc.*) should be made unnecessary wherever possible and innocuous when used.

(*vi*) Energy requirements should be recognized for their environmental and economic impacts and should be minimized. Synthetic methods should be conducted at ambient temperature and pressure.

(*vii*) A raw material or feedstock should be renewable rather than depleting whenever technically and economically practical.

(*viii*) Unnecessary derivatization (blocking group, protection/deprotection, temporary modification of physical/chemical processes) should be avoided whenever possible.

(*ix*) Catalytic reagents (as selective as possible) are superior to stoichiometric reagents.

(*x*) Chemical products should be designed in such a way that at the end of their function they do not persist in the environment and break down into innocuous degradation products.

(*xi*) Analytical methodologies need to be further developed to allow for real-time in-process monitoring and control prior to the formation of hazardous substances.

(*xii*) Substances and the form of a substance used in a chemical process should be chosen to minimize the potential for chemical accidents, including releases, explosions and fires.

---

*Paul T. Anastas and John C. Warner, Green Chemistry, Theory and Practice, Oxford University Press, New York, 1988.

1.3 PLANNING A GREEN SYNTHESIS

Following are given some of the points which should be followed for carrying out an organic synthesis.

1. **Percentage Atom Utilization:** There should be maximum incorporation of the starting materials and reagents in the final product. The extent of incorporation of the starting material into the product is given by the following formula [*R.A. Sheldon, Chem. Ind. (London), 1992, 903*]

$$\text{Atom economy} = \frac{\text{F.W. of atoms utilized}}{\text{F.W. of the reactants used in the reaction}} \times 100$$

2. **Evaluating the Type of the Reaction Involved:** The reaction involved must be evaluated with regard to its environmental impact or consequences. For this purpose, the nature of the starting material and the by products (if formed) must be examined. Following are given the different type of reactions which may be involved in a particular synthesis.

   (*a*) **Rearrangements:** As the name indicates, in these reactions the atoms that comprise a starting molecule change its orientation relative to one another including their connectivity and bonding pattern. Such reactions can be performed using a variety of procedures including thermal, photo and chemical means. From the point of green chemistry, in such reactions, both the starting material and the end products contain the same atoms and so there is no waste generated. In fact, a rearrangement reaction is 100% atom economical and fully efficient.

   (*b*) **Addition reactions:** In these reaction, a reagent adds to a substrate, all reagent, and the substrate are consumed during the reaction. No additional byproducts are generated and such reactions are very efficient and like rearrangement reactions are also 100% atom economical. Some typical addition reactions include the addition of bromine to an olefin, grignard reagent to a carbonyl compound and hydrogen cyanide to an α, β-unsaturated carbonyl compounds (Scheme–1).

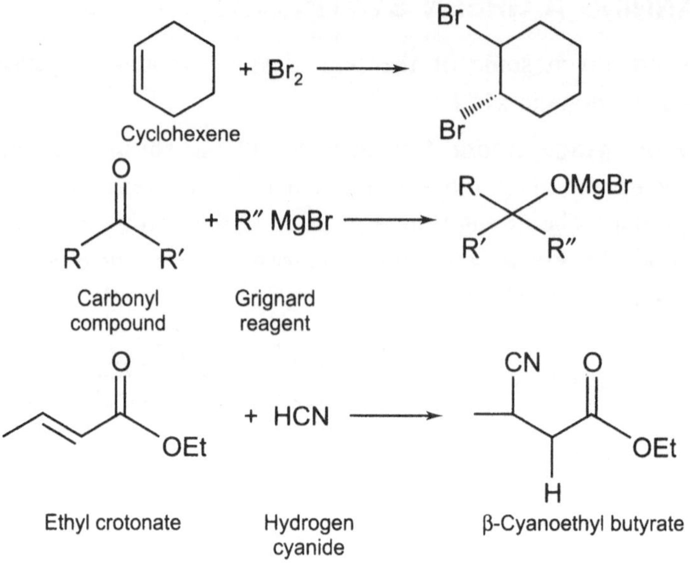

(Scheme–1)

(*c*) **Substitution reactions:** In these reactions the functional group of a substrate is replaced with another functional group. Typical examples include the well known S_N1 and S_N2 reactions. In these reactions, nucleophilic reagents displace a leaving group in an aliphatic carbon atom; the product formed incorporates the nucleophile with removal of the leaving group. Typical examples are given below (Scheme–2):

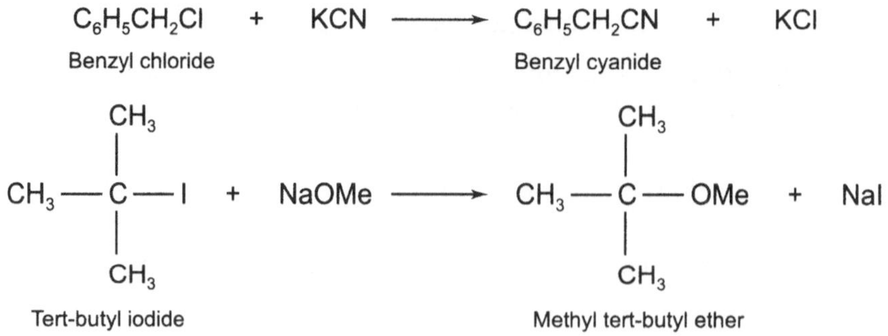

$$C_6H_5CH_2Cl \quad + \quad KCN \quad \longrightarrow \quad C_6H_5CH_2CN \quad + \quad KCl$$

Benzyl chloride   Benzyl cyanide

Tert-butyl iodide   Methyl tert-butyl ether

(Scheme–2)

In some cases, the leaving group is the desired product. For example, potassium iodide demethylation of a carboxylic acid methyl ester to give free carboxylate salt and methyl iodide (Scheme–3).

$$C_6H_5COOCH_3 \quad + \quad KI \longrightarrow C_6H_5COOK \quad + \quad CH_3I$$

Methyl benzoate   Pot-benzoate   Methyl iodide

(Scheme–3)

The usefulness of the methods depends on the nature of the leaving group generated. This pathway can be convenient and efficient if a substitution reaction sequence can be designed where the leaving group has been carefully selected.

(*d*) **Elimination reactions:** These are reverse of addition reactions and are procedures to generate unsaturation in the molecule. Examples of this type include dehydration of an alcohol to generate an olefin and loss of an alcohol from a hemiacetal to give an aldehyde (Scheme–4).

Propyl alcohol   Propene

Hemiacetal   Aldehyde

(Scheme–4)

As in the case of substitution reactions, the environmental implications of the leaving group should be examined, evaluated and controlled.

(*e*) **Pericyclic reactions:** These are concerted reactions and are characterised by the making or breaking of bonds in a single concerted step through a cyclic transition state involving π or σ electrons. Energy of activation for pericyclic reaction is supplied by heat in a thermally induced reaction and by ultraviolet light in a photo-induced reaction. Pericyclic reactions are highly stereospecific, often the thermal and photochemical processes yield products with different, but specific stereochemistry. Since pericyclic reactions do not involve ionic or free radical intermediates; solvents, and nucleophiles or electrophilic reagents have no effect on the course of the reaction. We normally come across three type of pericyclic reaction.

(*i*) **Cycloaddition reactions:** In these reactions, two molecules combine to form a ring; two π bonds being converted to two single bonds in the process. The most common example of a cycloaddition reaction is the Diels-Alder reaction (Scheme–5).

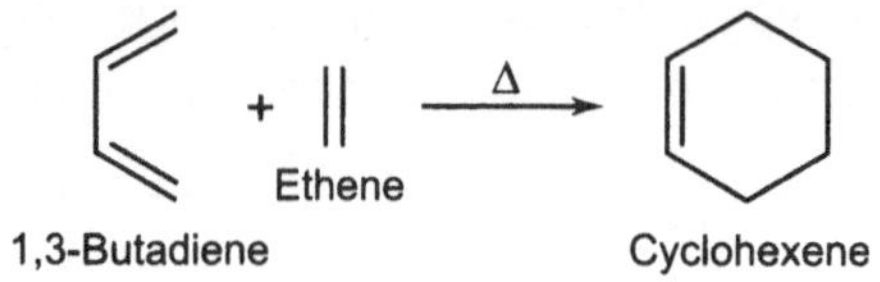
Δ
Ethene
1,3-Butadiene
Cyclohexene

(Scheme–5)

(*ii*) **Electrocyclic reaction:** These are reversible reactions in which a compound with two π electrons are used to form a sigma bond (Scheme–6).

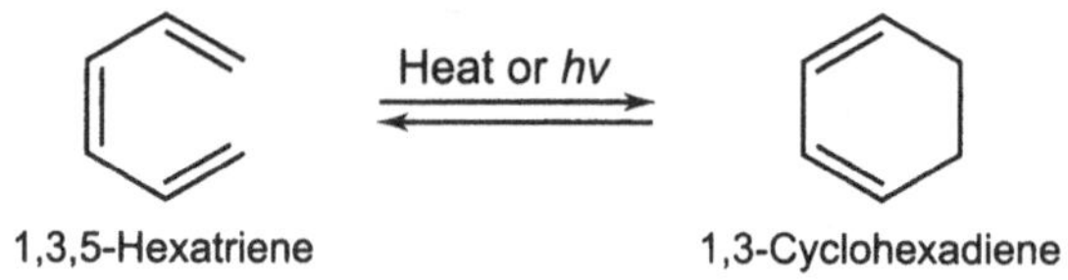
Heat or *hv*
1,3,5-Hexatriene
1,3-Cyclohexadiene

(Scheme–6)

(*iii*) **Sigmatropic rearrangements:** These are concerted intramolecular rearrangements in which an atom or a group of atoms shift from one position to another (Scheme–7).

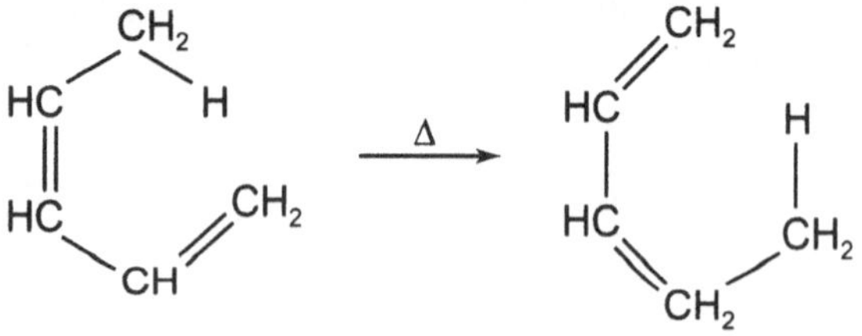
CH_2
HC
H
HC
CH_2
CH
Δ
CH_2
HC
H
HC
CH_2
CH_2

(Scheme–7)

Pericyclic reactions are very convenient from the point of view of environmental problems, since there is no byproduct obtained in these reactions.

3. **Selection of Appropriate Solvent:** The solvent selected for a particular reaction should not have any environmental pollution and health hazard. As far as possible, the reactions should be performed in aqueous phase or without the use of a solvent (solventless reaction). Recently, a novel type of solvent known as "Ionic liquids" have been developed and are used for various syntheses. Besides these, organic reactions have been carried out in supercritical water or in near water region (NCW), in supercritical CO_2, polyethylene glycol and its solutions.

(*i*) **Aqueous phase reaction:** A typical reaction which has been carried out in aqueous phase is the Diels-Alder reaction (Scheme–8).

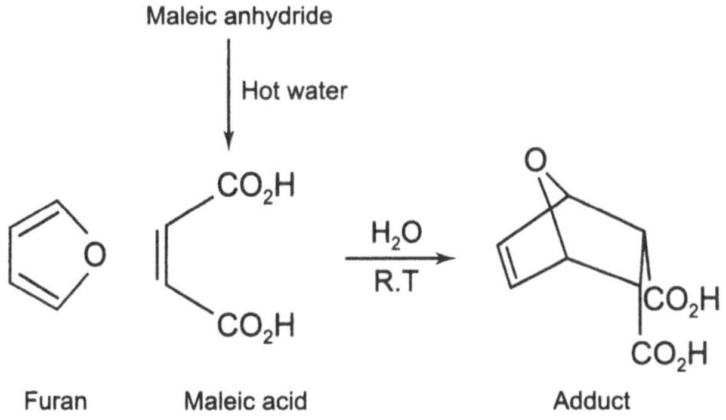

(Scheme–8)

Besides this a number of other reactions have also been performed in aqueous phase; these will be discussed subsequently.

(*ii*) **Reactions in ionic-liquid:** Ionic liquids are made up of at least two components in which the cation and anion can be varied. In these cases, the properties, such as melting point, viscosity, density and hydrophobicity can be varied by simple changes to the structure of ions. The ionic liquids are immiscible in water. By choosing the correct ionic liquids, higher product yield can be obtained and a reduced amount of waste is produced in a given reaction.

Ionic liquids are good solvents for a wide range of both inorganic and organic materials. They are also immiscible with a number of organic solvents and provide a non-aqueous, polar alternative for two phase system.

Ionic liquids find application in alkylations, allylations, hydroformylations, epoxidations, synthesis of ethers, Friedel craft reaction, Diel-Alder Reaction, Knoevenagel condensation and Wittig reaction.

(*iii*) **Organic synthesis in solid state:** The organic synthesis in solid state (*viz.*, solvent-free organic synthesis and transformations) are mostly green reactions. These are of two types:

(*a*) **Solid phase organic synthesis without any solvent:** The earliest record of an organic reaction in dry state is the Claisen rearrangement of allyl phenyl ether to o-allylphenol (Scheme–9).

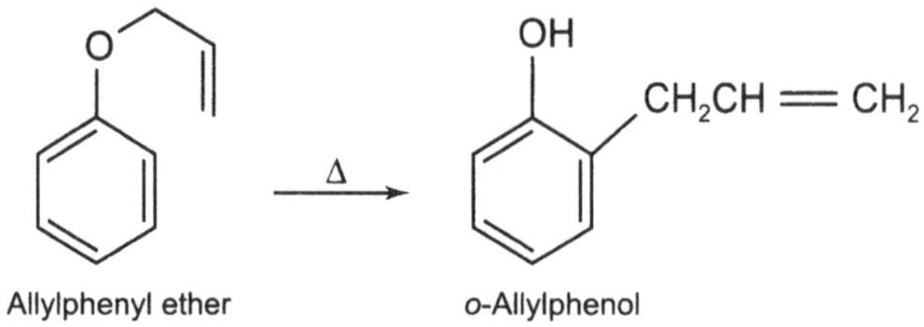

| Allylphenyl ether | o-Allylphenol |

(Scheme–9)

A large number of reactions have now been performed without any solvent. These will be discussed subsequently.

(*b*) **Solid supported organic synthesis:** In these reactions, the reactants are stirred in a suitable solvent (for example, water, alcohol, methylene chloride *etc.*,) with a suitable adsorbent or solid support like silica gel, alumina, phyllosilicate (m^{n+}–montomorillonite *etc.*,). After stirring, the solvent is removed in vacuo and the dried solid support on which the reactants have been absorbed are used for carrying out the reaction under microwave irradiation.

4. **Selection of Starting Materials:** As far as possible, the starting materials selected should be obtainable from renewable sources. The starting materials should not cause any harm (*e.g.*, allergy *etc.*,) to the person handling these.

5. **Use of Protecting Group** should be avoided as far as possible, since these generate wastes.

6. Use of Catalyst: We know that catalysts facilitate transformation and the conversions can be affected in short duration of time and consume less energy. Such reactions should be preferred. Use of phase transfer catalysts, crown ethers and biocatalysts is very well known.

7. Use of Microwaves, Sonication save lots of energy and time, and give much better yields.

1.4 INTRODUCTION TO ORGANIC SYNTHESIS

Organic synthesis basically involves bond forming reaction. This includes C—C bond formation, C—O bond formation, C—N bond formation, C—S bond formation, C-halogen bond formation, N—O bond formation and N=N bond formation. All these reactions have been discussed under normal conditions and also under benign condition, which is most important in the content of Green Chemistry; this is helpful for reducing environmental pollution.

Once, the bond formation has been affected, the next step is the transformations of the functional groups, so that the desired product could be obtained. Another important technique for organic synthesis is enantioselective synthesis and also the concept of retrosynthesis.

Carbon-Carbon Bond Formation

2.1 CARBON-CARBON SINGLE BOND FORMATION

2.1.1 Introduction

Carbon-carbon bond formation is most important for organic synthesis. In most of the cases, the organic molecules, due to the presence of an heteroatom renders the neighbouring carbon atom become electrophilic, as in the case of alkyl halides, alcohols and carbonyl compounds.

$$\overset{\delta+}{C}\overset{\delta-}{Cl} \qquad \overset{\delta+}{C}\overset{\delta-}{OH} \qquad \overset{\delta+}{C}=\overset{\delta-}{O}$$

In all such cases, reaction with a nucleophilic reagent (S_N2 reaction) can be represented as

$$CH_3\bar{S} + \overset{\delta+}{C}\overset{\delta-}{Cl} \longrightarrow CH_3S-C + Cl^-$$

Thus, for the formation of carbon-carbon bond, the substrates (alkyl halides *etc.*,) must react with a reagent containing a nucleophilic carbon.

$$C\bar{:} + \overset{\delta+}{C}Cl \longrightarrow C-C$$

A typical example is the nucleophilic carbon atom of a cyanide ion.

$$:C = \ddot{\underset{\cdot\cdot}{N}} \longleftrightarrow :C \equiv N$$

$$N \equiv C\ddot{\;} \;+\; \underset{}{\overset{\delta+}{C}} \underset{}{\overset{\delta-}{Cl}} \longrightarrow :N \equiv C \overset{}{\underset{}{\diagup}} \;+\; Cl^-$$

Following are given some of the common strategies that are important for the formation of carbon-carbon bond.

- The reaction of alkyl halides with various nucleophiles (for details see section 5.2.5.1).

- The reaction of alkyl halides into other groups via the formation of Grignard reagents (for details see section 5.2.5.2).

- The reaction of alkyl halides with lithium diethyl cuprate (Gilman's reagent) (for details see section [5.2.5.3 (*ii*)].

- The reaction of terminal alkynes with sodamide followed by treatment of the alkynide anion with alkyl halide [for details section 5.2.3 (*iv*)].

- Friedel crafts alkylation or acylation of aromatics (for details see section 5.2.4.1).

- The conversion of alcohols into the corresponding alkyl sulfonates followed by S_N2 reaction with various nucleophiles (for details see section 5.2.6.1).

- The conversion of aldehydes into ketones ($RCHO \longrightarrow RCOCH_2R'$) (for details see section 5.2.10 (*ix*)).

- Conversion of carboxylic acids into ketones [for details see section 5.2.12 (*v*)].

Besides the above procedures for carbon-carbon bond formation, a large number of other reactions have also been used. A discussion on these reactions form the subject matter of the following section.

It is also well known that compounds containing $C = C$ (alkenes) and $C \equiv C$ (alkynes) can be catalytically reduced to give compounds containing carbon–carbon single bond (alkanes).

2.1.2 Acetoacetic Ester Synthesis

Acetoacetic ester contains a reactive methylene group, which can be alkylated with alkyl halide in presence of potassium hydroxide, potassium carbonate and a PTC[1] (tetrabutyl ammonium chloride) and irradiation with microwaves (Scheme–1).

$$CH_3COCH_2CO_2Et \xrightarrow[\text{MW, 3 min}]{\text{RX, KOH, K}_2\text{CO}_3,\text{ PTC}} CH_3COCHCO_2Et \overset{R}{\underset{|}{}}$$

Ethyl acetoacetate

$$R = CH_3CH_2CH_2,\ PhCH_2,\ m\text{-}CH_3OC_6H_4CH_3,\ p\text{-}ClC_6H_4CH_2CH_2(CH_3)_2$$

(Scheme–1)

Normally, the alkylation is carried out with alkyl halide in presence of a strong base[2] under anhydrous conditions.

The formed substituted ethylacetate on hydrolysis with dilute alkali or hydrochloric acid followed by decarboxylation yield substituted acids or ketones (Scheme–2).

$$CH_3COCH-COOEt \overset{R}{\underset{|}{}} \begin{cases} \xrightarrow{H^+ \text{ or dil. KOH}} CH_3COCH_2R + CO_2 + EtOH \\ \xrightarrow{\text{conc. KOH}} CH_3COO^- K^+ + RCH_2COO^- K^+ + EtOH \end{cases}$$

(Scheme–2)

References

1. D. Runhua, W. Yuliang and Y. Yazhong, Synth. Commun., 1994, **24**, 111.
2. H.O. House, Modern Synthetic Reactions, New York: Benjamin.

2.1.3 Acyloin Condensation

It consists in treating a carboxylic ester with metallic sodium in large volume of xylene or toluene followed by protic solvents. The product obtained is α-hydroxy ketone, called acyloin[1] (Scheme–1).

(Scheme–1)

The acyloin condensation can be carried out under green conditions[2] using the coenzyme thiamine (Scheme–2).

$$CH_3CHO \xrightarrow{\text{Thiamine}} CH_3-CO-CH-CH_3$$
$$\underset{\underset{\text{Acyloin}}{OH}}{|}$$

(Scheme–2)

The acyloin condensations are easily affected by stirring aliphatic or aromatic aldehydes with a quaternary salt (*a* PTC)[3].

References

1. L. Bouveault and R. Loquin, Compt. Rend., 1905, **140**, 1593, S.M. McElvain, Organic Reactions, 1948, **4**, 256.

2. V.K. Ahluwalia, Green Chemistry, Environmentally Benign Reaction, Ane Books, India, 2006, page 20 and references cited therein.

3. W. Tagaki and H. Hara, J. Chem., Soc. Chem. Commun. 1973, 891.

2.1.4 Aldol Condensation

It is one of the most important carbon-carbon bond forming reactions and consist in self condensation of aldehydes (having α-hydrogen atom) on warming with dilute alkali to give β-hydroxyaldehydes (known as aldol)[1].

$$CH_3CHO + CH_3CHO \xrightarrow{^-OH} CH_3-\overset{\overset{\displaystyle OH}{|}}{C}-CH_2CHO$$

Acetaldehyde β-Hydroxybutyraldehyde (aldol)

(Scheme–1)

The aldol obtained above (under basic conditions) on heating gives α, β-unsaturated aldehyde (Scheme–2)

$$CH_3-CH=CH-\overset{\overset{\displaystyle O}{||}}{C}-H + H_2O + HO^-$$

Crotonaldehyde (2-Butenal)

(Scheme–2)

The aldol condensation can also be performed in presence of acid[2]. Thus, treatment of acetone with hydrogen chloride gives the aldol product, 4-methyl-3-penten-2-one (Scheme–3).

$$2CH_3-\overset{\overset{\displaystyle O}{\|}}{C}-CH_3 \xrightarrow{HCl} CH_3-\overset{\overset{\displaystyle O}{\|}}{C}-CH=\overset{\overset{\displaystyle CH_3}{|}}{C}-CH_3 + H_2O$$

Acetone   4-Methyl-3-penten-2-one

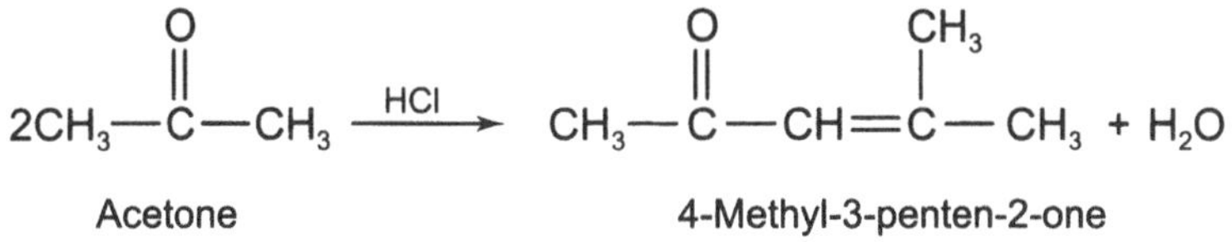

(Scheme–3)

The above reaction proceeds via the formation of acid catalysed formation of enol, which adds to the protonated carbonyl group of another molecule of acetone. Final step involves proton transfer and dehydration.

An interesting case of aldol condensation is **vinylogous aldol reaction**. Thus, the reaction of isophorone with benzaldehyde in water gives only vinylogous aldol product in low yield. However, in presence of CTACl, the condensation product, (E)-benzylidene isophorone is obtained in 80 per cent yield. Uses of tetrabutyl ammonium chloride (TBACl) gives a mixture of addition and condensation product (Scheme–4).

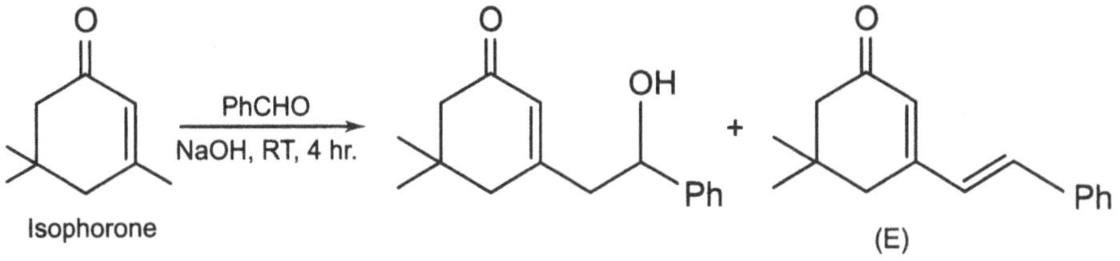

Isophorone		(E)
Water only	24%	
CTACl	–	80%
TBACl	27%	58%

(Scheme–4)

2.1.4.1 Aldol Condensation of Silyl Enol Ethers in Aqueous Media

The aldol condensation of silyl enol ethers with benzaldehyde, catalysed by titanium tetrachloride was with first reported[4] in 1973. However, these reactions are carried out in anhydrous solvents[5]. It has now been possible to perform aldol condensation of silyl enol ethers with aldehydes in aqueous phase[6] (Scheme–5).

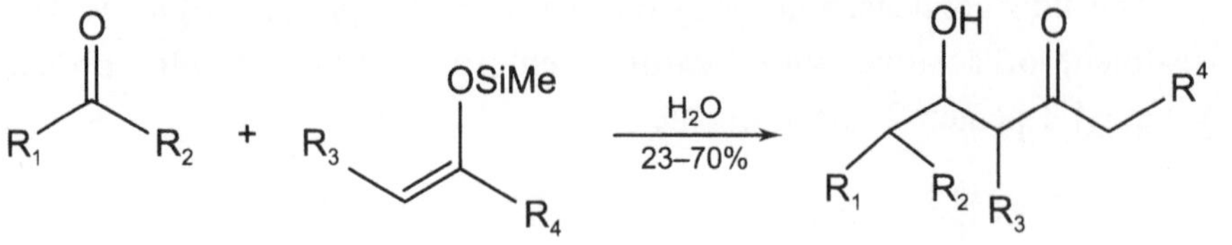

(Scheme–5)

The above reaction was carried out without any catalyst, but it took several days for completion, since water is a weak lewis acid. The addition of stronger lewis acid like ytterbium triflate greatly improved[7] the yield and rate (Scheme–6).

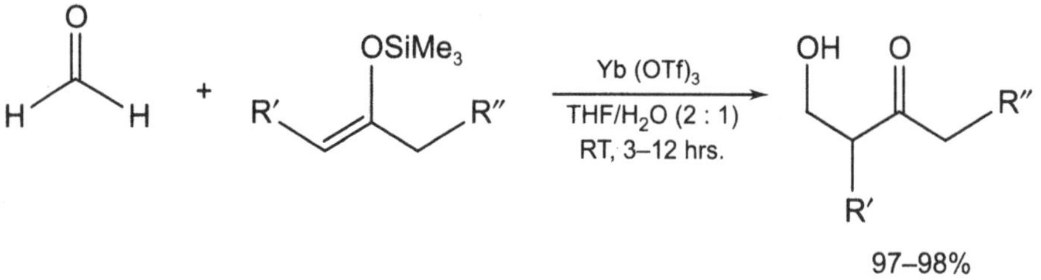

(Scheme–6)

The reaction of silyl enol ether of propiophenone with commercial formaldehyde in presence of ytterbium triflate gave good yields of the adduct[8] (Scheme–7).

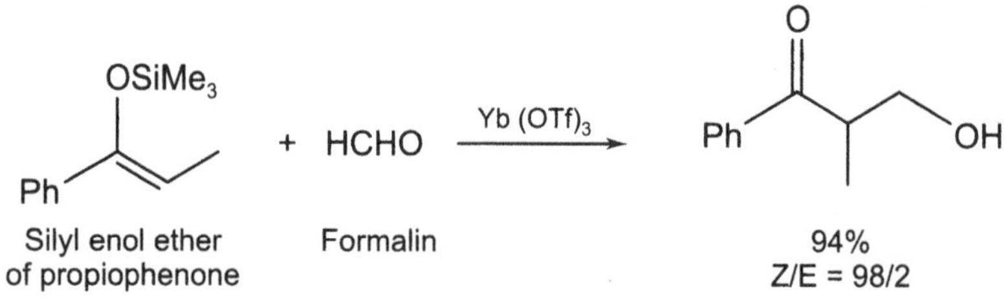

(Scheme–7)

In the above reactions, the catalyst could be recovered and used repetitively. This methodology has been extensively reviewed[3].

2.1.4.2 Aldol Condensation in Solid Phase

The aldol condensation of the lithium enolate of methyl 3,3-dimethylbutanoate with aromatic aldehydes yields[9] a mixture of syn and anti products (8:92 mixture) in 70 per cent yield (Scheme–8).

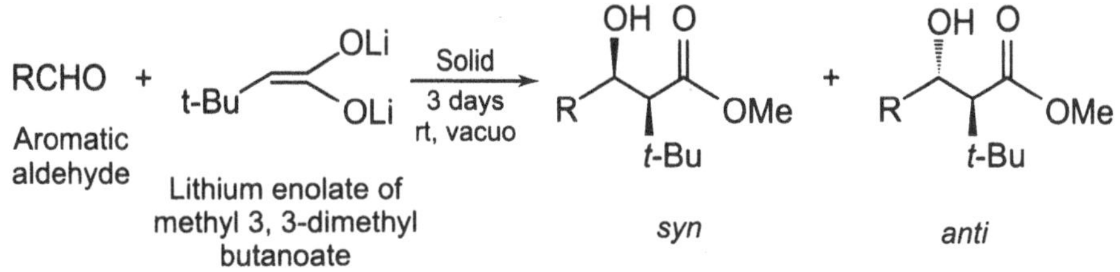

RCHO + [lithium enolate structure with OLi, t-Bu, OLi] $\xrightarrow[\text{rt, vacuo}]{\substack{\text{Solid} \\ \text{3 days}}}$ [product with OH O, R, OMe, t-Bu] + [product with OH O, R, OMe, t-Bu]

Aromatic aldehyde

Lithium enolate of methyl 3, 3-dimethyl butanoate

syn   *anti*

R = O – OMeC$_6$H$_4$ –, 4 – ClC$_6$H$_4$–, 4 – NO$_2$C$_6$H$_4$–, 3 – NO$_2$C$_6$H$_4$,
2 – NO$_2$C$_6$H$_4$ –, 4 – NO$_2$ – 2 – thenyl

(Scheme–8)

The above reaction could easily be carried out by mixing finely ground mixture of the starting materials in vacuum for 3 days at room temperature.

In the absence of any solvent, some aldol condensations proceed[10] more efficiently and stereoselectively. In this procedure, appropriate aldehyde and ketone, and sodium hydroxide is ground in a pestle and mortar at room temperature for 5 minutes. The product obtained is the corresponding chalcone; in this case, the initially formed aldol gets dehydrated to the chalcone (Scheme–9).

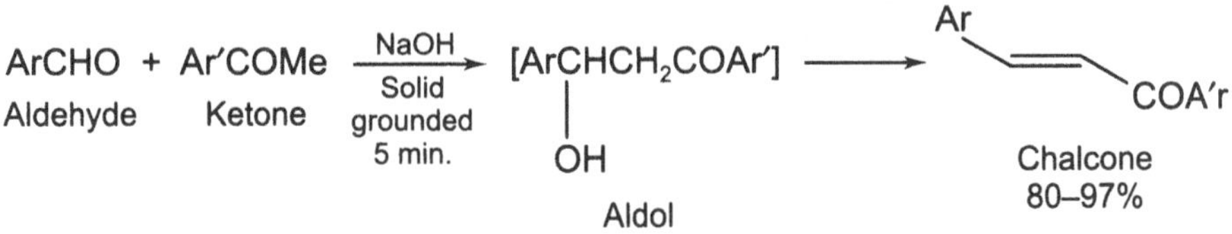

ArCHO + Ar′COMe $\xrightarrow[\substack{\text{Solid} \\ \text{grounded} \\ \text{5 min.}}]{\text{NaOH}}$ [ArCHCH$_2$COAr′] $\longrightarrow$ [chalcone structure: Ar, COA′r]

Aldehyde   Ketone

|
OH
Aldol

Chalcone
80–97%

Ar = Ar′ = Ph (30 min)
Ar = p-MeC$_6$H$_4$ – ; Ar′ = Ph (5 min) (10% yield)
Ar = Ar′ = p-MeC$_6$H$_4$ (5 min) (97% yield)
Ar = p-ClC$_6$H$_4$ – ; Ar′ = Ph (5 min) (98% yield)
Ar = p-ClC$_6$H$_4$ – ; Ar′ = p-MeC$_6$H$_4$ – (10 min) (79% yield)
Ar = p-ClC$_6$H$_4$ – ; Ar′ = p-BrC$_6$H$_4$ (10 min) (81% yield)

(Scheme–9)

The use of alcohol as a solvent in the above method using conventional procedure gives only the aldol in poor yields (10–25%).

2.1.4.3 Aldol Condensation in Supercritical Water

Supercritical water (SC H$_2$O), the critical temperature and pressure of which are 374°C and 22.1 MPA has been used as a solvent due to its unique physical and chemical properties that are quite different from those of ambient water[11]. High temperature water behaves like many other organic solvents in which several organic compounds are soluble.

Aldol condensation reactions can be accomplished in high temperature water. Though 2,5-hexanedione is unreactive in pure water, but it undergoes intramolecular aldol condensation in presence of small amount of base (NaOH) to form 3-methylcyclopentene-2-enone in 81 per cent yield[12].

2.1.4.4 Aldol Condensation in Ionic Liquids

Aldol condensation of propanal to form 2-methylpent-2-enal has been carried out in non-coordinating imidazolium ionic liquid[13]. The reaction is believed to proceed through an aldol intermediate and yielded unsaturated aldehyde under the reaction conditions. In aldol condensation, highest product selectivity was found in [b$_{min}$][PF$_6$] (Scheme–10).

$$CH_3CH_2CHO + \underset{\substack{| \\ CH_3}}{CH_2CHO} \xrightarrow{[b_{min}][PF_6]} \left[\underset{\substack{| \\ CH_3}}{CH_3CH_2\overset{\overset{\textstyle OH}{|}}{CH}-CH-CHO} \right]$$

Propanal $\qquad$ Propanal $\qquad\qquad$ Aldol product

$$\downarrow -H_2O$$

$$CH_3CH_2CH=\underset{\substack{| \\ CH_3}}{C}-CHO$$

2-Methylpent-2-enal

(Scheme–10)

2.1.4.5 Asymmetric Aldol Condensations

Asymmetric version of aldol condensation has been utilized for enantioselective carbon–carbon bond formation. The proline catalysed asymmetric direct aldol reaction of different aromatic aldehyde with acetone and other ketones in ionic liquid [b$_{min}$] [PF$_6$] gave good yield of the aldol product with reasonable enantioselectives[14] (Scheme–11).

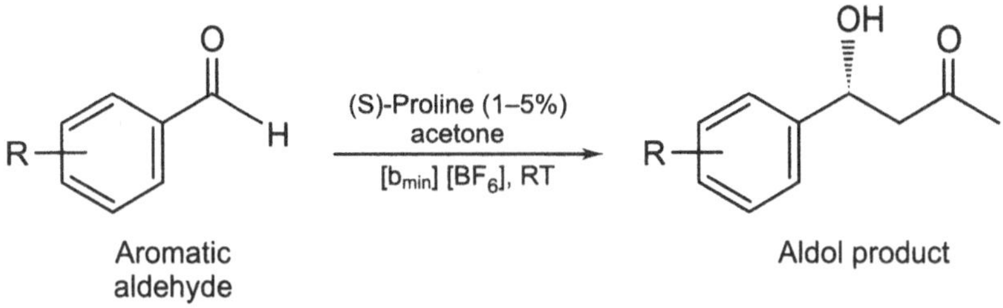

Aromatic aldehyde → Aldol product

(Scheme–11)

L-Proline catalysed direct asymmetric aldol reaction of acetone with various aromatic and aliphatic aldehydes in polyethylene glycol (PEG–400) has been reported[15] to give asymmetric aldol products (Scheme–12).

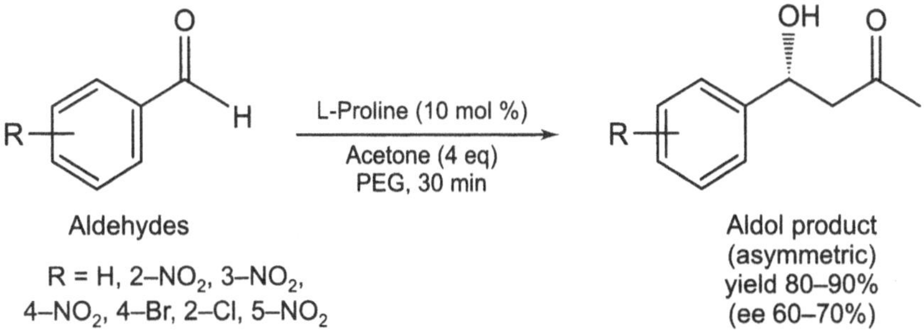

Aldehydes

R = H, 2–NO$_2$, 3–NO$_2$,
4–NO$_2$, 4–Br, 2–Cl, 5–NO$_2$

Aldol product
(asymmetric)
yield 80–90%
(ee 60–70%)

(Scheme–12)

In a similar way, aliphatic aldehydes like isobutyraldehyde and cyclohexane carboxaldehyde have been used as substrates[15] and the asymmetric aldol products obtained in 90 and 65 per cent yields, respectively.

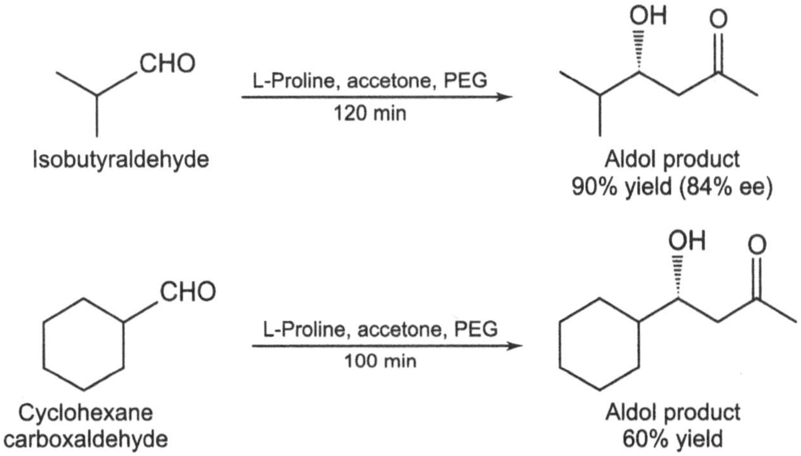

Isobutyraldehyde

Aldol product
90% yield (84% ee)

Cyclohexane carboxaldehyde

Aldol product
60% yield

(Scheme–13)

In the above condensations, polyethylene glycol (PEG), [HO— (CH_2 $CH_2O)n$ — H] used as a solvent is a biologically compatible product and is a green solvent, and is commercially available and is reusable.

See also Mukaiyama reaction (2.1.27), a stereoselective aldol condensation and Henery reaction (2.2.16.5).

References

1. A.T. Nielsen and W.J. Houlihan, Organic Reactions, 1968, **16**, 1; H.O. House, Modern Synthetic Reaction, W.A. Benjamin, California, 2nd ed. 1972, pp. 629–682.

2. V.K. Ahluwalia, Green Chemistry: Environmentally Benign Reactions, Ane Books, India, 2006, Pages 23–24.

3. F. Fringuelli, G. Pani, O. Piematti and F. Pizzo, Tetrahedron, 1994, **50**, 149.

4. T. Mukaiyama, K. Narasaka and T. Banno, Chem., Lett., 1973, 1011; T. Mukaiyama, K. Banno and K. Narasaka, J. Am. Chem., Soc., 1974, **96**, 7503; T. Mukaiyama, Org. React., 1982, **28**, 2303.

5. K. Takai and C.M. Heathcock, J. Org. Chem., 1985, **50**, 3247; A.E. Vougioskas and H.B. Kagan, Tetrahedron Lett., 1987, **28**, 5513.

6. A Lubineau, J. Org. Chem., 1986, **51**, 2142; A. Lubineau and E. Meyer, Tetrahedron, 1988, **44**, 6065.

7. S. Kobayashi and I. Hachiya, J. Org. Chem., 1994, **59**, 3590; For a review on lanthanide catalyzed organic reactions in aqueous media, see S. Kobayashi, Synlett., 1994, 589.

8. S. Kobayashi, Chem., Lett., 1991, 2187; S. Kobayashi and I. Hachiya, Tetrahedron Lett., 1992, 1625.

9. W. Weft and R. Bakthavatechalam, Tetrahedron Lett., 1991, 32, 1535.

10. F. Toda, K. Tanaka and K. Hamai, J. Chem. Soc. Perkin Trans-I, 1990, 3207.

11. O. Kajmoto, Chem., Rev., 1999, **99**, 355; P.E. Savage, Chem., Rev., 1999, **99**, 603. D. Broli, C. Kaul, A. Krammer, P. Krammer, T. Richter, M. Jung, H. Vogal and P. Zehner, Angew. Chem. Int. Edn., 1999, **38**, 2998; M. Siskin and M. Katrizzky, Chem., Rev., 2001, **101**, 825.

12. J. An, L. Bagnell, T. Cablewski, C.R. Strauss and R.W. Trainer, J. Org., Chem., 1997, **62(8)**, 2505.

13. C.P. Mchnert, N.C. Dispenziere and R.A. Cook, Chem., Commun; 2006, 1610.

14. P. Kotrusz, I. Kmentova, S. Gotav, S. Toma and E. Solacaniova, Chem., Commun., 2002, 2510.

15. S. Chandrasekhar, N. Ramakrishna Reddy, S. Shmeem Sultana, Ch. Narsihmulu and K. Venkataraman Reddy, Tetrahedron Lett., 2006, 338.

2.1.5 Alkylations

2.1.5.1 Alkylation of Compounds Containing Reactive Methylene Group

Compounds containing reactive methylene group can be alkylated[1] in a microwave oven with alkyl halide in presence of KOH, K_2CO_3 and a PTC, tetrabutyl ammonium chloride. The alkylation of ethylacetoacetate has already been discussed in section 2.1.2. Another example is given[1] below (Scheme–1).

$$C_6H_5SCH_2CO_2Et \xrightarrow[\text{Microwave}]{\text{RX, KOH–K}_2\text{CO}_3\text{, PTC}} \underset{\underset{R}{|}}{C_6H_5SCH} - CO_2Et$$

Ethyl phenyl
mercaptoacetate

(Scheme–1)

Alkylation of active methylene compounds can also be carried out in ionic liquid, N-butylpyridinium tetrafluoroborate, [bpy] [BF₄], (a recyclable solvent). One such example is given below[2] (Scheme–2).

Meldrum's acid   Dialkylated product

(Scheme–2)

See also alkylation of aromatics, Friedel-Crafts alkylation and acylation, claisen rearrangement, enamine reaction.

Active methylene groups can be alkylated using microwave irradiation. Thus, treatment of phenylsulphonylacetate with alkyl halide in presence of potassium carbonate and a phase transfer catalyst leads to the alkylated product in good yield[2a] (Scheme–3).

Phenylsulphonylacetate   Alkylated product
                                        83%

(Scheme–3)

2.1.5.2 Alkylation of Nitriles

Nitriles having α-Hydrogen(s) can be mono alkylated[3] with alkylhalides in aqueous alkali in presence of a PTC (benzyl triethylammonium chloride). The usual conditions of such alkylation is the use of metal hydrides (like potassium tertiary butoxide) in anhydrous organic solvents. Some examples are given below (Scheme–4).

$$C_6H_5CH_2CN \ + \ C_2H_5Cl \ \xrightarrow[\text{aq. NaOH, 50\%}]{C_6H_5CH_2N^+(C_2H_5)_3Cl^-} \ C_6H_5\overset{\displaystyle |}{\underset{\displaystyle C_2H_5}{C}}HCN \ + \ NaCl$$

$$80\text{--}85\%$$

$$C_6H_5CH_2CN \ + \ (C_6H_5)_2\,CHCl \ \xrightarrow[\text{aq. NaOH}]{\text{PTC}} \ (C_6H_5)_2\,CH\overset{\displaystyle |}{\underset{\displaystyle C_6H_5}{C}}HCN$$

$$94\%$$

$$C_6H_5\overset{\displaystyle |}{\underset{\displaystyle C_2H_5}{C}}HCN \ + \ C_6H_5CH_2Cl \ \xrightarrow[\text{aq. NaOH}]{\text{PTC}} \ C_6H_5CH_2\,\overset{\displaystyle |}{\underset{\displaystyle C_2H_5}{C}}(C_6H_5)CN$$

$$94\%$$

(Scheme–4)

2.1.5.3 Alkylation of Aldehydes

Aldehydes containing only α-hydrogen, such as isobutyraldehyde can be alkylated[4] in good yield with alkyl halides in presence of 50 per cent aqueous NaOH solution in presence of catalytic amount of a PTC (tetrabutyl ammonium iodide) (Scheme–5).

$$(CH_3)_2CHCHO \ + \ C_6H_5CH_2Cl \ + \ NaOH \ \xrightarrow{(Bu_4N^+I^-)} \ C_6H_5CH_2\overset{\displaystyle CH_3}{\underset{\displaystyle CH_3}{\overset{\displaystyle |}{\underset{\displaystyle |}{C}}}}\!\!-CHO$$

Iso butyraldehyde Benzyl   aq
                  chloride

2,2-Dimethyl-3-phenyl
propionaldehyde
75%

(Scheme–5)

2.1.5.4 Alkylation of Ketones

Acidic ketones like phenyl acetone on reaction with chloroacetonitrile in presence of aq. alkali give C-alkylated product[5] under Darzen conditions; in this case glycidic esters, which are normal Darzen condensation products are not obtained (Scheme–6).

$$C_6H_5CH_2COCH_3 \ + \ ClCH_2CN \ + \ NOH \ \xrightarrow[\overset{+}{Q}\overset{-}{X}]{PTC} \ C_6H_5\underset{\underset{CH_2CN}{|}}{CH}COCH_3$$

Phenylacetone   Chloro   alkali
acetonitrile

(Scheme–6)

See also Darzen reaction in presence of PTC (2.1.12.1).

2.1.5.5 C–Alkylations

C-Alkylation of isoquinoline derivatives has been affected[6] by treating with alkyl halide using PTC as catalyst and sonication (Scheme–7).

$$\xrightarrow[\text{NaOH,)))), RT, 20–25 min}]{RX, \ R_4\overset{+}{N}\overset{-}{Br}}$$

R=PhCH₂
60% yield

R = p-ClC₆H₄CH₂
50% yield

(Scheme–7)

2.1.5.6 Alkylation of Terminal Alkynes

The hydrogen attached to a triply bonded carbon of a terminal alkyne (called acetylenic hydrogen) can be replaced by an alkyl group. This is achieved by treating the terminal alkyne with a strong base like sodium amide, the formed alkynide anion on treatment with an appropriate alkyl halide gives the alkyl-substituted alkyne. The sequence of reactions in given below (Scheme–8).

$$CH_3-C\equiv C-H \ \xrightarrow[-NH_3]{NaNH_2} \ CH_3-C\equiv \overset{-}{C}\overset{+}{Na} \ \xrightarrow[-NaX]{CH_3X} \ CH_3-C\equiv C-CH_3$$

Propyne   Sod. amide   Alkynide ion   84%
2-Butyne

(Scheme–8)

In the above reaction, the alkyl halide used to react with alkynide anion must be methyl or primary and also unbranched at its second (beta) carbon. This is essential in order to avoid the formation of other products, predominantly by elimination.

The newly formed alkyne (2-Butyne) can be reduced catalytically to give *n*-butane, which is not possible to obtain in by using ethane in the above reaction.

References

1. R. Runhua, W. Yuliang and J. Yaozhong, Synth. Commun., 1994, **24**, 111, 1917.
2. S. Su, Z. – C. Chen and Q. G. Zhen, Synth. Commun. 2003, **33**, 817.
2a. W. Yuliang and J. Yaozhong, Synth.Commun., 1992, **22**, 2287.
3. M. Makosza, Tetrahedron, 1968, **24**, 175.
4. H. Dietl and K.C. Brannock, Tetrahedron Lett., 1973, 1273.
5. A. Jonck, M. Fedorynski and M. Makosa, Tetrahedron Lett., 1972, 2394.
6. J. Ezgerra and J. Alvorez – Builla, J. Chem. Sol., Chem. Commun., 1984, 54.

2.1.6 Arndt-Eistert Synthesis

It is a simple general method[1] for the carbon-carbon bond formation. The method consists in converting an acid into its next higher homologous acid. In this procedure, the acid is first converted to its acid chloride, which is then treated with diazomethane resulting in the formation of the diazoketone, which on treatment with silver oxide gets converted to a ketene. The ketene gets esterified under the conditions of the reaction. Subsequent saponification gives the homologues acid. The various steps involved are (Scheme–1):

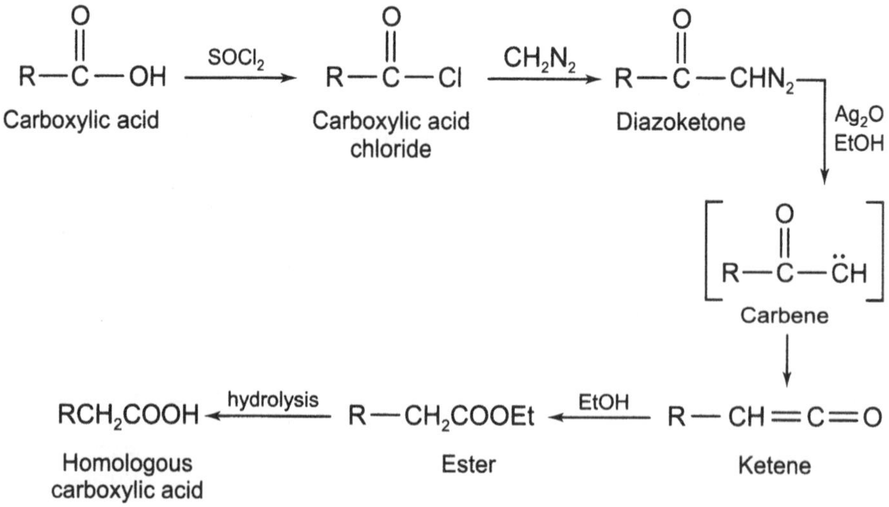

$$R-\overset{\overset{O}{\|}}{C}-OH \xrightarrow{SOCl_2} R-\overset{\overset{O}{\|}}{C}-Cl \xrightarrow{CH_2N_2} R-\overset{\overset{O}{\|}}{C}-CHN_2$$

Carboxylic acid   Carboxylic acid chloride   Diazoketone $\begin{matrix} Ag_2O \\ EtOH \end{matrix}$

$$\left[R-\overset{\overset{O}{\|}}{C}-\ddot{C}H \right]$$

Carbene

$$RCH_2COOH \xleftarrow{hydrolysis} R-CH_2COOEt \xleftarrow{EtOH} R-CH=C=O$$

Homologous carboxylic acid   Ester   Ketene

(Scheme–1)

In the above sequence of reactions, the conversion of diazoketene into the ketone involves a rearrangement known as **Wolffs rearrangement** under the catalytic influence of silver oxide[2].

2.1.6.1 Photochemical Arndt-Eistert Reaction

The conversion of diazoketone into carboxylic ester can be affected under photochemical conditions[3]. Some examples are given below:

(*i*) Synthesis of methyl γ-cyclohexyl butyrate (Scheme–2).

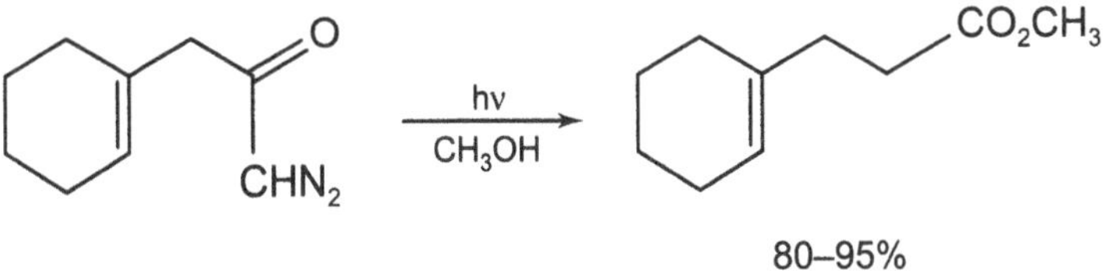

80–95%

(Scheme–2)

(*ii*) Synthesis of δ-furanyl glutarate (Scheme–3).

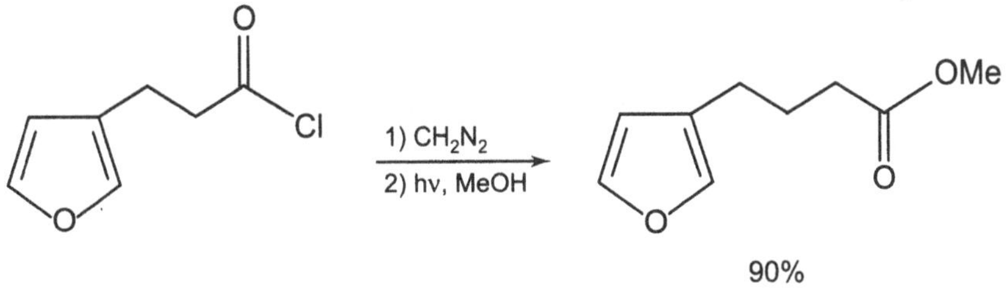

90%

(Scheme–3)

(*iii*) Synthesis of methyl cyclopentene carboxylate (Scheme–4).

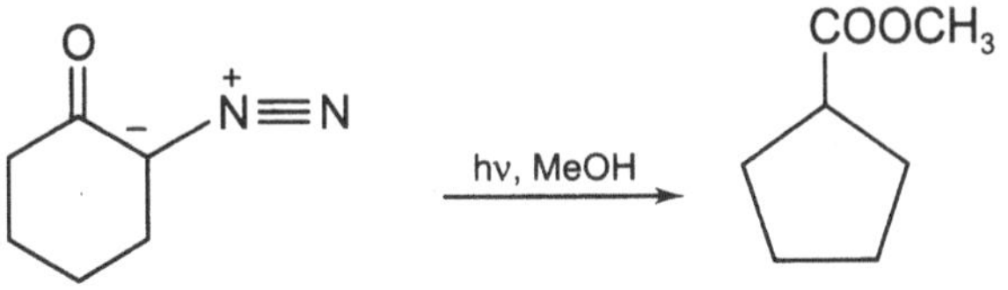

(Scheme–4)

(*iv*) A variation of Arndt-Eistert reaction is that in cyclic diazoketones, the rearrangement leads to ring contraction. This reaction has been extensively used for the preparation of strained small ring compounds, such as bicyclo [2.1.1]-hexane and benzocyclobutane (Scheme–5).

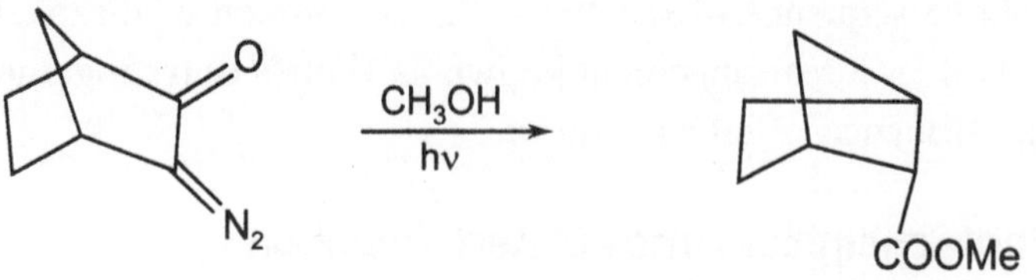

54%
Methylcyclo [2.1.1] hexane
carboxylate

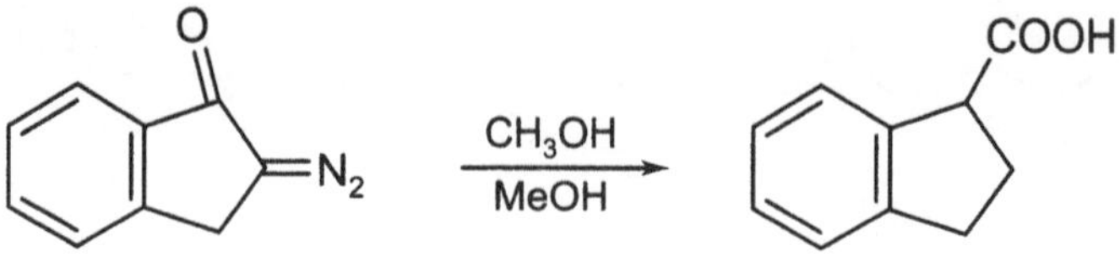

Benzobicyclobutane carboxylic
acid

(Scheme–5)

References

1. F. Arndt and B. Eistert, Ber., 1935, **68**, 200.
2. V.K. Ahluwalia and R.K. Parashar, Organic Reaction Mechanisms, Narosa Publishing House, 2011, Page 275.
3. A.B. Smith, Chem., Commun., 1974, 695.
4. J. Meinwald and Y.C. Meinwald, in Advances in Aliyclic Chemistry Ed. H. Hart and G. J. Karabatsos, Vol. 1, P.I (New York, Academic Press).

2.1.7 Baker-Venkataraman Rearrangement

The base catalysed rearrangement of o-acyloxy (preferably o-benzyloxy) ketones yields β-diketones (which are important intermediates for the synthesis of flavones or chromones) is known as Baker-Venkataraman rearrangement[1] (Scheme–1).

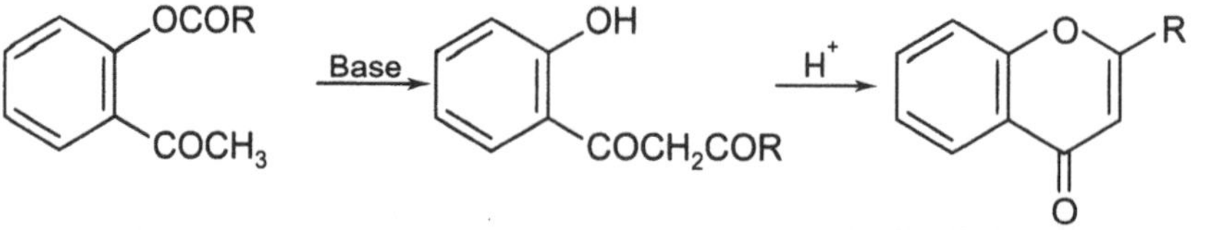

o-Acyloxyacetophenone   β-Diketone   2-Methyl chromone R = CH_3
R = CH_3 or C_6H_5   Flavone R = C_6H_5

(Scheme–2)

2.1.7.1 Baker-Venkataraman Rearrangement using PTC

A convenient one step process has been developed[2] for the synthesis of β-diketone. In this procedure, benzoyl chloride is added to a stirred mixture of 2-hydroxyacetophenone, tetrabutylammonium hydrogen sulphate and aqueous potassium carbonate or potassum hydroxide (2 hr. stirring). Working up of the reaction mixture gives the β-diketone (Scheme–2).

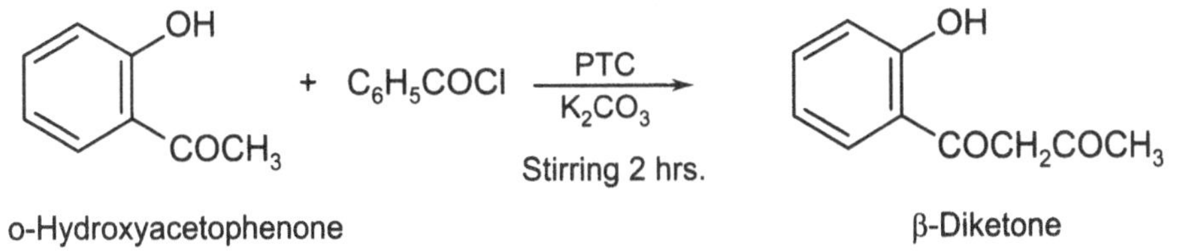

o-Hydroxyacetophenone

β-Diketone

(Scheme–2)

References

1. W. Baker, J. Chem., Soc., 1933, 1381; H.S. Mahal and K. Venkataraman, J. Chem., Soc., 1934; 1767.

2. V.K. Ahluwalia et.al., unpublished results.

2.1.8 Benzidine Rearrangement

The acid catalysed rearrangement of hydrazobenzene to 4,4′-diaminobiphenyl (*p*-benzidine) is known as benzidine rearrangement[1] (Scheme–1).

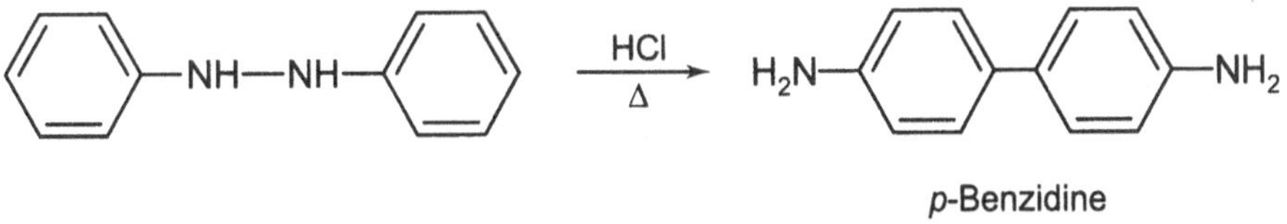

p-Benzidine

(Scheme–1)

In the above rearrangement, *p*-benzidine is obtained as the major product.

When both para positions are blocked, the major product is 2,2′-isomer (Scheme–2).

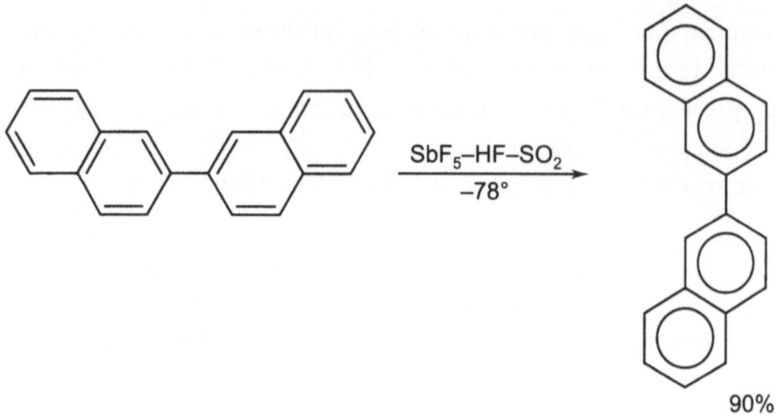

90%

(Scheme–2)

The mechanism is believed to be intramolecular rearrangement[2].

References

1. A.W. Hoffmann, Proc. Roy. Soc. London, 1863, **12**, 576; T. Shlradsky and S. Auramovki-Grisau, J. Het. Chem., 1980, **17**, 189.
2. V.K. Ahluwalia and R.K. Parashar, Organic Reaction Mechanism, Narosa Publishing House, New Delhi, 2011, page 286.

2.1.9 Benzoin Condensation

Aromatic aldehydes (having no α-hydrogen) on treatment with sodium or potassium cyanide undergo self condensation to give α-hydroxy ketones, known as Benzoin condensation[1]. This condensation does not take place with aliphatic aldehydes[2] (Scheme–1).

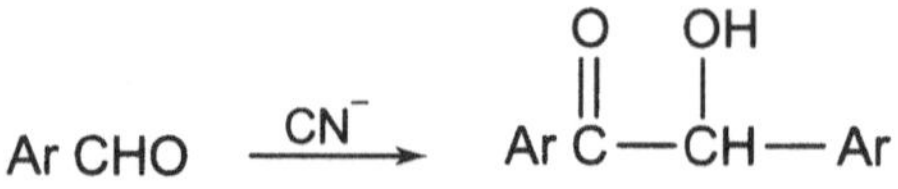

(Scheme–1)

2.1.9.1 Benzoin Condensation under Catalytic Conditions

Benzoin condensation of aldehydes are strongly catalysed by quaternary ammonium chloride in a two phase system[3] (Scheme–2).

$$Ar\ CHO \xrightarrow[\text{NaCN}]{\overset{+}{Q}\overset{-}{X}} Ar-\underset{\underset{OH}{|}}{CH}-\underset{\overset{\|}{O}}{C}-Ar$$

$$C_6H_5CHO \xrightarrow{Bu_4\overset{+}{N}\overset{-}{CN}} C_6H_5-\underset{\underset{OH}{|}}{CH}-\underset{\overset{\|}{O}}{C}-C_6H_5$$

$$63\%$$

(Scheme–2)

Benzoin condensation can also be carried out with either aqueous KCN/ neat aromatic aldehyde or solid KCN/aldehyde dissolved in benzone or acetonitrile at 25–60° using 18-crown 6 or dibenzo-18-crown-6-as catalalyst.

Both aliphatic and aromatic aldehydes undergo benzoin condensation[4] with solid potassium hydroxide using 3-benzyl-4-methylthiazolium chloride as catalyst (Scheme–3).

$$CH_3CHO \xrightarrow[\text{KOH}]{\text{3-Benzyl-4-methylthiazoliumchloride}} CH_3-\underset{\underset{OH}{|}}{CH}-\underset{\overset{\|}{O}}{C}-CH_3$$

Acetaldehyde

Acetoin (100%)

(Scheme–3)

Aromatic aldehydes react within few minutes, but aliphatic aldehydes require 5–10 hr.

In a similar way, benzoin condensation with both aromatic or aliphatic aldehydes proceed remarkably well using N-laurylthiazolium bromide as catalyst with aqueous phosphate solution[5] (Scheme–4).

$$R-CHO \xrightarrow[\text{aq. phosphate}]{\text{N-Laurylthiazolium Chloride}} R-\underset{\underset{OH}{|}}{CH}-\underset{\overset{\|}{O}}{C}-R$$

$$10–95\%$$

(Scheme–4)

Benzoin condensation also takes place in aqueous media using inorganic salts (*e.g.*, LiCl); the reaction is 200 times faster[6] than in ethanol without any salt.

Benzoin condensation is also brought about by coenzyme[3] Thiamine.

References

1. A.J. Lapworth, J. Chem. Soc., 1903, **83**, 995; 1904, **85**, 1206.
2. For a review, see Ide and Buek, org. React., 1948, **4**, 269–304.
3. J. Solodar, Tetrahedron Lett., 1971, 287.

4. C.M. Starks and C. Liotta, Phase Transfer Catalysts, Principles and Techniques, Academic Press, Inc., N. Y. 1978, page 343.

5. W. Tagaki and H. Hara, J. Chem., Soc., Chem. Commun., 1973, 891.

6. E.T. Kool and R. Breslow, J. Am. Chem., Soc., 1988, **110**, 1596.

2.1.10 Bouveault Reaction

The action of Grignard reagents on N, N′-disubstituted formamides yield aldehyde. This is knoun as Bouveault reaction[1] or Bauveault aldehyde synthesis (Scheme–1).

R′ \>N—CHO   + R′MgX   ⟶   [R′ \>N—CHOMgX with R′ on N]

| N,N-Disubstituted formamides | Grignard reagent |

$\downarrow H_2O$

R′—CHO

Aldehyde

(Scheme–1)

The required grignard reagent is prepared by the reaction alkyl or aryl halides on magnesium in ether on sonication[2]. In this case magnesium on sonication gets activated.

2.1.10.1 Bouveault Reactions under Sonication

In place of Grignard reagents, organolithium reagents give better yields[3]. The organolithium reagents are obtained by sonication of aryl halide with lithium using low intensity ultra sonic. These reagents are used in Bouveault reaction and give higher yields than the traditional methods[4] (Scheme–2).

$RX \xrightarrow[))))]{Li} R^-Li^+ \xrightarrow{HC(O)NMe_2}$ [$R\,CH \overset{O^-Li^+}{\underset{NMe_2}{}}$]

$\downarrow H_2O^+$

$R\,CHO + Me_2NH$

(Scheme–2)

In non-ultrasonic reaction, a number of side products are obtained. The method is improved when DMF is replaced by more elaborate and expensive formamide, $Me_2NCH_2CH_2N(Me)CHO$.

A simplification of this method is the sonication of an aryl halide and DMF with excess lithium for 15 min. followed by dropwise addition of 1-bromobutane, sonication for 30 min. more; the producted obtained is *o*-substituted aldehyde.

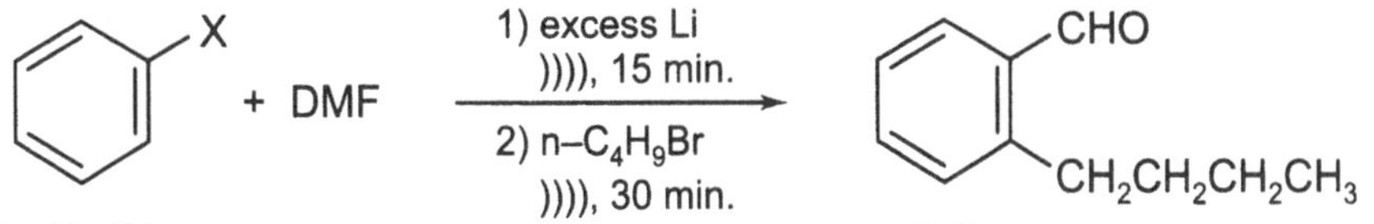

Aryl halide 70%

(Scheme–3)

Use of iodomethane in place of *n*-butylbromide in the above reaction gives[5] *o*-tolualdehyde.

References

1. L. Bouveault, Bull. Soc. Chem. France, 1904, 31, 1306, 1322; N. Smith, J. Org. Chem., 1941, **6**, 489; Sice′, J. Am Chem. Soc., 1953, **75**, 3697; Jones et.al., J. Chem. Soc., 1958, 1054.
2. J.L. Luche and J.C. Damiano, J. Am. Chem. Soc., **102**, 7926; H. Bonnermann, R. Bogdanovic, D.W. Hl. Brinkman and B. spliletoff, Angew. Chem., Int. Ed. Eng., 1983, **22**, 728; J.D. Sprich and G.S. Lewandos, Inorg. Chem. Acta, 1982, **76**, 1241.
3. Evans, Chem. Ind. (London), 1957, 1956.
4. C. Petrier, A.L. Gemal and J.L. Luche, Tetrahedron Lett., 1982, **23**, 3361.
5. J.L. Luche, Ultrasonics, 1987, **25**, 40.

2.1.11 Claisen Rearrangement

Allyl phenyl ether on heating at 200° undergoes an intramolecular reaction known as claisen rearrangement[1] (Scheme–1).

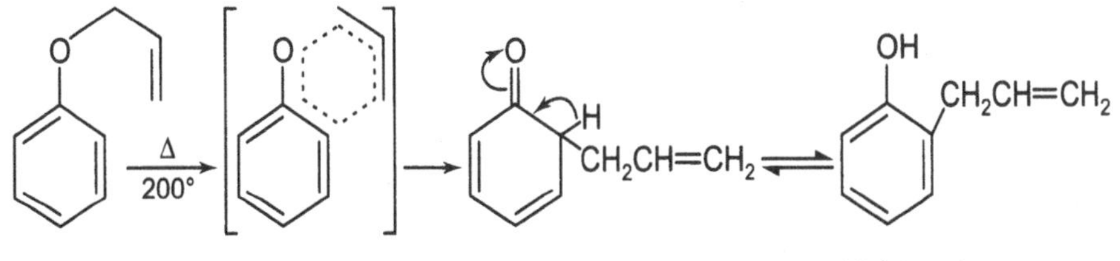

Allylphenyl ether T.S. *o*-Allylphenol

(Scheme–1)

The above claisen rearrangement could also be carried out using microwaves[2].

2.1.11.1 Claisen Rearrangement in Water

The ortho-claisen rearrangement of allyl phenyl ether in aqueous media was effected using microwaves when 2-allylphenol was obtained exclusively after 10 min. (Scheme–2)[2].

Allylphenyl ether	H$_2$O, MW 10 min	2-Allylphenol (100%)

(Scheme–2)

2.1.11.2 Claisen Rearrangement in Near Critical Water

Reactions conducted in a microwave oven in water above 200° exist due to the pressure limit of about 20 bar for most of the commercially available MW instruments. Pure water reaches an autogenic pressure of 50 bar at 250°. This condition is called near critical water (NCW) region between 200 and 300°C[3].

The Claisen rearrangement of allylphenylether to 2-allyl phenol was successfully performed at 240° to give the product (2-allylphenol) in 84 per cent yield. However, at 200°, only 10 per cent conversion was achieved. It was found that by increasing the temperature and time to 250° and 1 hr., dihydrobenzofuran was produced in 72 per cent conversion due to the involvement of water into the reaction pathway (Scheme–3).

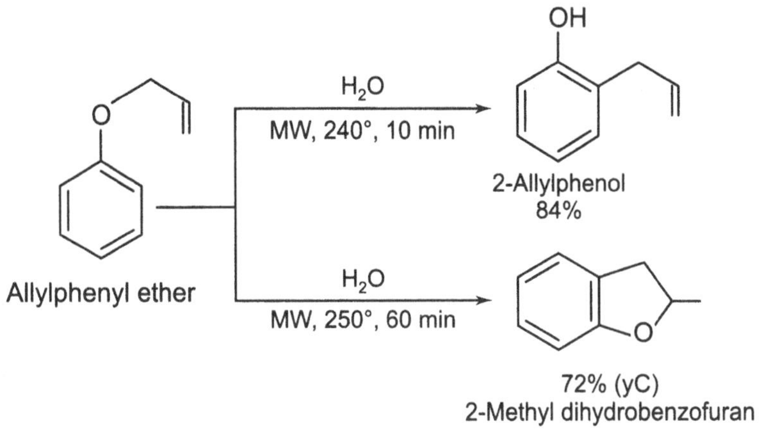

2-Allylphenol
84%

72% (yC)
2-Methyl dihydrobenzofuran

(Scheme–3)

References

1. L. Claisen, Ber., 1912, **45**, 3157; L. Claisen and E. Tietze, Ber., 1925, 58, 275; Org. React., 1944, **2**.

2. R.J. Giguere, A.M. Namen, O.Lopez, A. Arepally, D.E. Ramos, G. Majetich and J. Defaw, Tetrahedron Lett., 1987, **28**, 6553.

3. K.D. Raner, C.R. Strauss, R.W. Trainer and J.S. Thern, J. Org. Chem., 1995, **60**, 2456.

4. L. Bagnell, T. Cablewski, C.R. Strauss and K.W. Trainor, J. Org. Chem., 1996, **61**, 7355.

2.1.12 Darzen Reaction

The condensation of aldehydes or ketones with α-haloester in presence of a base (like potassium tertiary butoxide, a hindered ketone to avoid S_N2 displacement of the cloride) gives α, β-epoxy esters (glycidic esters) is known as Darzen Reaction[1] (Scheme–1).

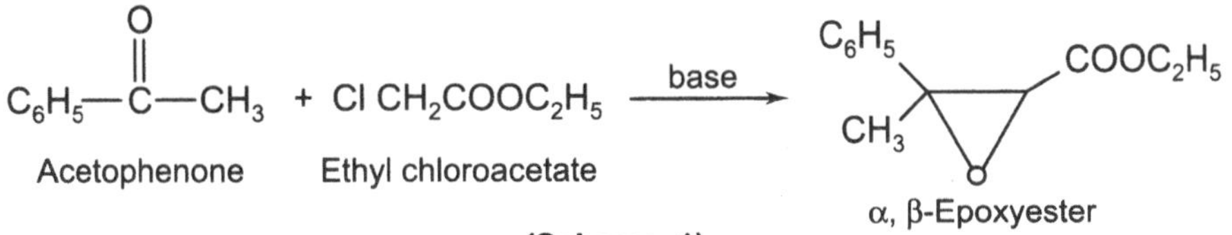

Acetophenone   Ethyl chloroacetate

α, β-Epoxyester

(Scheme–1)

In place of α-haloester, α-halonitriles (*e.g.*, chloroacetonitrile, $ClCH_2CN$) can also be used.

2.1.12.1 Darzen Reaction in Presence of Phase Transfer Catalyst

Darzen reaction has been found to occur in alkaline solution in presence of a PTC (benzyltriethyl ammonium chloride)[2] (Scheme–2).

$$\underset{\substack{R' \\ \text{Aldehyde} \\ \text{or ketone}}}{\overset{R}{>}}C=O \ + \ \underset{\substack{\text{Chloroaceto} \\ \text{nitrile}}}{ClCH_2CN} \ + \ NaOH \xrightarrow[\text{aq.}]{C_6H_5CH_2\overset{+}{N}Et_3\overset{-}{Cl}} \underset{\substack{R' \\ }}{\overset{R}{>}}\underset{O}{\overset{}{C}}-CH-CN$$

55–80%

R	R'
Ph	H
CH₃	CH₃
(CH₂)₃	H
(CH₂)₅	H
Ph	Ph

Cyclohexanone + Cl CH₂CN $\xrightarrow{Bu_4N^+I^-, 15-20°}$

Chloro
acetonitrile

(Scheme–2)

With aldehydes and unsymmetrical ketones, both possible stereoisomers are obtained. However, with more acidic ketones, such as phenylacetone, the ketone carbanion is formed rather than from the nitrile (which is normally the case), leading to alkylation of the ketone (Scheme–3).

$$\underset{\text{Phenylacetone}}{C_6H_5CH_2COCH_3} \ + \ \underset{\substack{\text{Chloro} \\ \text{acetonitrile}}}{ClCH_2CN} \ + \ NaOH \xrightarrow[\text{aq.}]{PTC} \ \underset{\substack{| \\ CH_2CN}}{C_6H_5\overset{}{C}HCOCH_3}$$

(Scheme–3)

In place of quaternery ammonium salts (PTC), crown ethers (*e.g.*, dibenzo-18-crown-6) (1 mol%) can also be used[3] (Scheme–4).

$$\underset{R'}{\overset{R}{>}}C=O \ + \ CH_3CN \xrightarrow[KOH]{\text{[18]-crown-6}} \underset{R'}{\overset{R}{>}}C=CHCN$$

Benzaldehyde + Cl CH₂CN $\xrightarrow[\substack{50° \\ \text{Dibenzo-18-crown-6}}]{\text{aq. NaOH}}$ 78%

Chloro
acetonitrile

(Scheme–4)

In a similar way, phenyl acetone can be alkylated[4] with *n*-butylbromide in 50 per cent aq. sodium hydroxide in presence of dicyclohexyl-18-crown-6 in better than 90 per cent yield (Scheme–5).

$$C_6H_5CH_2COCH_3 \ + \ n\text{-BuBr} \ \xrightarrow[\substack{50\% \text{ aq. NaOH} \\ \text{Dicyclohexyl-18-crown-6}}]{80°} \ C_6H_5CHCOCH_3$$

| Phenylacetone | Butyl bromide | | Bu(*n*) |

(Scheme–5)

The PTC catalysed condensation of α-thiocarbonyl compound with 2-chloroacrylonitrile yielded[5] 2-cyano-2, 3-epoxytetrahydrothiophenes. The reaction probably involved first the thiol addition followed by an internal Darzen reaction (Scheme–6).

α-Thiol cyclohexanone

2-Chloro acrylonitrile

2-Cyano-2, 3-epoxy tetrahydro thiophene (73%)

(Scheme–6)

References

1. C. Darzen, Compt. Rend., 1904, **139**, 1214; 1905, **141**, 766; 1906, **142**, 214; M. Ballester, Chem. Rev., 1995, **55**, 283.
2. A. Jonczky, M. Fedorynski and M. Makosaza, Tetrahedron Lett., 1972, 2395.
3. G.W. Gokel, S.A. DiBiase and B.A. Lipiska, Tetrahedron Lett., 1976, 3495.
4. M. Cinquini, F. Montanori and P. Tundo, Chem. Commun., 1974, 878.
5. J.M. Melntosh and H. Khalili, J. Org. Chem., 1977, **42**, 2123.

2.1.13 Dieckmann Condensation

Diesters of C_6 and C_7 dibasic acids undergo intramolecular claisen condensation in presence of base to give good yields of cyclic β-ketoesters. This is known as Dieckmann condensation[1] (Scheme–1).

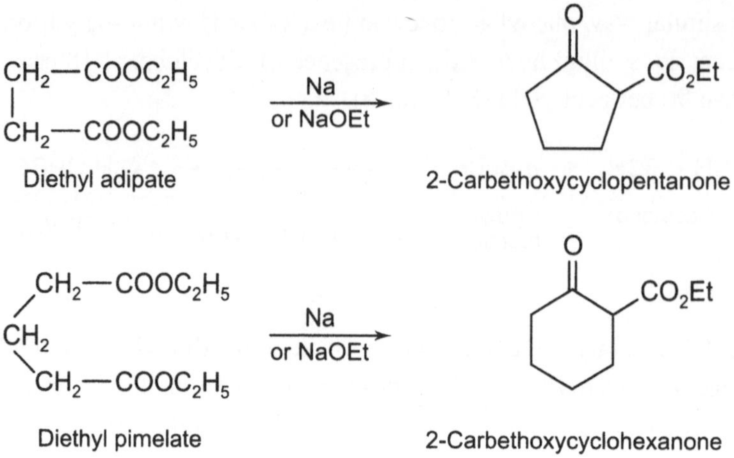
CH2—COOC2H5
CH2—COOC2H5
Diethyl adipate
Na
or NaOEt
O
CO2Et
2-Carbethoxycyclopentanone
CH2—COOC2H5
CH2
CH2—COOC2H5
Diethyl pimelate
Na
or NaOEt
O
CO2Et
2-Carbethoxycyclohexanone

(Scheme–1)

2.1.13.1 Dieckmann Condensation in Solid State

Dieckmann condensation has also been achieved in solid state[2] by direct distillation of the powered reaction mixture, which was neutralised with *p*-TsOH . H_2O (Scheme–2).

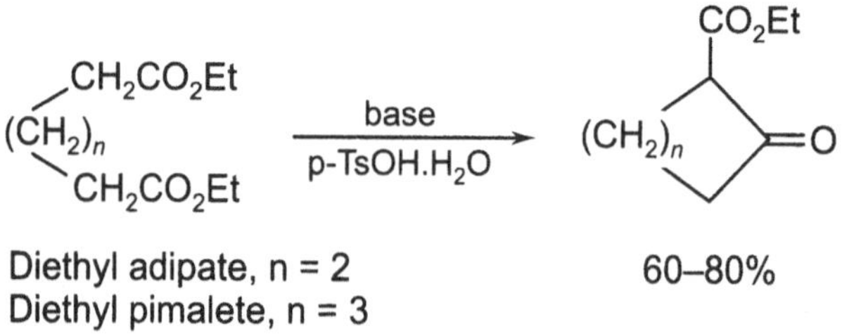
CH2CO2Et
(CH2)n
CH2CO2Et
base
p-TsOH.H2O
CO2Et
(CH2)n
=O
Diethyl adipate, n = 2
Diethyl pimalete, n = 3
60–80%

(Scheme–2)

2.1.13.2 Dieckmann Condensation under Sonication

Dieckmann condensation proceeds very well on sonication in a short time[3]. On sonication, potassium is easily transformed to a silvery blue suspension in toluene[4] (Scheme–3).

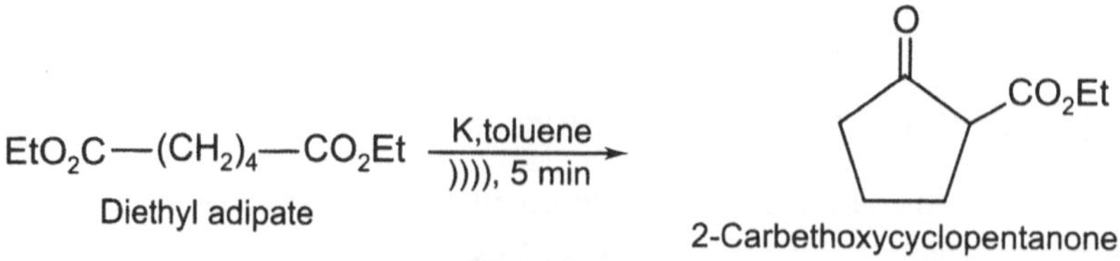
EtO2C—(CH2)4—CO2Et
Diethyl adipate
K,toluene
)))), 5 min
O
CO2Et
2-Carbethoxycyclopentanone

(Scheme–3)

2.1.13.3 Dieckmann Condensation using Polymer Support Technique

In Dieckmann cyclisation, it is necessary to control the relative rates of the two competing reactions, *viz.*, intramolecular and intermolecular cyclisations; the former must be faster. Normally, this is achieved by having the compound in larger volume of the solvent (dilution technique). Alternatively, one carboxyl group is anchored on to a polymeric matrix so that the intermolecular reaction is prevented. Using this technique, mixed esters of dicarboxylic acids can be cyclised[5-8] (Scheme–4).

(Scheme–4)

References

1. W. Dieckmann, Ber., 1894, **27**, 102, 965; 1900, **33**, 2670; Ann, 1901, **317**, 53, 93; C.R. Housen and B.E. Hudson Organic Reactions, 1942, **I**, 274.
2. F. Toda, T. Suzuki and S. Higa, J. Chem., Soc., Perkin Trans., 1998, 3521.
3. Cited in Review by J.M. Khurana, Chem., Edn., 1990 (Oct, Dec), p 27.
4. J.L. Luche, C. Petrier and C. Dupty, Tetrahedron Lett., 1985, **26**, 753.
5. J. Schalffer, J.J. Bloomfield, Org. React., 1967, 15.
6. J.L. Crowley and H. Rapport, J. Chem., Soc., 1970, **92**, 6363.
7. H.L. Lochte and A.G. Pinman, J. Org. Chem., 1960, **25**, 1462.
8. O.S. Broun, Synthesis, 1975, **5**, 326.

2.1.14 Diels-Alder Reaction

The [4 + 2] cycloaddition reaction between a conjugated diene (4π-electron system) and a compound containing a double or triple bond, called the dienophile (2π-electron system) to form an adduct is known as Diels Alder Reaction[1]. It is a typical carbon-carbon bond forming reaction (Scheme–1).

Dienes   Dienophiles

(Scheme–1)

2.1.14.1 Diels-Alder Reaction using Microwaves

Anthracene and maleic anhydride (in the form of finely ground mixture) on heating in a beaker using diglyme in microwave[2] oven gives an adduct (Scheme–2).

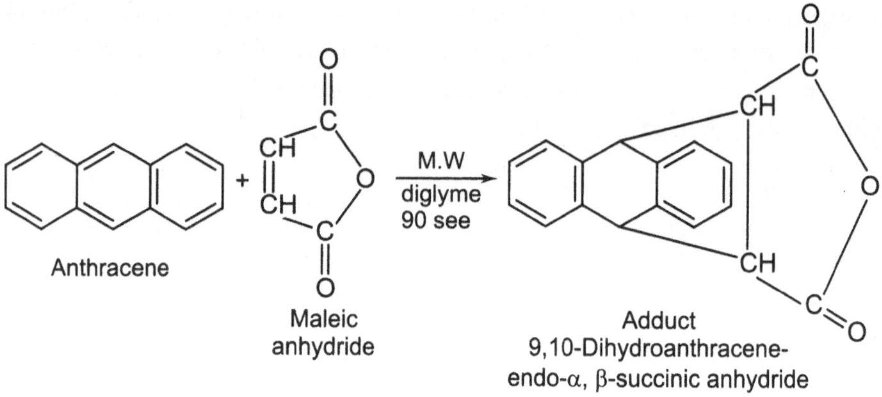

Anthracene   Maleic anhydride

M.W
diglyme
90 see

Adduct
9,10-Dihydroanthracene-
endo-α, β-succinic anhydride

(Scheme–2)

2.1.14.2 Diels-Alder Reaction in Water

Furan on treatment with maleic acid in water at room temperature gives endo-cis-1, 4-edoexo-Δ^5-cyclohexene-2, 3-dicarboxylic acid[3] (Scheme–3).

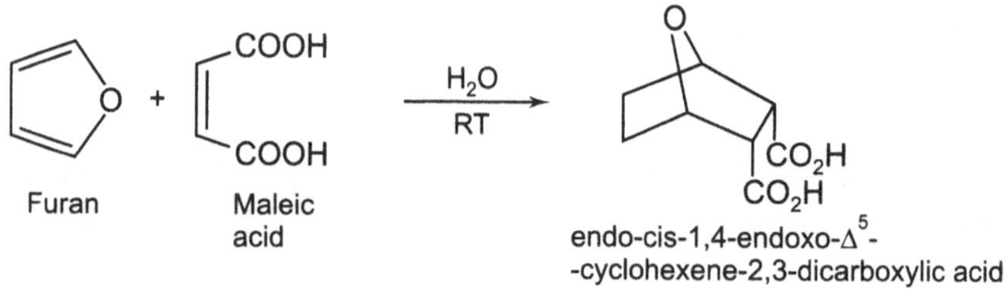

Furan   Maleic acid

H_2O
RT

endo-cis-1,4-endoxo-Δ^5-
-cyclohexene-2,3-dicarboxylic acid

(Scheme–3)

2.1.14.3 Diels-Alder Reaction using Sonication

Diels-Alder reaction is facilitated by sonication[4]. A typical example is given below (Scheme–4).

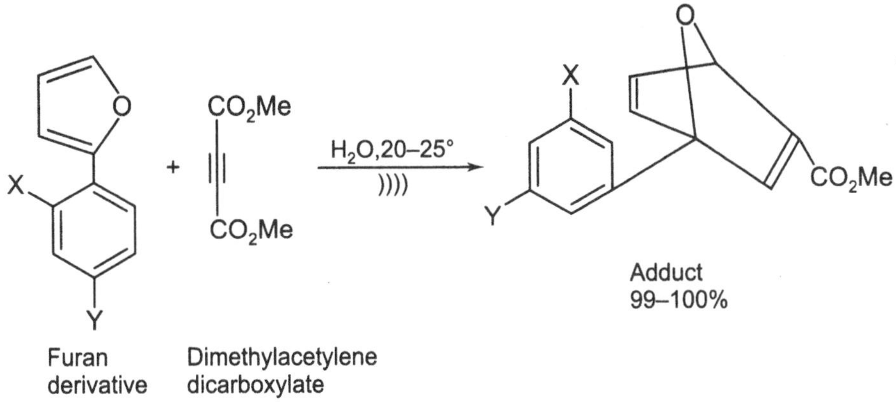

Furan derivative + Dimethylacetylene dicarboxylate → Adduct 99–100%

(Scheme–4)

2.1.14.4 Diels-Alder Reaction in Supercritical Water

Diels-Alder cycloaddition reactions have been conducted in SC – H_2O. A number of different dienes/dienophiles have been used[5].

For details about super critical water, see Aldol condensation in supercritical water (2.1.4.3).

2.1.14.5 Diels-Alder Reaction in Ionic Liquids

Ionic liquids, such as [b_{min}] [BF_4], [B_{min}] [ClO_4] have been used for Diels-Alder Reaction. There is rate enhancement, high yields and strong eno selectivities in these reactions[6].

2.1.14.6 Diels-Alder Reaction in Supercritical Carbon Dioxide

A large large number of Diels-Alder reactions have been conducted in SC — CO_2. For details see ref[7].

2.1.14.7 Diels-Alder Reaction in Polyethylene Glycol

The Diels-Alder reaction of 2, 3-dimethyl-1, 3-butadiene with acrolein gave[8] the adduct in good yield (Scheme–5).

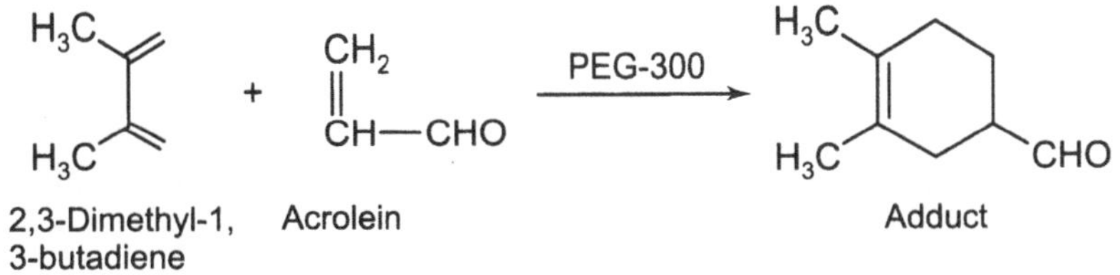

2,3-Dimethyl-1, 3-butadiene + Acrolein → Adduct

(Scheme–5)

2.1.14.8 Diels-Alder Reaction under Photochemical Conditions

A typical example is the irradiation of butadiene in presence of ketonic sensitisers. The reaction affords a mixture of cis- and trans- divinyl-cyclobutane along with normal Diels-Alder adduct, vinylcyclohexene.

1,3-Butadiene   *trans*   *cis*   4-Vinylcyclohexene
                                   Divinylcyclobutane

(Scheme–6)

References

1. O. Diels and K. Alder, Ann., 1928, **460**, 98; 1929, **470**, 62; Ber., 1929, **62**, 2081, 2087; S.A. Norton., Chem., Revs., 1942, **31**, 319; D.A. Oppolzer, Ang. Chem. Int. Ed., 1977, **16**, 10.

2. S.S. Bari, A.K. Bose, A.G. Choudhary, M.S. Manhas, V.S. Raju and E.W. Robb, J. Chem., Ed., 1992, **69(11)**, 938.

3. R.B. Woodward and Harold Baer, J. Am. Chem. Soc., 1948, **70**, 1161; R. Breslow and D. Rideout, J. Am. Chem. Soc., 1980, **102**, 7816.

4. J. Lee and J.K. Sayder, J. Am. Chem., Soc., 1989, **111**, 1522.

5. M.B. Kovzenski and J.W. Kolis, Tetrahedron Lett., 1997, **38**, 5611.

6. T. Fischer, T. Sethi, T. Welton and J. Woolf, Tetrahedron Lett., 1999, **40**, 793; M.J. Earle; P.B. McMormal and K. R. Seddon, Green Chem., 1999, **1**, 23.

7. V.K. Ahluwalia and R.S. Varma, Green Solvents for Organic Synthesis, Narosa Publishing House, New Delhi, 2009, page 5.9 to 5.12 and the references cited therein.

8. N.F. Leininger, R. Clontz, J.L. Gainer and D.V. Kirwan in Clean Solvents, Alternative Media for Chemical Reactions and Processing, ed. M.A. Abraham and L. Moems, ACS symposium series 819, American Chemical Society, Washington DC, 2002, p. 208.

2.1.15 Enamine Reaction[1]

Enamines, obtained by the reaction[2] of an aldehyde or a ketone having α-hydrogen with a secondary amine (such as pyrrolidine, morpholine and piperidine) in presence of a dehydrating agent (like catalytic amount of toluene-*p*-sulfonic acid) can be alkylated with alkyl halide to give the alkylated product or acylated with acyl chloride to give the corresponding acylated products. These products on hydrolysis give 2-alkylcyclohexanone and 2-acylcyclohexanone, respectively. This synthesis is known as Stork Enamine Synthesis (Scheme–1).

(Scheme–1)

References

1. G. Stork, R.Terrell and J. Szmuszkovics, J. Am. Chem. Soc., 1954, **76**, 2029; J.W.Whitesell and M.A. Whitesell, Synthesis, 1983, 517.
2. G. Stork, A. Brizzoland, H. Landsman, J. Szmuszkovicz J. Am. Chem., Soc., 1963, **85**, 207.

2.1.16 Ene Reaction

The addition of an olefin having an allylic hydrogen (ene) to a compound having a double bond (C $=$ C, C $=$ O, N $=$ N, called the enophile) gives a new compound with the formation of a new σ bond to the terminal carbon of the allyl group. This is followed by the 1, 5-migration of the allylic hydrogen and subsequent change in the position of the allylic double bond. This reaction is commonly known as ene reaction[1] (Scheme–1).

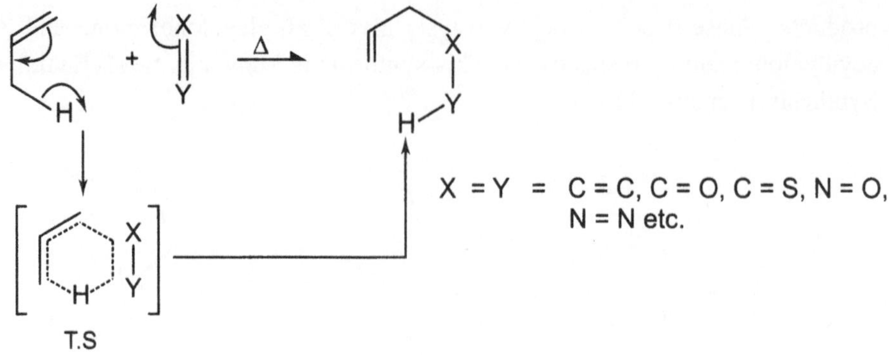

(Scheme–1)

In case of 1,7-diene, a cyclic product is obtained. This is called **intramolecular ene reaction**[2] (Scheme–2).

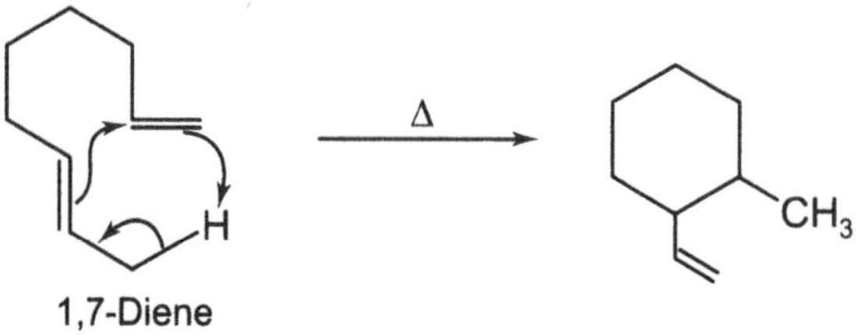

1,7-Diene

(Scheme–2)

A typical example of ene reaction is the reaction of propene with maleic anhydride at 200° to give the product (A) and the cyclohexene forms (B) (Scheme–3).

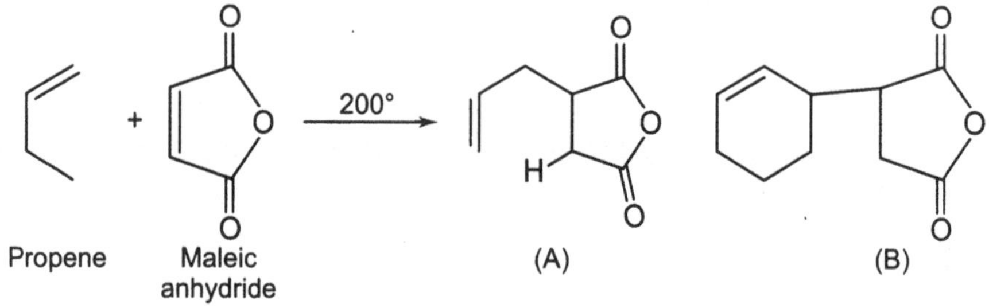

Propene   Maleic   (A)   (B)
            anhydride

(Scheme–3)

The reaction can be catalysed by Lewis acids ($AlCl_3$, $SnCl_4$, $TiCl_4$); the reaction proceeds at lower temperature with much improved selectivity[3].

References

1. K. Alder, F. Pascher and A. Smith, Ber., 1943, **76**, 27.

2. W.D. Huntsman, V.C. Solman and D. Eros, J. Am. Chem., Soc., 1958, **80**, 5455.

3. B.B. Snider and E.A. Deutsch. J. Org. Chem., 1982, **47**, 745.

2.1.17 Friedel-Crafts Reaction

Friedel-Crafts alkylation and acylation are two important carbon-carbon bond forming reactions[1] (Scheme 1).

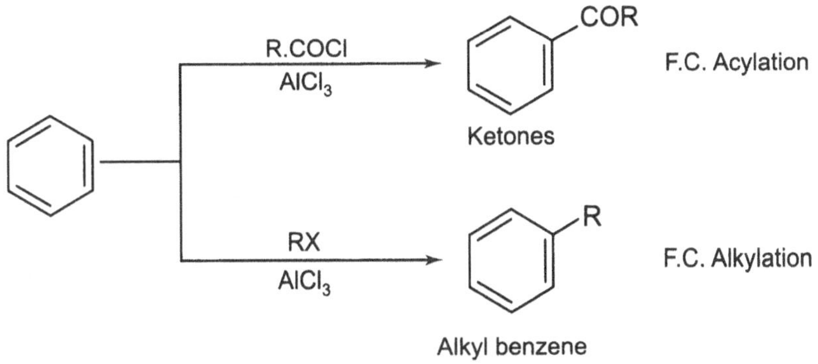

(Scheme–1)

2.1.17.1 Friedel-Crafts Alkylation

Friedel-Crafts alkylation of aromatic hydrocarbon can be affected by using a mixture of alkyl halide and aluminium trihalide (Lewis acid). The reaction occurs as described below (Scheme–1).

$$R{-}CH_2{-}Cl + AlCl_3 \longrightarrow R\overset{+}{C}H_2 \cdots [AlCl_4]^- \longrightarrow R\overset{+}{C}H_2 + Al\overset{-}{C}l_4$$

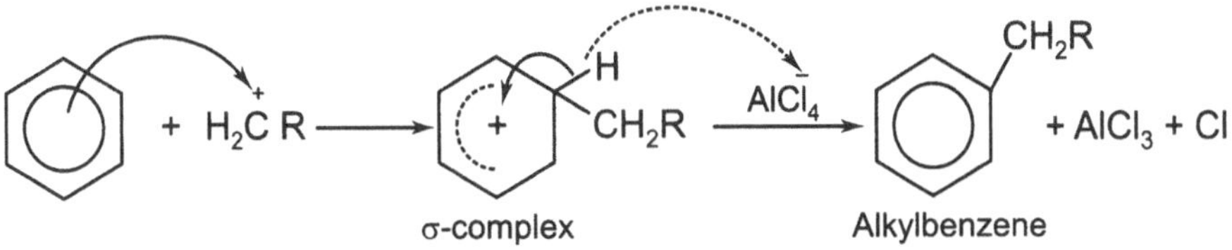

(Scheme–2)

The alkyl halides can be 1°, 2° or 3°. As seen, carbocations are generated as intermediates. The carbocations can also be generated from alkene or alcohols and so both alcohols and olefins can be used in place of alkyl halids (Scheme–3).

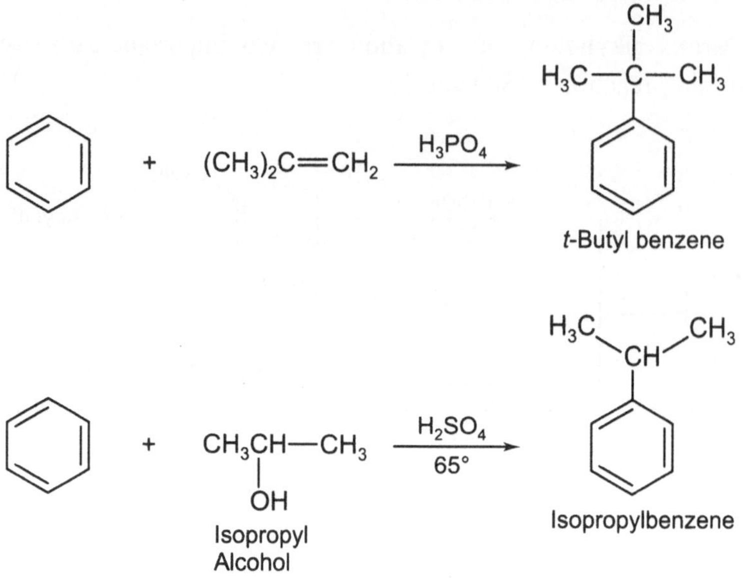

(Scheme–3)

In Friedel-Crafts alkylation, there can be isomerisation, depending on the structure of the alkyl halide. Two examples are given below (Scheme–4).

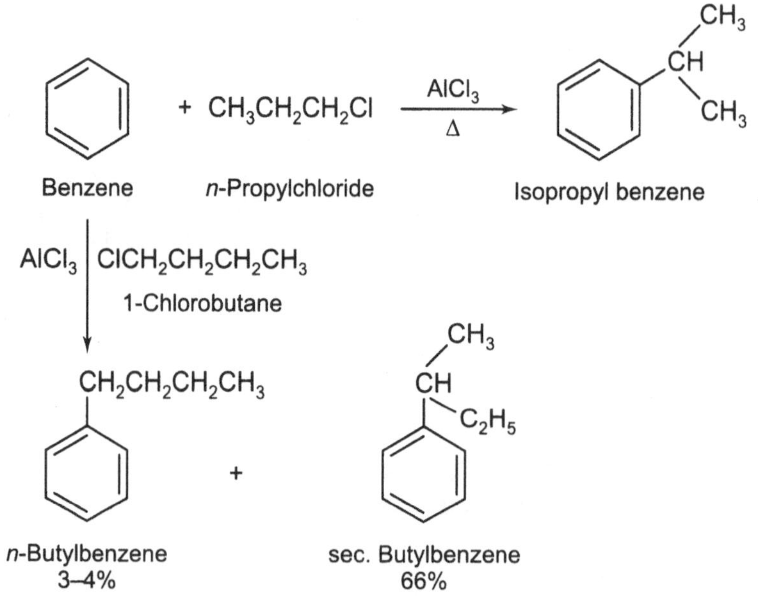

(Scheme–4)

2.1.17.2 Friedel-Crafts Alkylation in Water

Friedel-Crafts reaction of aromatic compounds with methyl trifluoropyruvate in water yielded[2] various α-hydroxy esters (Scheme–5).

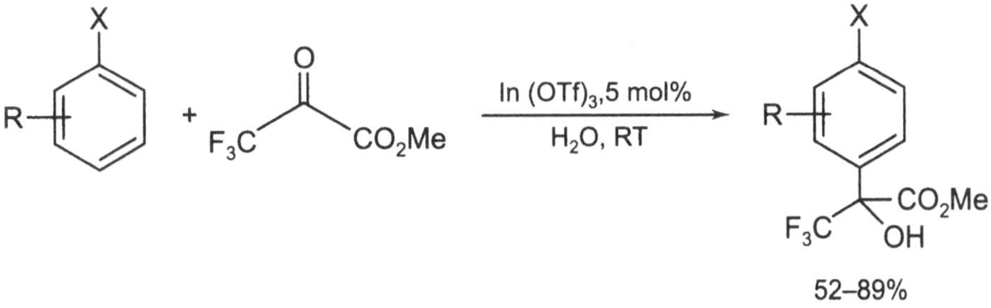

(Scheme–5)

2.1.17.3 Friedel-Crafts Alkylation in Supercritical Carbon Dioxide

Friedel-Crafts alkylation of aromatic compounds has been performed in supercritical fluids (SCF) media[3]. Thus, alkylation of mesitylene in SC-propane (T_C = 91.9°, P_C = 46.0 bar), which acted both as solvent and alkylating agent, yielded a mixture of mono-, di- and tri-alkylated products in 25, 6 and minor percentage, respectively. However, use of SC — CO_2 and propan-2-ol as the alkylating agent gave only the monoalkylated product with conversion of 42 per cent (Scheme–6).

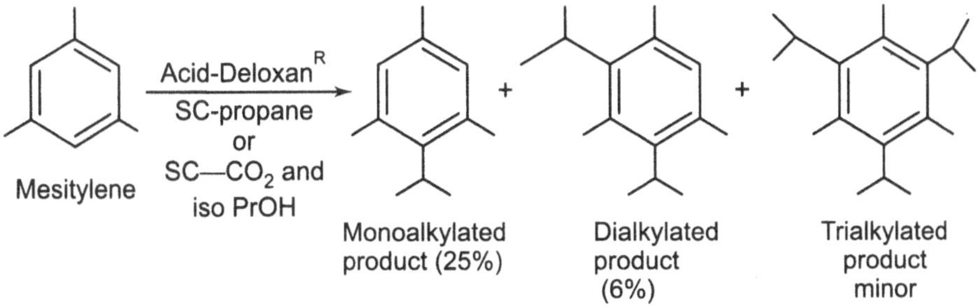

(Scheme–6)

2.1.17.4 Friedel-Crafts Alkylation in Ionic Liquids

Friedel-Crafts alkylation of aromatic compounds with alkenes using $Sc(OTf)_3$-ionic liquid system gave the benefit of simple procedure, easy recovery and reuse of catalyst. This contributed to the development of environmentally benign and waste free processes[4].

The use of ionic liquid, [e$_{min}$] Cl — AlCl$_3$ in place of solid AlCl$_3$ in Friedel-Crafts alkylation enhances the reaction rates and selectivity[5-10], and it also acts as a solvent for the reaction.

2.1.17.5 Friedel-Crafts Alkylation using Polymer Supported Aluminium Chloride

Use of polymer supported aluminium chloride (prepared[11] by the reaction of polystyrene with aluminium chloride, (P)—〈benzene〉—AlCl$_3$) gives much better yields in the Friedel-Crafts alkylation of aromatic compounds. This also constitutes an environmentally benign and waste free process.

2.1.17.6 Friedel-Crafts Acylation

Friedel-Crafts acylation of aromatic compounds is affected by reaction with acylhalides in presence of anhydrous aluminium chloride. It is a general method for the synthesis of ketones and involves the following steps (Scheme–7).

(Scheme–7)

2.1.17.7 Friedel-Craft Acylation in Ionic Liquids

Acetylation of naphthalene in chloroaluminiate ionic liquid gives[12] 89 per cent of 1-acetyl naphthalene (the thermodynamically unfavoured 1-isomer is obtained as the major product) (Scheme–8).

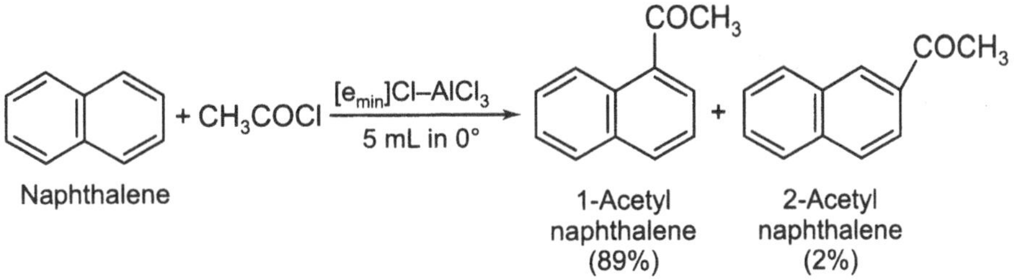

(Scheme–8)

In a similar way acetylation of anthracene, pyrene and phenanthrene could be affected.

The chloroaluminate (III) ionic liquids are powerful lewis acids and are prepared[13] by mixing the appropriate organic halide salts with aluminium (III) chloride, the two solids melt on mixing to form ionic liquid.

Polymer supported aluminium chloride, (P)—⟨benzene⟩—$AlCl_3$, can also be used with advantage for the Friedel-Crafts acylation as in the case of Friedel crafts alkylation (discussed earlier).

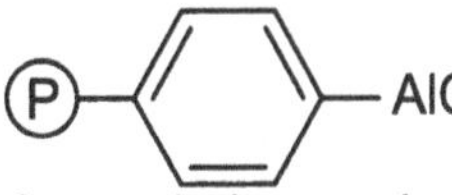

2.1.17.8 Friedel-Crafts Acylation using Ultrasound

The Friedel-Crafts acylation of aromatics is facilated[14] by ultrasound (Scheme–9).

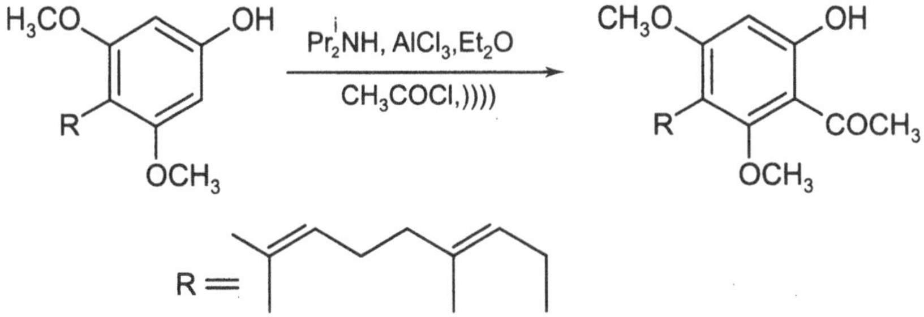

(Scheme–9)

References

1. C. Friedel and J.M. Crafts, Compt. Rend., 1877, **84**, 1392.
2. R. Ding, H. B. Zhang, Y.J. Chen., L. Liu, D. Wang and C. –J. Li, Synlett., 2004, 555.

3. M. Hitzler, F.R. Smail, S.K. Ross and M. Poliakoff, J. Chem. Soc. Chem. Commun., 1998, 359.
4. C.E. Song, W.H. Shim, E.J. Roh and J.H. Chori, Chem. Commun., 2000, 1695.
5. J.A. Boon, J.A. Levisky, J.L. Pflug and J.S. Wilkes, J. Org. Chem., 1986, **51**, 480.
6. A Stark, B.L. McClean and R.D. Singer, J. Am. Chem. Soc., Dalton Trans. 1999, 63.
7. C.W. Lee, Tetrahedron Lett., 1999, **40**, 2461.
8. E. Ota, J. Electrochem, Soc., 1987, **134**, 512.
9. S. Kobayashi, J. Synth., Org. Chem., Jpn, 1995, **53**, 370.
10. C.J. Adams, M.J. Earle, G. Roberts and K.R. Seddon, Chem. Commun., 1998, 2097.
11. D.C.Neckers, D.A. Kooistra and G.W. Green, J. Am. Chem. Soc., 1972, 9284.
12. C.J. Adams, M.J. Earle, G. Roberts and K.R. Seddon, Chem. Commun., 1998, 2097.
13. M.J. Earle and K.K. Seddon, Pure Appl. Chem., 2000, **72**, 1391 and the references cited therein.

2.1.18 Fries Rearrangement

It invovles[1] rearrangement of phenolic esters to o- and or p-phenolic ketones on heating with aluminium chloride or other acid catalysts (Scheme–1).

| Phenylacetate | o-Hydroxy acetophenone | p-Hydroxy acetophenone |

(Scheme–1)

In place of aluminium chloride, it is advantageous to use polymer supported aluminium chloride [(P)—⟨benzene ring⟩— $AlCl_3$), which can be recycled] or the chloroaluminate ionic liquid, [e_{min}] Cl — $AlCl_3$, as in the case of Friedel crafts alkylation and acylation.

There is considerable rate enhancement[2] of Fries migration by conventional microwave ovens (Scheme–2).

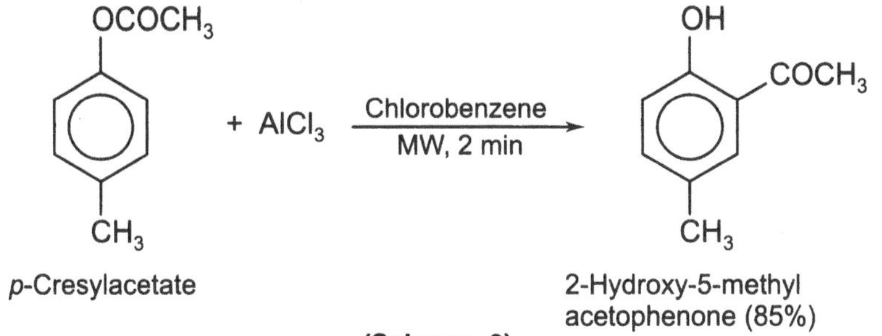

p-Cresylacetate   2-Hydroxy-5-methyl
                              acetophenone (85%)

(Scheme–2)

References

1. K. Fries and G. Fink, Ber., 1908, **41**, 4271; K. Fries, and W. Pfaffendorf, Ber. 1910, **43**, 212.

2. D.L. Clive, Chem. Commun., 1970, 1014.

2.1.19 Hofmann-Martius Rearrangement

The rearrangement[1] of N-alkylarylamine hydrochloride salts on heating at 200–300° gives ortho and para alkyl substituted amine (Scheme–1).

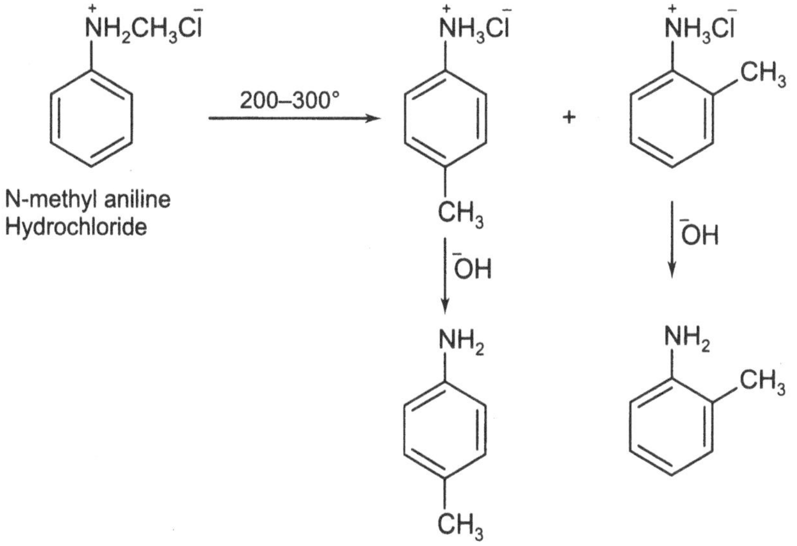

N-methyl aniline
Hydrochloride

(Scheme–1)

In case para position is occupied, the ortho isomer is formed.

The above rearrangement is facilitated by heating in a microwave oven.

Reference

1. A.W. Hofmann and C.A. Martius, 1871, **4**, 742; A.W. Hofmann, Ber, 1972, **5**, 720.

2.1.20 Houben-Hoesch Reaction

The reaction of activated phenols with nitriles in presence of hydrogen chloride gas and zinc chloride gives ketones. The reaction is known as Houben-Hoesch reaction[1,2] (Scheme–1).

(Scheme–1)

See also Nencki Reaction (2.1.28).

References

1. K. Hoesch, Ber, 1915, **48**, 1122; J. Houben, Ber., 1926, **50**, 2878; R. Roger, Chem., Rev., 1961, **61**, 184.

2. J. Trueare, J. Org. Chem., 1963, **28**, 3206.

2.1.21 Kiliani-Fischer Synthesis

It is the most useful method for extending the carbon chain in sugars. The method consists[1] in addition of hydrocyanic acid (HCN) to a carbonyl group of aldoses or ketoses to form cyanohydrins. Subsequent steps are shown as follows (Scheme–1).

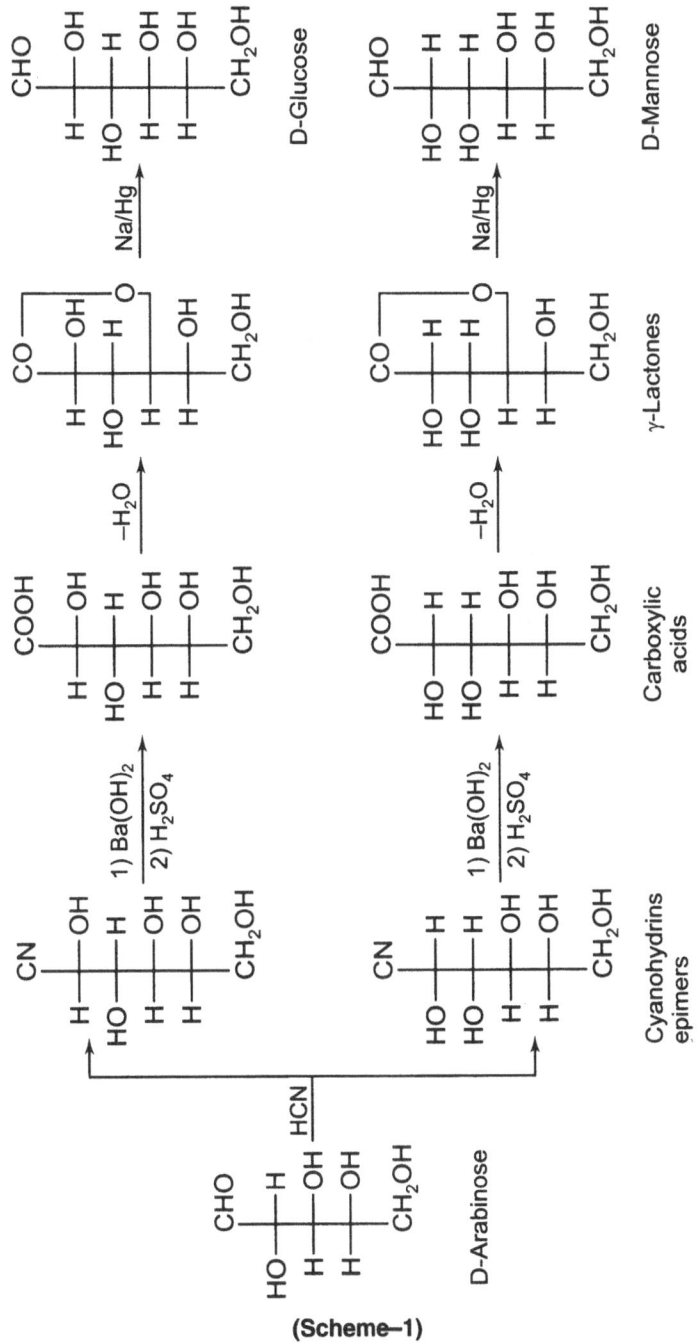

(Scheme–1)

Reference

1. H. Kiliani, Ber., 1885, **18**, 3066; E. Fischer, Ber., 1889, **22**, 2204; C.S. Hudson, Advances in Carbohydrate Chemistry, 1945, **1**, 2; D.T. Moury, Chem. Rev., 1948, **42**, 234.

2.1.22 Kolbe Electrolytic Reaction

It involves[1,2] electrolysis of sodium or potassium salt of a carboxylic acid (Scheme–1).

$$R\text{—}COO^- \xrightarrow{\text{electrolyte}} R\text{—}R + 2CO_2$$

$$2RCH_2COOR \xrightarrow{\text{electrolyte}} RCH_2CH_2CH_2R + 2CO_2$$

(Scheme–1)

The mechanism involves free radical formation (Scheme–2).

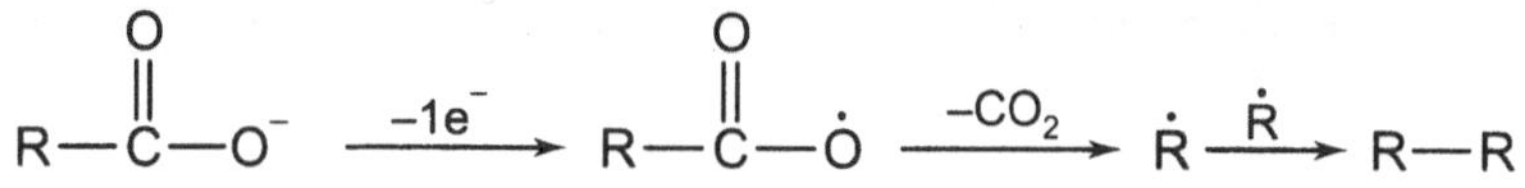

(Scheme–2)

References

1. H. Kolbe, Ann; 1948, **69**, 257; B.C.L. Weedon, Quart. Rev., 1952, **6**, 380.
2. M. Finkelstein, J. Org. Chem., 1969, **25**, 156.

2.1.23 Kolbe-Schmitt Reaction

It involves[1] the reaction of sodium salt of phenols with carbondioxide (Scheme–1).

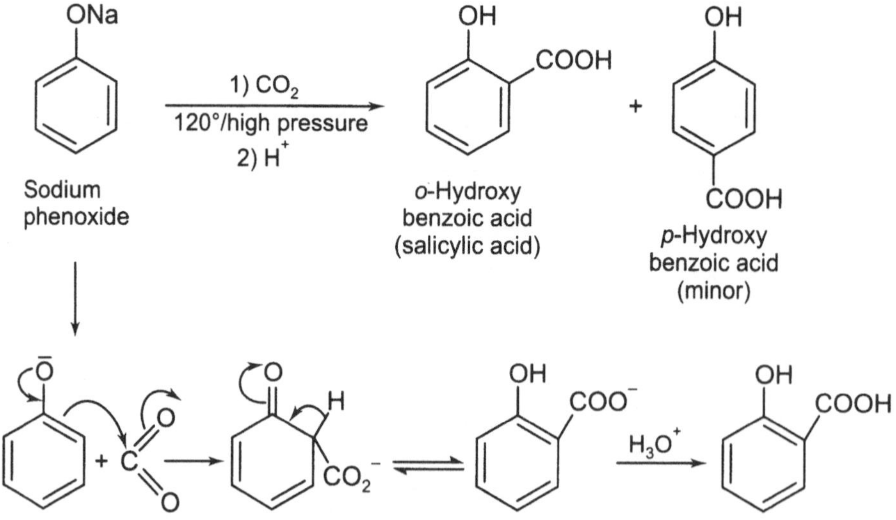
ONa
1) CO_2
120°/high pressure
2) H^+
Sodium phenoxide
OH
COOH
o-Hydroxy benzoic acid (salicylic acid)
+
OH
COOH
p-Hydroxy benzoic acid (minor)
O
C
O
H
CO_2^-
OH
COO^-
H_3O^+
OH
COOH

(Scheme–1)

This reaction is used industrially for the preparation of salicylic acid. The reaction is more facile with *di* or trihydroxy phenols (Scheme–2).

R = R' = H (Resorcinol)
R = OH; R' = H (Pyrogallol)
R = H; R' = OH (Phloroglucinol)

Resorcylic acid R = R' = H
Pyrogallol carboxylic acid
R = OH; R' = H
(2,3,4-Trihydroxybenzoic acid)
Phloroglucinol carboxylic acid
R = H; R' = OH
(2,4,6-Trihydroxybenzoic acid)

(Scheme–2)

Direct carboxylation of phenols can be accomplished by reaction in supercritical carbondioxide[2,3].

References

1. H. Kolbe, Ann., 1860, **113**, 125; R. Schmidt. J. Prak. Chem., 1985, **31**, 397; A.S. Lindsey and H. Leskey, Chem., Rev., 1957, **57**, 583.
2. A.R. Renslo, R.D. Weinstein, J.W. Tester and R.L. Danheiser, J. Org. Chem., 1997, **62**, 4530.
3. .R.D. Weinstein, A.R. Renslo, R.L. Danheisen, J.G. Harris and J.W. Tester, J. Phys. Chem., 1966, **100**, 12337.

2.1.24 Hydroboration

Hydroboration of olefins and acetylenes give organoboranes, which can be used to make carbon-carbon bonds[1]. Many of the reactions involve the migration of a group from boron to carbon. Normally these reactions give high yields under mild conditions.

Hydroboration is considerably enhanced by ultrasound particularly in heterogeneous system. Thus, tricyclohexyl borane is obtained in 1 hr. by the reaction of cyclohexene with $BH_3 . SMe_2$ (traditionally 24 hrs. are required at 25° for this reaction to be completed) (Scheme–1).

Cyclohexene

BH_3SMe_2
THF, 1 hr.,))))
(50Hz, 150W)

Tricyclohexylborane

(Scheme–1)

The above procedure involving sonication provides an interesting alternative and provides excellent yields in short periods. Even chiral reagents can be obtained conveniently from α-pinene and 9-borobicyclononane (9-BBN) (Scheme–2).

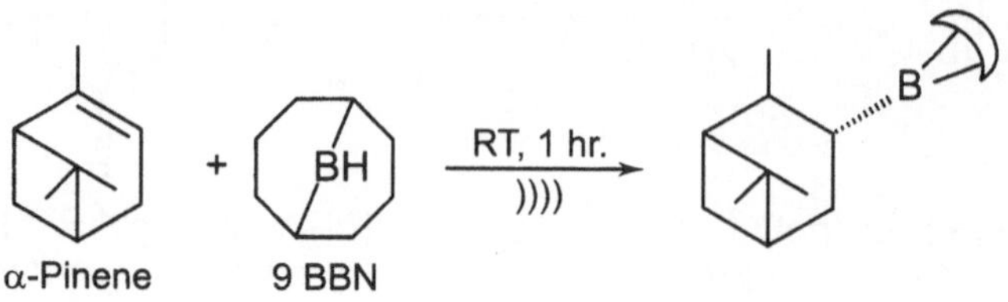

α-Pinene   9 BBN

(Scheme–2)

Hydroborating agents, frequently used for hydroboration are disiamylborane[1], thexylborane[2] and 9-borabicyclo [3.3.1] nonane[3]. These are obtained[1,3] as given below (Scheme–3).

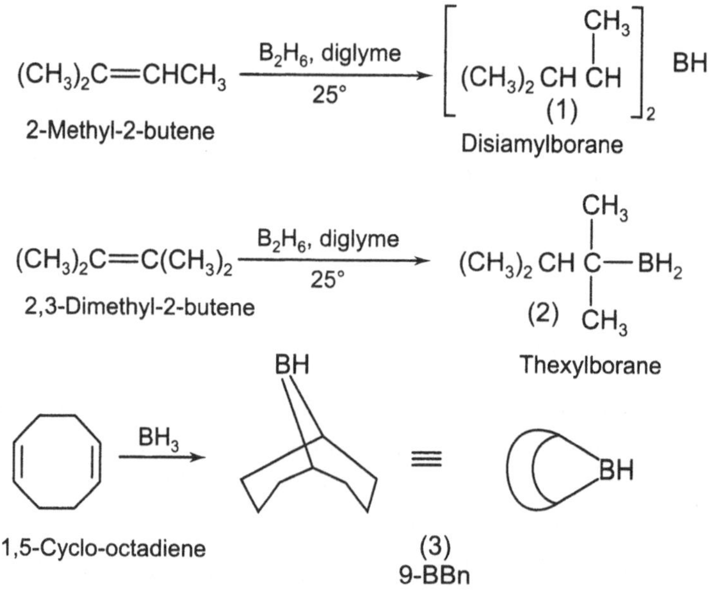

(Scheme–3)

Two important reactions of organo boranes are oxidation and protonolysis. In oxidation, the organoborane is oxidised with H_2O_2 in presence of alkali to give an alcohol. The net addition of hydrogen is as per Markownikoff's rule. An example is given as follows (Scheme–4).

(Scheme–4)

Thus, the addition of borane to alkene follows the anti Markownikoff' rule.

In protonolysis, the organoborane is treated with acid. The product obtained is an alkane. An example is given below (Scheme–5).

(Scheme–5)

Besides oxidation and protonolysis, the organoboranes are particularly important for carbon-carbon bond formation. This is achieved by carbonylation of organoboranes[4] followed by subsequent workup. Under appropriate conditions, the carbonylation of organoboranes can be used to give primary, secondary and tertiary alcohols and open chain, cyclic and polycyclic ketones[5].

Thus, if an organoborane is heated with carbon monoxide to 100–125°, all the three groups migrate from boron to carbon; the formed product on oxidation with hydrogen peroxide and alkali gives a tertiary alcohol. However, in presence of water, the reaction is stopped after migration of two groups from boron to carbon. Oxidation of the reaction mixture at this stage gives a ketone[6]. In case, carbonylation of organoborane is carried out in presence of sodium or lithium borohydride, only one group migrates from boron to carbon. In fact NaBH$_4$ reduces the product of first migration. Subsequent hydrolysis gives a primary alcohol. All these are transformations are represented as follows (Scheme–6).

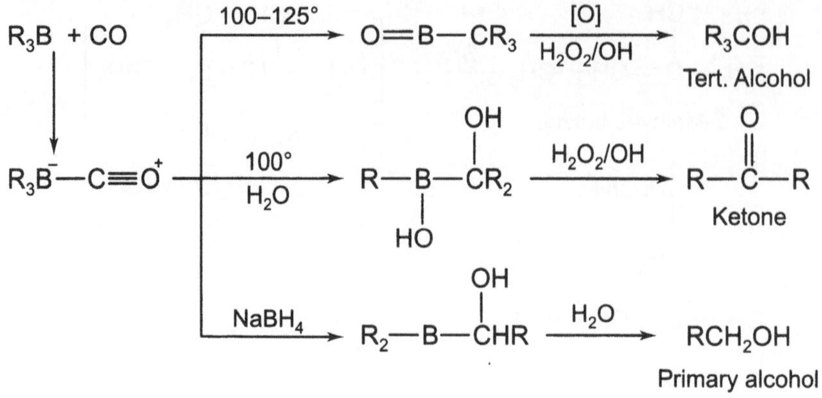

(Scheme–6)

The above procedure can be extended for the synthesis of unsymmetrical ketones by using the appropriate mixed organoborane, which is prepared from thexylborane or thexylchoroborane. A typical example is given below (Scheme–7).

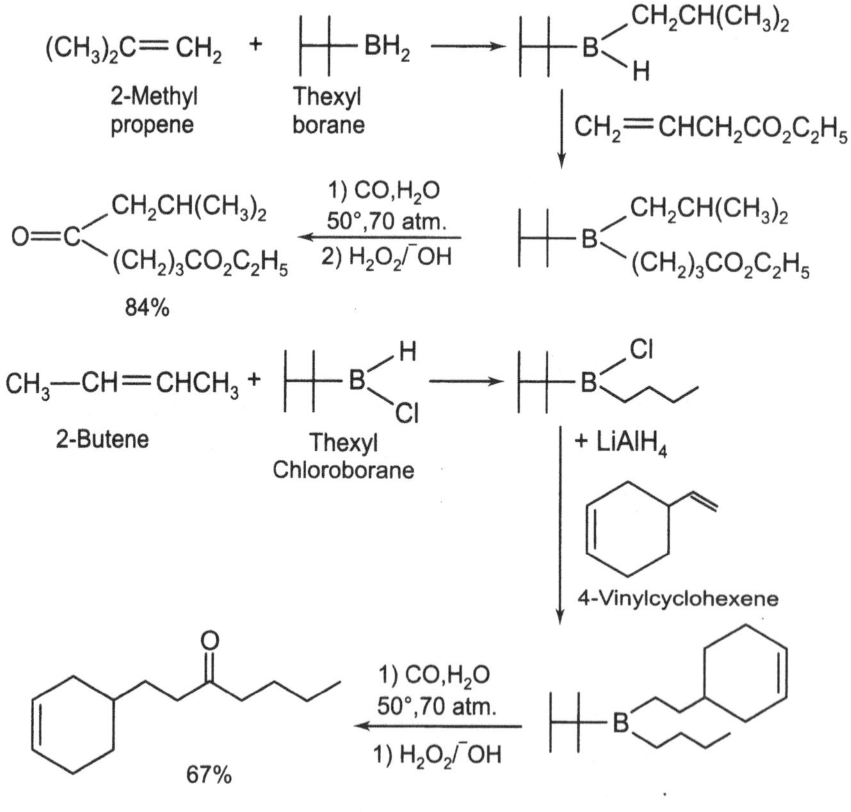

(Scheme–7)

In an alternative procedure, the organoborane is reacted with cyanide ion and trifluoroacetic anhydride. In this reaction, the borane initially forms an adduct with cyanide ion. Subsequent steps are shown below (Scheme–8).

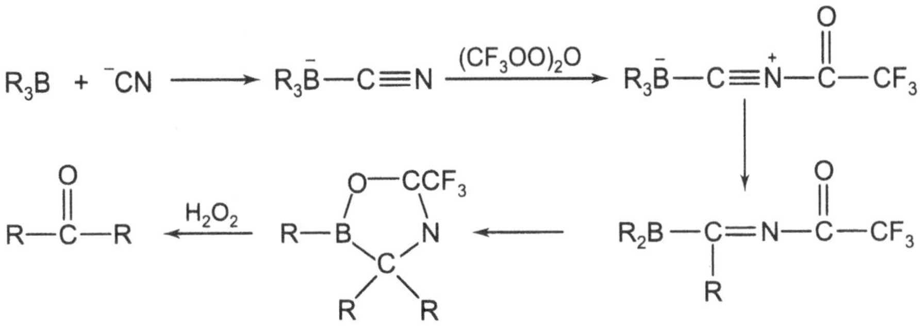

$$R_3B \ + \ ^-CN \longrightarrow R_3\bar{B}-C{\equiv}N \xrightarrow{(CF_3OO)_2O} R_3\bar{B}-C{\equiv}\overset{+}{N}-\overset{\overset{O}{\|}}{C}-CF_3$$

$$R-\overset{\overset{O}{\|}}{C}-R \xleftarrow{H_2O_2} R-\underset{\underset{R \quad R}{C}}{B}\underset{}{\overset{O-CCF_3}{\diagdown}}N \longleftarrow R_2B-\underset{R}{\overset{\|}{C}}{=}N-\overset{\overset{O}{\|}}{C}-CF_3$$

(Scheme–8)

The above sequence of reactions are highly stereoselective, the original boron-carbon bond is replaced by a carbon-carbon bond with retention of configuration, for example, 1-methylcyclopentene is converted into *trans*-2-methylcyclopentylmethanol (Scheme–9).

1-Methylcyclopentene   trans-2-Methylcyclopentylmethanol

(Scheme–9)

It is also possible to affect enantioselective synthesis by hydroboration of alkenes with enantiomerically pure hydroborating agent, monoisocamphenyl borane (IpcBH$_2$)[8]. The hydroboration of alkene establishes a new stereo centre. A third alkyl group can be introduced by a second step of hydroboration. The formed trialkyl borane on treatment with acetaldehyde releases the original α-pinene. Final application of carbonylation gives a chiral ketone[9]. Various steps involved are given ahead (Scheme–10).

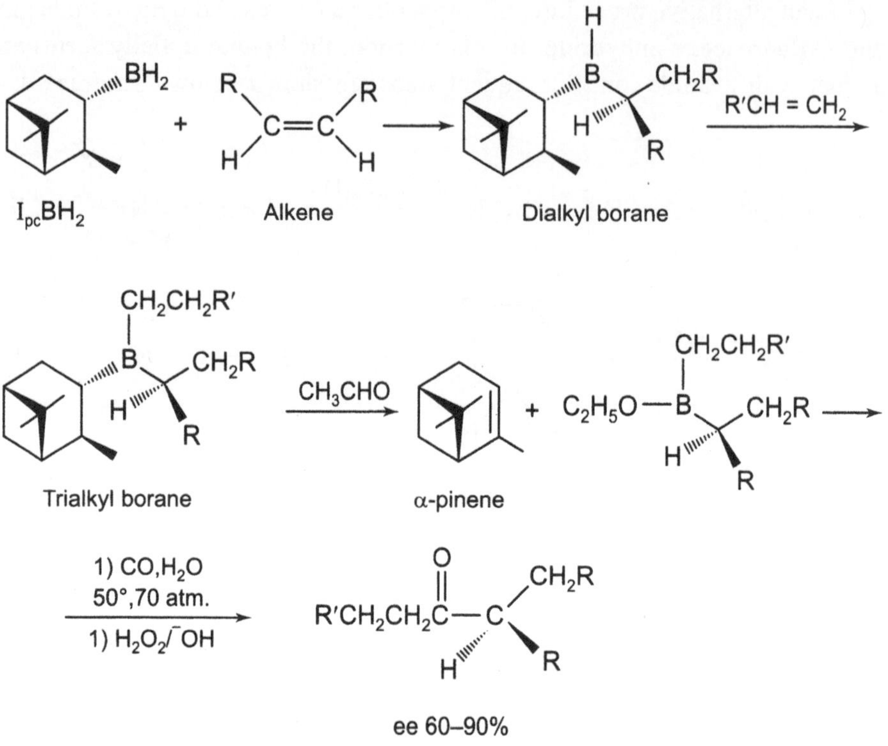

(Scheme–10)

A useful method[10] for the formation of carbon-carbon bond involves the treatment of organoboranes with basic AgNO₃, resulting in the coupling of the alkyl groups (Scheme–11).

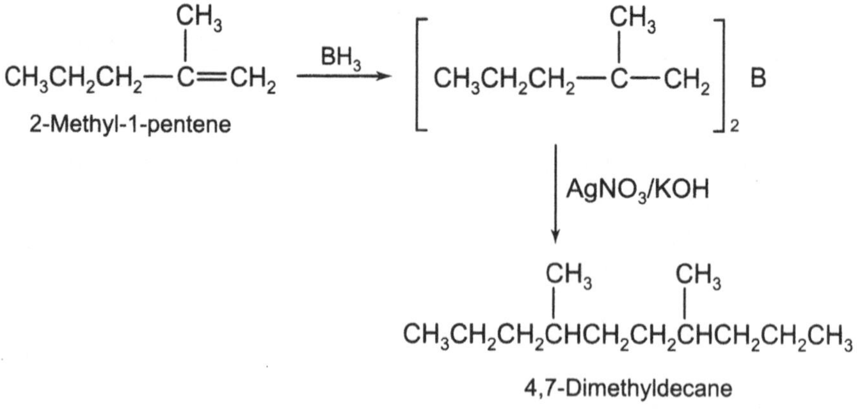

(Scheme–11)

References

1. H.C. Brown, K.W. Kramer, A.B. Levy and M.M. Mialand, Organic Synthesis in Boranes, 1975, New York, Wiley; R. Criegee, Agnew. Chem. Internal edn., 1975, **14**, 745.

2. H.C. Brown and U.S. Racherla, Tetrahedron Lett., 1985, **26**, 2187.

3. H.C. Brown, Organic Synthesis Today and Tomorrow, ed. B.M. Trost and C.R. Hutchinson, Oxford, Pergman, 1981, P. 121; J.A. Soderquist and H.C. Brown, J. Org. Chem., 1981, **44**, 4599.

4. For a review on carbon-carbon bond forming reactions of organoboranes see E. Negishi and M. Idacavage, Org. React., 1985, **33**, 1.

5. H.C. Brown, M.M. Midland and A.B.Levy, J. Am. Chem. Soc., 1972, **94**, 3662.

6. H.C. Brown and M.W. Rathke, J. Am. Chem. Soc., 1967, **89**, 2738.

7. M.W. Rathke and H.C. Brown, J. Am. Chem. Soc., 1967, **89**, 2740.

8. H.C. Brown, P.K. Jadhav and A.K. Mandal, J. Org. Chem., 1982, **47**, 5074.

9. H.C. Brown, P.K. Jadhav and M.C. Desai, Tetrahedron, 1984, **40**, 1325.

10. H.C. Brown and C.H. Snyder, J.Am. Chem. Soc., 1961, **83**, 1002.

2.1.25 Malonic Ester Synthesis

Malonic ester on alkylation with an alkylhalide in presence of sodium ethoxide gives alkyl malonic ester. Its hydrolysis and acidification gives substituted malonic acid, which on heating looses carbon dioxide to give substituted acetic acid (Scheme–1).

$$CH_2(CO_2Et)_2 \xrightarrow[\text{2) RX}]{\text{1) } C_2H_5ONa} R\,CH(CO_2Et)_2 \xrightarrow[\text{2) H}^+]{\text{1) hydrolysis}}$$

Malonic ester   Alkyl malonic ester

$$\longrightarrow RCH(CO_2H)_2 \xrightarrow[-CO_2]{\Delta} R\,CH_2CO_2H$$

Alkylmalonic acid   Alkylsubstituted acetic acid

(Scheme–1)

As already stated (see alkylations, section 2.1.5.1), alkylation of the reactive methylene group can be affected by heating with alkyl halide in presence of potassium carbonate on heating in microwave oven for 3 min.

Malonic ester can also be used for the synthesis of disubstituted acetic acids, dicarboxylic acids and cyclopropane carboxylic acid[1–3] (Scheme–2).

$$CH_2(CO_2Et)_2 \xrightarrow[\text{2) RX}]{\text{1) NaOEt}} R\,CH(CO_2Et)_2 \xrightarrow[\text{2) R'X}]{\text{1) NaOEt}} \underset{R'}{\overset{R}{\diagdown}} C \overset{CO_2Et}{\underset{CO_2Et}{\diagup}}$$

$$\xrightarrow[\text{hydrolysis}]{\text{HCl}} \underset{R'}{\overset{R}{\diagdown}} C \overset{CO_2H}{\underset{CO_2H}{\diagup}} \xrightarrow[-CO_2]{\Delta} \underset{R'}{\overset{R}{\diagdown}} CH{-}COOH$$

Disubstituted acetic acids

$$NaCH(CO_2Et)_2 \; + \; Cl\,CH_2CO_2Et \longrightarrow (EtO_2C)_2\,CHCH_2CO_2Et$$

$$\xrightarrow[\substack{\text{2) H}^+/\text{H}_2\text{O} \\ \text{3) }\Delta,-CO_2}]{\text{1) KOH/H}_2\text{O}} HOOC(CH_2)_2COOH$$

Succinic acid
Dicarboxylic acid

$$\underset{CH_2Br}{\overset{CH_2Br}{|}} + \overset{+}{Na}\overset{-}{C}H(CO_2Et)_2 \longrightarrow \underset{CH_2Br}{\overset{CH_2{-}CH(CO_2Et)_2}{|}} \xrightarrow{\text{NaOEt}}$$

$$\longrightarrow \underset{CH_2Br}{\overset{CH_2{-}\overset{-}{C}H(CO_2Et)_2}{|}} \longrightarrow \underset{}{\triangleright\!\!\triangleleft}\overset{CO_2Et}{\underset{CO_2Et}{}} \xrightarrow[\substack{\text{2) HCl} \\ \text{3) }\Delta,\,-CO_2}]{\text{1) KOH/H}_2\text{O}} \triangleright\!\!-COOH$$

Cyclopropane
carboxylic acid

(Scheme–2)

References

1. H.O. House, Modern Synthetic Reactions, W.A. Benjamin, 1972, 510–518; 756–761.
2. R.G. Riley and R.M. Silverstein, Tetrahedron, 1974, **30**, 1171.
3. J.E. McMurry and J.M. Musser, J. Org. Chem., 1975, **40**, 2556.

2.1.26 Michael Addition

It is one of the most useful carbon-carbon bond forming reaction and has wide applications in organic synthesis. The base catalysed reaction between α, β-unsaturated carbonyl compounds *e.g.*, cinnamaldehyde, $C_6H_5CH = CHCHO$, benzylidene acetone, $C_6H_5CH = CHCOCH_3$, mesityloxide, $(CH_3)_2C = CHCOCH_3$) and a compound with active methylene group (*e.g.*, malonic ester, acetoacetic ester, cyanoacetic esters, nitro paraffins) is known as Michael addition[1]. The base employed is sodium ethoxide or a secondary amine (usually piperidine). An example is given as follows (Scheme–1).

$$H_2C=CH-\overset{\overset{\displaystyle O}{\|}}{C}-CH_3 \;+\; CH_2(CO_2Et)_2 \;\xrightarrow{\text{NaOEt}}\; (EtO_2C)_2CHCH_2CH_2\overset{\overset{\displaystyle O}{\|}}{C}CH_3$$

Methyl vinyl ketone   Diethyl   Adduct
                           malonate

(Scheme–1)

2.1.26.1 Michael Addition under PTC Conditions

The Michael addition of active nitrites to acetylenes can be carried out[2,3] in presence of quaternary ammonium salts (Scheme–2).

$$C_6H_5-\underset{\underset{\displaystyle R}{|}}{CH}-CN \;+\; HC\equiv CR' \;\xrightarrow[\substack{\text{DMSO}\\ \text{solid NaOH}}]{C_6H_5CH_2\overset{+}{N}Et_3\overset{-}{Cl}}\; C_6H_5-\underset{\underset{\displaystyle R}{|}}{\overset{\overset{\displaystyle CN}{|}}{C}}-CH=CHR'$$

R	R'	% yield
Me	H	83
Et	H	80
isoPr	H	82
C5H11	H	88
Et	Ph	94
isoPr	Ph	83
PhCH2	Ph	98

(Scheme–2)

2.1.26.2 Michael Addition in Aqueous Solution

The Michael addition of 2-methyl-cyclopentane-1, 3-dione with methyl vinyl ketone in water gave[4] an adduct using a basic catalyst (pH > 7). The adduct further cyclises to give fused ring systems (Scheme–3).

2-Methyl   Methyl   Adduct
cyclopentane   vinyl
1,3-dione   ketone

(Scheme–3)

Michael addition of 2-methyl cyclohexane-1, 3-dione with methyl vinyl ketone in water gave[5] an adduct, which cyclised to Wieland-Miescher[5] ketone (Scheme–4).

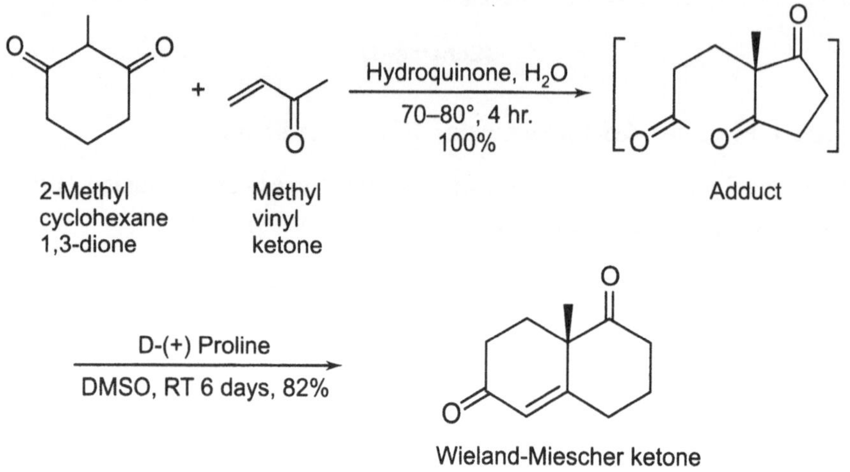

| 2-Methyl cyclohexane 1,3-dione | Methyl vinyl ketone | | Adduct |

Wieland-Miescher ketone

(Scheme–4)

The Michael addition of nitromethane to methyl vinyl ketone in water in the absence of a catalyst gave[6] 4 : 1 mixture of adducts A and B (Scheme–5).

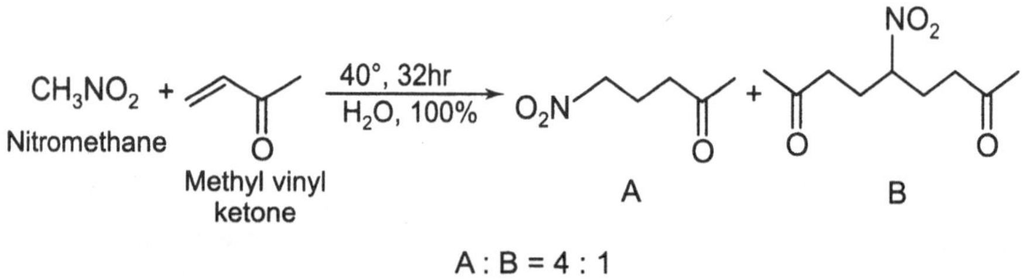

A : B = 4 : 1

(Scheme–5)

Use of methyl alcohol as solvent (in place of H_2O) gave 1 : 1 mixture of adducts A and B. The above reaction does not occur in neat conditions or in solvents like THF, PhMe *etc.*, in the absence of catalyst.

2.1.26.3 Michael Addition in Solid State

2′-Hydroxychalcones undergo a solid state Michael type addition to yield[7] the corresponding flavonones (Scheme–6).

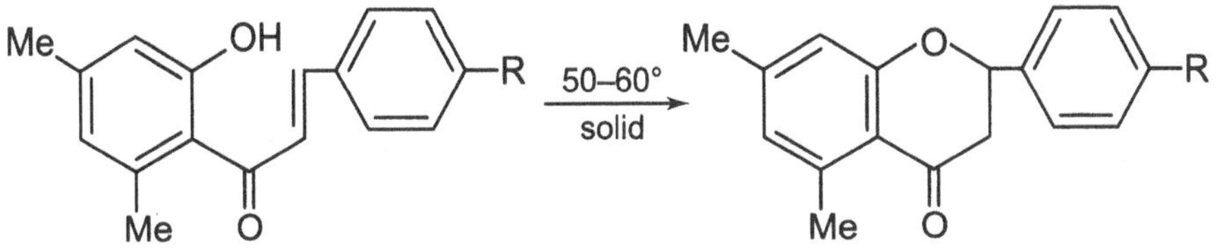

2'-Hydroxy-4',6'-dimethyl Chalcones
R = H, Cl or Br

5,7-Dimethyl flavonone

(Scheme–6)

2.1.26.4 Michael Addition in Presence of PTC

The Michael addition of chalcone to 2-phenylcyclohexanone in presence of PTC gave[8] 2,6-disubstituted cyclohexanone derivative in high distereoselectivity (90% ee) (Scheme–7).

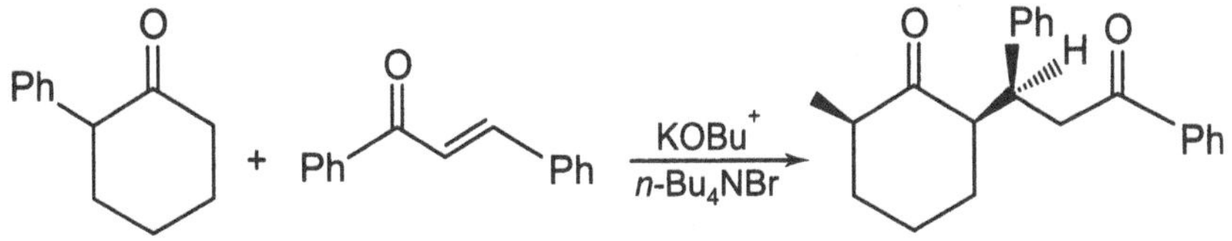

2-Phenyl
Cyclohexanone

Chalcone

99%
(90% ee)

(Scheme–7)

Michael addition of diethyl (acetamido) malonate to chalcone using asymmetric phase transfer catalyst (ephedrinium salt) in presence of KOH in the solid state gave[9] the adduct in 56% yield with ee of 50% (Scheme–8).

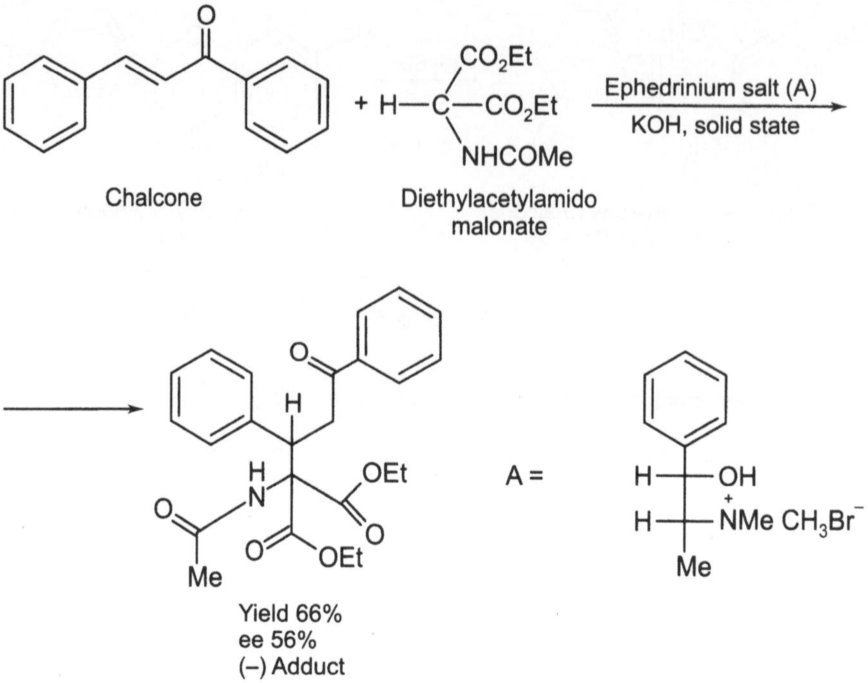

Chalcone   Diethylacetylamido malonate

Yield 66%
ee 56%
(−) Adduct

(Scheme–8)

2.1.26.5 Michael Addition in Ionic Liquids

Michael addition of acetyl acetone to methyl vinyl ketone in presence of catalyst $Ni(acac)_2$ in ionic liquid [b_{min}] [BF_4] yielded[10] the adduct in high selectivity. The ionic liquid could be recycled (Scheme–9).

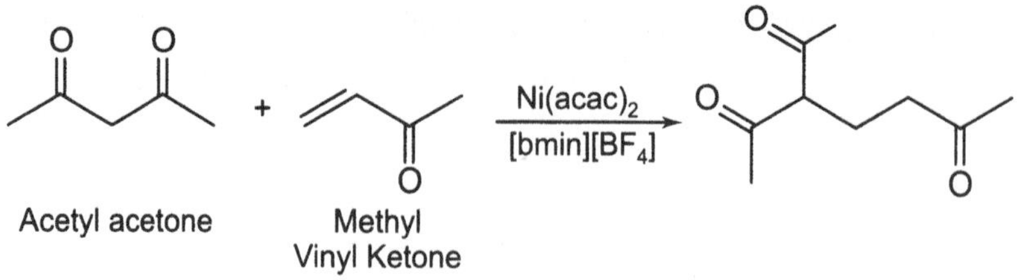

Acetyl acetone   Methyl Vinyl Ketone

(Scheme–9)

References

1. A. Michael, J. Prakt. Chem., 1987(2), **35**, 349; E.D. Bergmann, D. Ginsburg and R. Pappo, Org. Reactions, 1959, **10**, 179.
2. M. Makosza, Tetrahedron Lett., 1966, 5489; Polish Patent, 55113 (1968); CA. 1969, **70**, 106006.

3. M. Maskosza, J. Czyzewski and M. Jawdosiak, Org. Synth., 1976, 55, 99.

4. Z.G. Hajos and D.R. Parrish, J. Org. Chem., 1974, **39**, 1612; U. Elder, G. Sauer and R. Wiechert, Angew. Chem. Int. Edn. Engl., 1971, **10**, 496.

5. N. Harada, T. Sugioka, U. Uda and T. Kuriki, Synthesis, 1990, 53.

6. A Lubineau and J. Auge, Tetrahedron Lett., 1992, **33**, 8073.

7. B. Satish, K. Panneesel-Vam, D. Zacharids and G.R. Desivaju, J. Chem., Soc. Perkin Trans., 1995, **2**, 325.

8. E. Diez-Barra, A. de la Hoz, Merino and P. Sanchez-Verdu, Tetrahedron Lett., 1997, **38**, 2359.

9. A. Loupy, J. Sansoulet, A. Zaparucha and C. Merinne, Tetrahedron Lett., 1989, **30**, 333.

2.1.27 Mukaiyama Reaction in Aqueous Phase

It is a stereoselective aldol condensation[1] of silyl enol ethers of ketones with aldehydes in presence of titanium tetrachloride. As an example, condensation of silyl enol ether of 3-pentanone with 2-methylbutyraldehyde in presence of TiCl$_4$ gave the aldolate product, which on hydrolysis yielded aldol product, Manicone, an alarm pheromone (Scheme–1).

(Scheme–1)

In place of TiCl$_4$, other Lewis acids like tin tetrachloride and borontrifluoride etherate can also be used.

The first water promoted Mukaiyama reaction of silyl enol ethers with aldehydes was reported[2] in 1986 (Scheme–2).

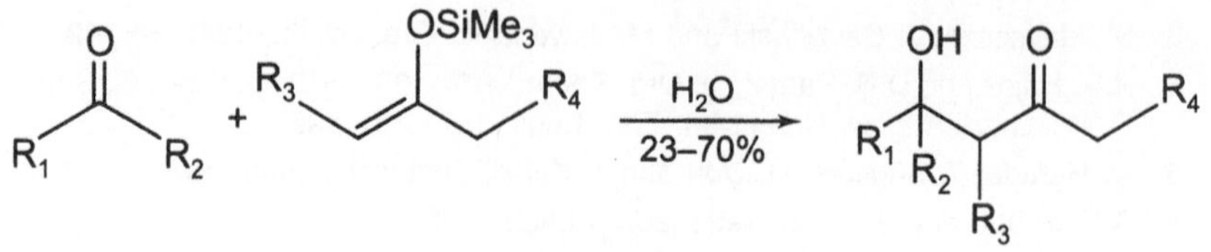

(Scheme–2)

Since water serves as a weak Lewis acid, the above reaction took several days for completion. The addition of stronger Lewis acid (*e.g.*, ytterbium triflate) greatly improved the yield and also the rate of the reaction[3] (Scheme–3).

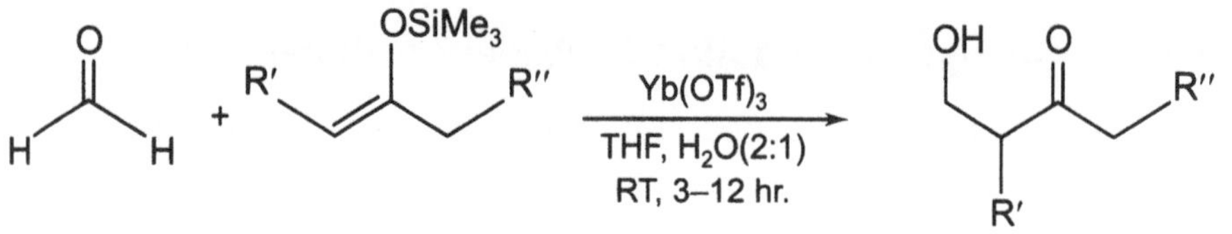

(Scheme–3)

The reaction of trimethyl-silyl ether of cyclohexanone with benzaldehyde occurs[2] in water in presence of $TiCl_4$ in heterogeneous phase at room temperature (Scheme–4).

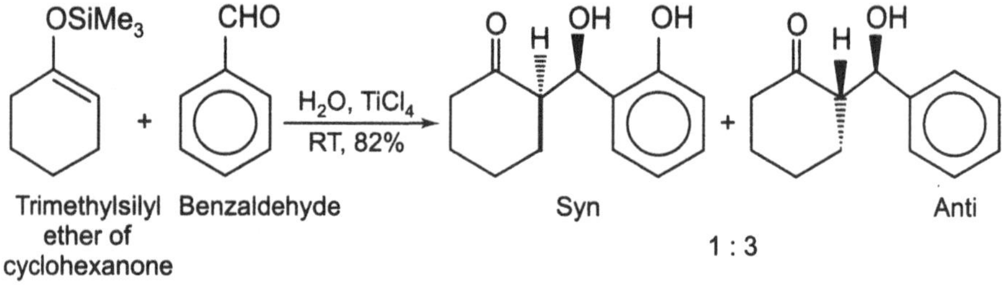

(Scheme–4)

Much better yields are obtained on sonication. The reaction is favoured by an electron withdrawing substituent in the para position of the phenyl ring in benzaldehyde.

References

1. T. Mukaiyama, Chem., Lett., 1982, 353; J. Am. Chem., Soc.; 1973, **95**, 967; Chem., Lett., 1986, 187; T. Mukaiyama, Organic Reactions, 1982, **28**, 187.
2. A. Lubineau, J. Org. Chem., 1986, **51**, 2142; A. Lubineau, E. Meyer, Tetrahedron, 1988, **44**, 6065.
3. S. Kobayashi and I. Hachiya, J. Org. Chem., 1994, **59**, 3590. For a review on lanthanide catalyzed organic reactions in aqueous media, see S. Kobayashi, Synlett., 1994, 589.

2.1.28 Nencki Reaction

Dihydroxy and trihydroxy benzene on heating with acetic acid in presence of anhyd. zinc chloride give[1] the corresponding ketones.

Resorcinol R = H
Pyrogallol R = OH

CH₃COOH
anhyd. ZnCl₂, Δ

β-Resacetophenone, R–H
Gallacetophenone, R = OH

See also Houben-Hoesch reaction (2.1.20).

Reference

1. M. Nencki, N. Sieber, J. Prakt. Chem., 1881, **23**, 147; M. Nencki and W. Schmid, J. Prakt. Chem., 1881, **23**, 546; M. Nencki, J. Prakt. Chem., 1882, **25**, 273.

2.1.29 Organometallic Compounds

The organometallic compounds are very important in synthetic organic chemistry, since they are useful for the formation of carbon-carbon bonds. These are of two types, *viz.*, simple and mixed or complex. In simple organometallic compounds, the metal is linked only to the R group (s), *e.g.*, dimethyl cadmium (Me₂ Cd). These have carbon-metal bond with ionic character. On the other hand, in mixed or complex organometallic compounds, the metal is attached to the R group along with X (halide) group. Examples include organo magnesium, Lithium reagents and organosilanes. These possess carbon metal bond with ionic character. The ionic character of the carbon-metal bond depends on the nature of the metal and decreases in the order.

K > Na > Li > Mg > Al > Zn, Cd > Hg. The present discussion is restricted to organometallic compounds of lithium, magnesium, zinc, copper, and silicon.

2.1.29.1 Organo-Magnesium Reagents

The organomagnesium halides, commonly known as Grignard reagents were discovered by Victor Grignard, who was awarded Nobel Prize in 1912 for his work on the tremendous synthetic potential of Grignard reagents. These have the general formula RMgX and are prepared by the reaction of alkyl halide with magnesium in dry ether (Scheme–1).

$$R-X + Mg \xrightarrow{\text{ether}} R\,MgX$$

Grignard
reagent

(Scheme–1)

For an alkyl halide, the ease of formation of a Grignard reagent is of the order RI > RBr > RCl. In case, the reaction is sluggish, small amount of iodine is added to start the reaction. Different types of organohalogen compounds can be used as starting materials. These include alkyl, aryl and vinyl chlorides, bromides and iodides, although organo chlorides are the least reactive (Scheme–2).

$$R-Br + Mg \xrightarrow{\text{ether}} RMgBr$$
$$R = CH_3,\ C_2H_5,\ C_3H_7 \text{ etc.}$$

$$C_6H_5Br + Mg \xrightarrow{\text{ether}} C_6H_5MgBr$$

(Scheme–2)

Organo halides that also have an acidic proton (alcohol, phenol, carboxylic acid or amine) or a reactive functional group (carbonyl, nitrile, epoxide or nitro group) cannot be used to prepare a Grignard reagent. This problem can, however, be circumvented by protection of the functional group, making a Grignard reagent and final deprotection.

Besides, the method described above, the Grignard reagents can also be prepared from substrates that have acidic hydrogens, *e.g.*, terminal alkynes, cyclopentadine *etc.*, by reacting with a preformed Grignard reagent (Scheme–3).

$$R-C\equiv CH + RMgBr \xrightarrow{\text{ether}} R-C\equiv C-MgBr + RH$$

(Scheme–3)

Grignard reagents can be very conveniently obtained by sonication of a suspension of magnesium in alkyl halide using ether as solvent. Sonication activates the magnesium and finds applications in the synthesis of Grignard reagents[2-4] without the use of activators (Scheme–4).

$$R\!-\!X + Mg \xrightarrow[))))]{\text{ether}} RMg\,X$$
$$(90\%)$$

(Scheme–4)

Following are given some of the applications of Grignard reagents in carbon-carbon bond formations.

(*i*) **Reaction with carbondioxide:** The carbon atom of a Grignard reagent reacts as a nucleophile towards carbon atom of CO_2 giving a halomagnesium salt of a carboxylic acid, which on treatment with water gives a carboxylic acid (Scheme–5).

Phenyl Magnesium chloride   Halomagnesium salt of the carboxylic acid

Carboxylic acid

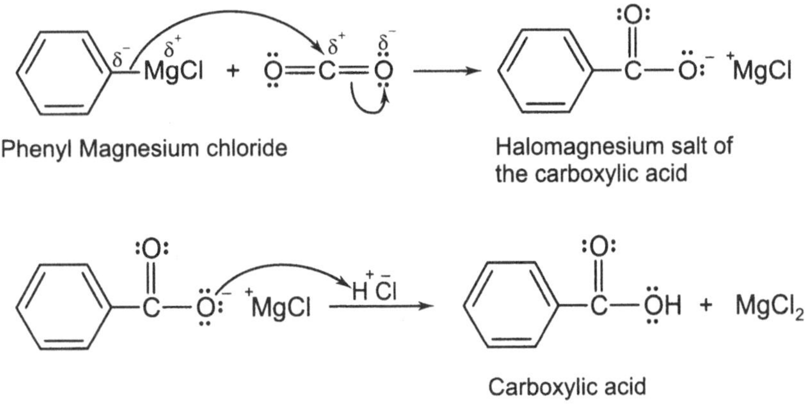

(Scheme–5)

(*ii*) **Reaction with carbonyl compounds:** Depending on the nature of carbonyl compounds, 1°, 2° or 3° alcohols are obtained. As an example, formaldehyde, acetaldehyde and a ketone give 1°, 2° and 3° alcohols, respectively (Scheme–6).

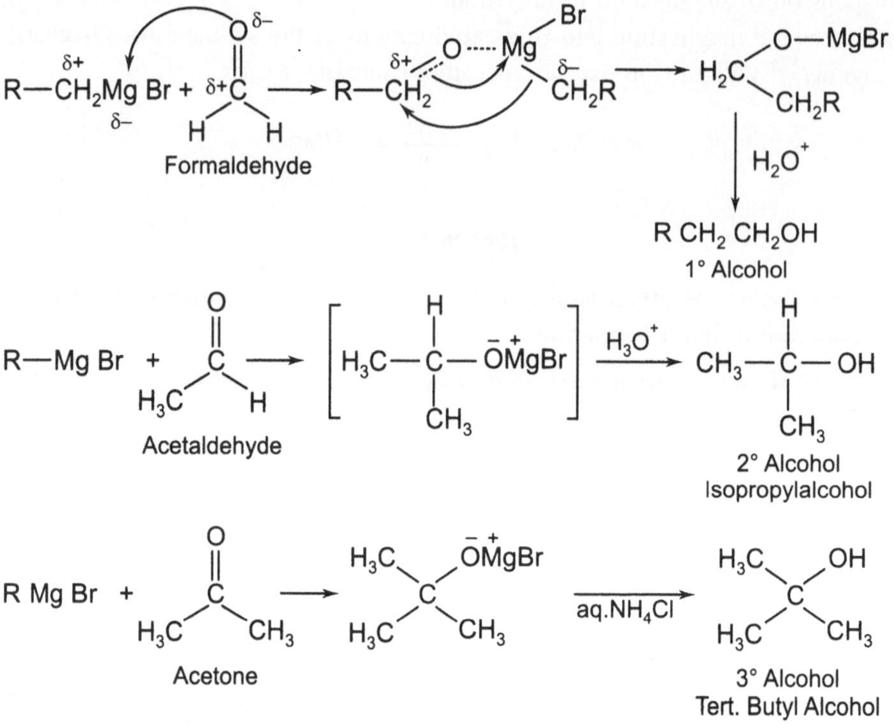

Formaldehyde

R CH$_2$ CH$_2$OH
1° Alcohol

Acetaldehyde

2° Alcohol
Isopropylalcohol

Acetone

3° Alcohol
Tert. Butyl Alcohol

(Scheme–6)

In all the above transformation, the final product contains one more carbon atom.

(*iii*) **Reaction with epoxides:** In this case the product obtained (1° alcohol) contains two more carbon atoms than the starting material (Scheme–7).

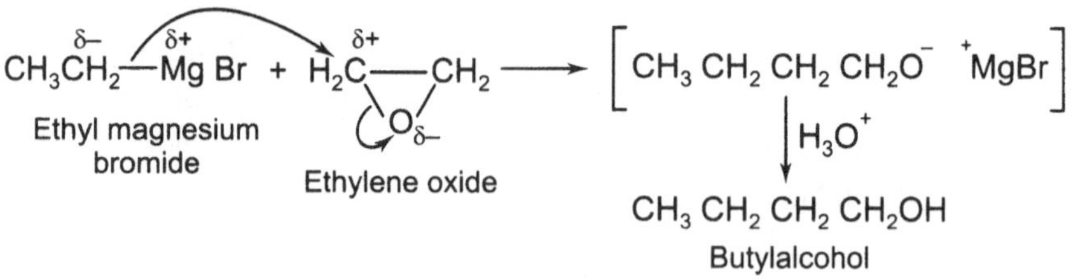

Ethyl magnesium bromide

Ethylene oxide

CH$_3$ CH$_2$ CH$_2$ CH$_2$OH
Butylalcohol

(Scheme–7)

Unsymmetrical epoxides react with Grignard reagent at the less hindered carbon atom of the ring. So, 2° or 3° alcohols are obtained (Scheme–8).

$$CH_3CH_2\overset{\delta-\ \ \delta+}{—}MgBr + \overset{\delta+}{CH_2}—\underset{\underset{\delta-}{O}}{\overset{R}{\underset{|}{C}}}—R' \longrightarrow CH_3\,CH_2\,CH_2—\underset{\underset{O^-}{|}}{\overset{R}{\underset{|}{C}}}—R'$$

Ethyl magnesium
bromide

$$\downarrow H_3O^+$$

$$CH_3\,CH_2\,CH_2—\underset{\underset{OH}{|}}{\overset{R}{\underset{|}{C}}}—R'$$

R = H; R' = CH₃   2° Alcohol

R = R' = CH₃   3° Alcohol

(Scheme–8)

Tertiary alcohols can also be formed by reacting Grignard reagent with esters, acid chlorides and acid anhydrides.

(*iv*) **Reaction with alkyl cyanides:** Reaction of Grignard reagent with alkyl cyanides and subsequent workup give ketones (Scheme–9).

$$R\,MgX + R'C\!\equiv\!N \longrightarrow RR'C\!=\!NMgX \xrightarrow{H_2O} \left[RR'C\!=\!NH\right]$$

$$\downarrow H_3O^+$$

$$RCOR'$$

(Scheme–9)

(*v*) **Reaction with alkyl halides:** Grignard reagents couple with less reactive alkyl halides in presence of transition metal catalyst, cobalt chloride to give the coupled product (Scheme–10).

$$C_6H_5MgBr + C_6H_5CH_2Br \xrightarrow{CoCl_2} C_6H_5CH_2C_6H_5$$

Phenyl magnesium   Benzyl   Diphenyl
bromide   bromide   methane
                                     (87%)

(Scheme–10)

This transition metal catalysed reaction is known as **Kharasch reaction.**

α-Haloketones react with Grignard reagent without a metal catayst to give α-arylketones (Scheme–11).

$$C_6H_5COCH_2Cl \quad + \quad C_6H_5MgBr \quad \xrightarrow[\text{2) Ice/HCl}]{\text{1) Ether, }-30°} \quad C_6H_5CO\ CH_2C_6H_5$$

(Scheme–11)

References

1. V. Grignard, Compt. Rend., 1900, **130**, 1322; D.A. Shirley, Organic Reactions, 1954, 8, 28.
2. J. L. Luche and J.C. Damiano, J. Am. Chem., Soc., 1980, **102**, 7926.
3. J.D. Sprich and G.S. Lewadson, Inorg. Chem., Actd, 1982, **76**, 1241.
4. W. Oppolzer and A. Nakao, Tetrahedron Lett., 1986, **27**, 5471.

2.1.29.2 Organolithium Reagents

The organo lithium reagents, characterised by a C — Li bond are excellent reagents in organic synthesis, particularly in carbon-carbon bond formation. These are generally prepared[1] by the reaction of an alkyl or aryl halide with lithium metal in a hydrocarbon solvent in inert atmosphere (Scheme–1).

$$R—X + 2Li \longrightarrow R—Li \quad + \quad LiX$$

(Scheme–1)

Some commercially available lithium reagents are (Scheme–2).

$$CH_3—Li$$
Methyl lithium

$$CH_3\ CH_2CH_2CH_2—Li$$
Butyl lithium

$$CH_3\ CH_2—\overset{\overset{\displaystyle Li}{|}}{CH}—CH_3$$
Sec. Butyl lithium

$$CH_5—\overset{\overset{\displaystyle CH_3}{|}}{\underset{\underset{\displaystyle CH_3}{|}}{C}}—Li$$
tert. Butyl lithium

(Scheme–2)

Phenyl lithium reagents are easily obtained (Scheme–3).

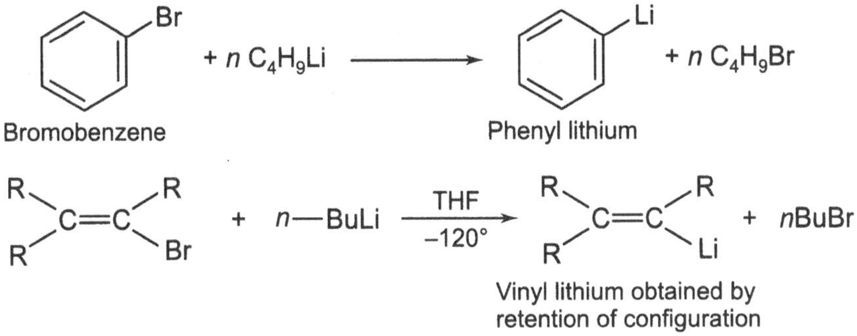

(Scheme–3)

The vinyl lithium reagents are obtained by retention of configuration[2]. The organo lithium reagents are prepared[3,4] in > 90% yield by the reaction of appropriate bromide with lithium wire on sonication (Scheme–4).

$$R\text{—}X \ + \ Li \ \xrightarrow{\ ))))\ } \ R\text{—}Li \ (90\%)$$

(Scheme–4)

Organo lithium reagents are very reactive, powerful nucleophiles and strong bases. These are better than Grignard reagents. Some of the synthetic applications particularly carbon-carbon bond formations are given below.

    (*i*) **Reaction with carbonyl compounds:** Depending on the carbonyl compounds, primary, secondary and tertiary alcohols are obtained (Scheme–5).

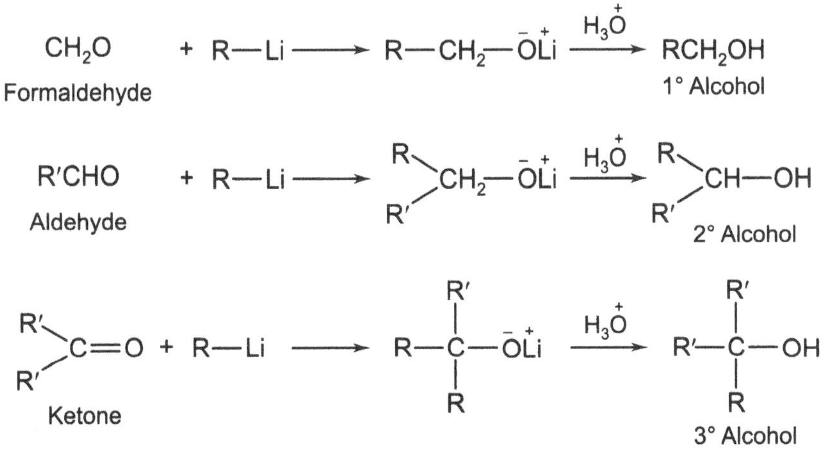

(Scheme–5)

Unlike Grignard reagents, the lithium reagents react with hindered ketones giving tertiary alcohols (Scheme–6).

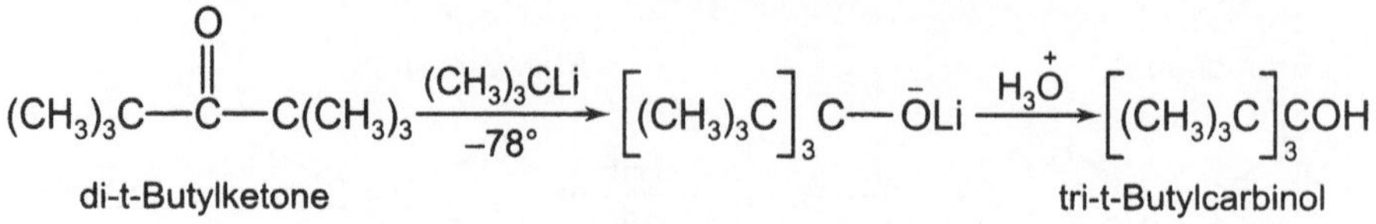

$$(CH_3)_3C-\overset{\overset{\displaystyle O}{\|}}{C}-C(CH_3)_3 \xrightarrow[-78°]{(CH_3)_3CLi} \left[(CH_3)_3C\right]_3 C-\bar{O}Li \xrightarrow{H_3\overset{+}{O}} \left[(CH_3)_3C\right]_3 COH$$

di-t-Butylketone   tri-t-Butylcarbinol

(Scheme–6)

See also Barbier reaction, section 3.1.5.

(*ii*) Reaction with alkyl cyanides (Scheme–7)

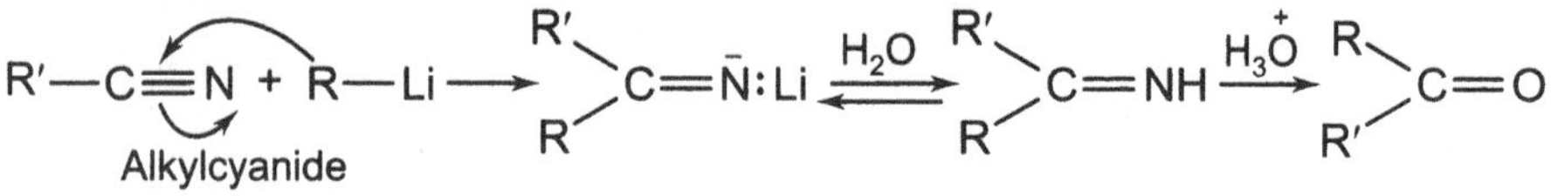

$$R'-C\equiv N + R-Li \longrightarrow \overset{R'}{\underset{R}{\diagdown}}C=\bar{N}:Li \underset{}{\overset{H_2O}{\rightleftharpoons}} \overset{R'}{\underset{R}{\diagdown}}C=NH \xrightarrow{H_3\overset{+}{O}} \overset{R}{\underset{R'}{\diagdown}}C=O$$

Alkylcyanide

(Scheme–8)

(*iii*) Reaction with carboxylic acids (Scheme–9)

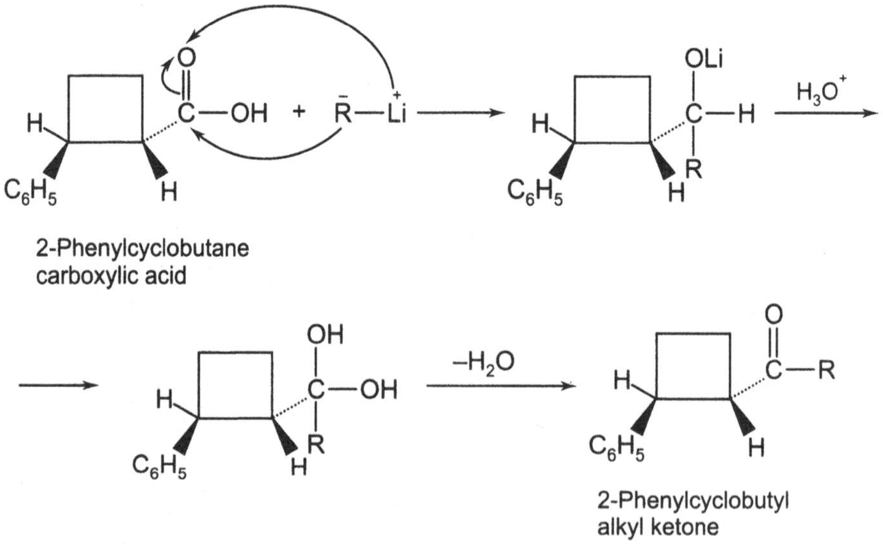

2-Phenylcyclobutane
carboxylic acid

2-Phenylcyclobutyl
alkyl ketone

(Scheme–9)

(*iv*) **Reaction with alkenes** (Scheme–10)

$$CH_2\!\!=\!\!CH_2 + Li\!-\!R \longrightarrow R\!-\!CH_2\!-\!CH_2\!-\!Li \xrightarrow{\ CH_2=CH_2\ }$$

Ethylene   Alkyl lithium

$$\longrightarrow R\!-\!CH_2\!-\!CH_2\!-\!CH_2\!-\!CH_2\!-\!Li \xrightarrow{\ CH_2=CH_2\ }$$

$$\longrightarrow R\!-\!CH_2\!-\!CH_2\!-\!CH_2\!-\!CH_2\!-\!CH_2\!-\!CH_2\!-\!Li$$

(Scheme–10)

The elongation of chain continues till the original alkene (ethylene) lasts.

It is appropriate to state that Grignard reagents have no action on ethylene.

(*v*) **Nucleophilic displacement:** The halogen of the alkyl halides can be substituted with alkyl group of the organo lithium reagent to give hydrocarbons (Scheme–11).

$$R\!-\!X \ + \ R'\!-\!Li \longrightarrow R\!-\!R' \ + \ LiX$$

Alkylhalide   Alkyl
              lithium

(Scheme–11)

This takes place by S_N2 mechanism as in the case of Wurtz reaction (Scheme–12).

$$\bar{R}'\!-\!\overset{+}{Li} \ + \ R\!-\!X \longrightarrow R'\!-\!R \ + \ LiX$$

(Scheme–12)

Allyllithium and benzyllithium reagents can be alkylated by secondary alkyl bromides and a high degree of inversion of configuration is observed[5] (Scheme–13).

Ph CH_2 Li   +   (sec. Butylbromide) $PhCH_2\!-\!\overset{CH_2CH_3}{\underset{H}{\overset{|}{C}}}\!\cdots\!CH_3$

Benzyllithium

58% yield
100% inversion

(Scheme–13)

(*vi*) **Reaction of aromatic compounds with aryllithium** (Scheme–14)

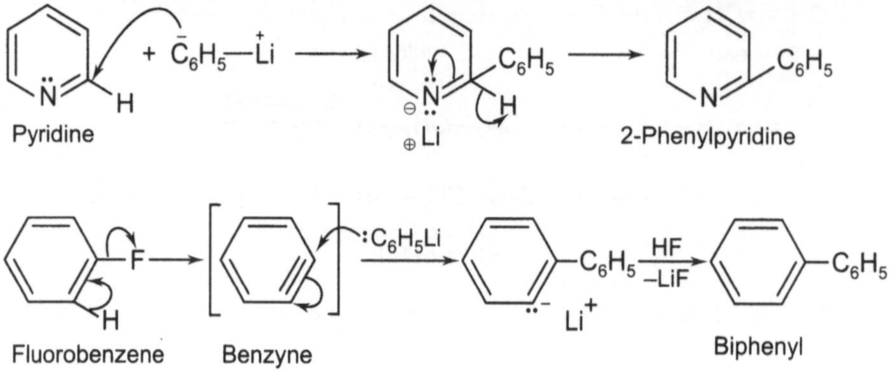

Pyridine   ⊕Li   2-Phenylpyridine

Fluorobenzene   Benzyne   Biphenyl

(Scheme–14)

Alkylation by allylic halides with phenyl lithium proceeds via a cyclic mechanism[6]. This is well understood by using labelled allyl chloride.

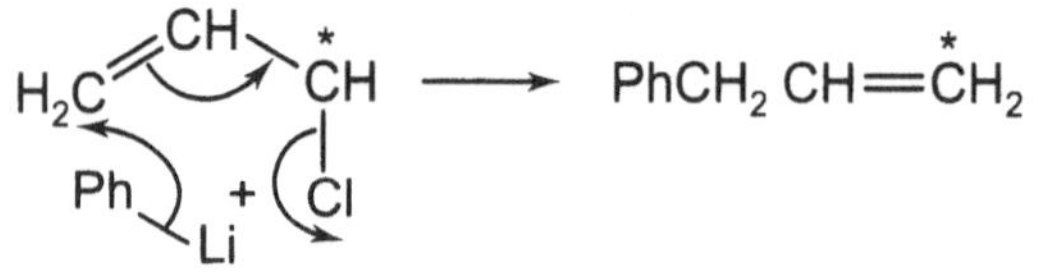

(Scheme–15)

References

1. W.H. Glaze and C.H. Freeman, J. Am. Chem. Soc., 1969, **91**, 7198.
2. R.B. Millen and G. McGravey, J. Org. Chem., 1979, **44**, 4623.
3. J. L. Luche and J.C. Damiano, J. Am. Chem. Soc., 1980, **102**, 7926.
4. J.D. Sprich and G.L. Lewadon, Inorg. Chem. Acta, 1982, **76**, 1241.
5. A.I. Meyers, P.D. Edwards, W.F. Ricker and T.R. Bailey, J. Am. Chem. Soc., 1984, **106**, 3270; A.I. Meyers and G. Milot, J. Am. Chem. Soc., 1993, **115**, 6652.

2.1.29.3 Organosilanes

These are mostly prepared by the nucleophilic displacement of halogen from a halosilane by an organometallic reagent[1].

$$RCH=CHMgBr + (CH_3)_3SiCl \longrightarrow RCH=CHSi(CH_3)_3$$

(Scheme–1)

These reagents react with acyl chlorides to give the corresponding ketones. The raction is catalysed by aluminium chloride or stannic chloride[2] (Scheme–2).

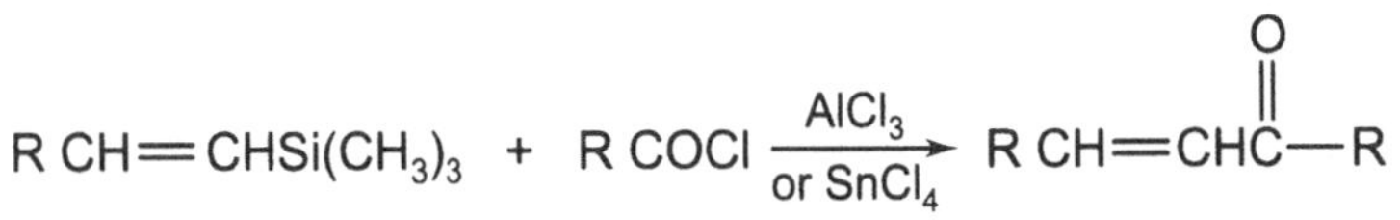

$$R\,CH{=}CHSi(CH_3)_3 \;+\; R\,COCl \xrightarrow[\text{or } SnCl_4]{AlCl_3} R\,CH{=}CH\overset{\displaystyle O}{\overset{\|}{C}}{-}R$$

(Scheme–2)

These reagents also react with dichloromethyl methyl ether, the reaction is catalysed by titanium tetrachloride[3].

Allylic silanes (obtained the reaction of allyl Grignard reagent with trimethyl silyl chloride) also react with acylhalides to give the corresponding ketones[4] (Scheme–3).

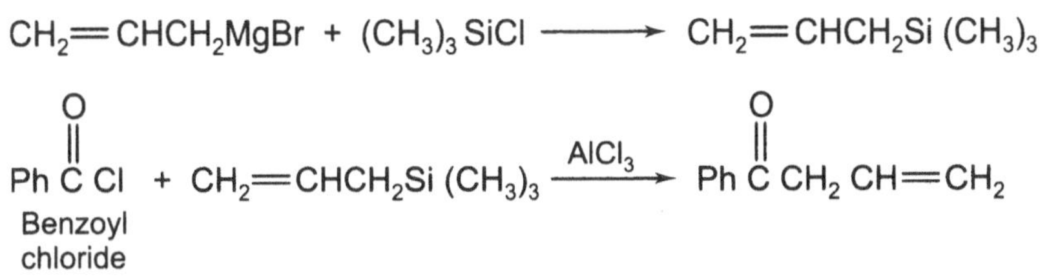

$$CH_2{=}CHCH_2MgBr \;+\; (CH_3)_3\,SiCl \longrightarrow CH_2{=}CHCH_2Si\,(CH_3)_3$$

$$\underset{\substack{\text{Benzoyl}\\ \text{chloride}}}{Ph\,\overset{\displaystyle O}{\overset{\|}{C}}\,Cl} \;+\; CH_2{=}CHCH_2Si\,(CH_3)_3 \xrightarrow{AlCl_3} Ph\,\overset{\displaystyle O}{\overset{\|}{C}}\,CH_2\,CH{=}CH_2$$

(Scheme–3)

Allylic silanes also react with carbonyl compounds[5] as shown below (Scheme–4).

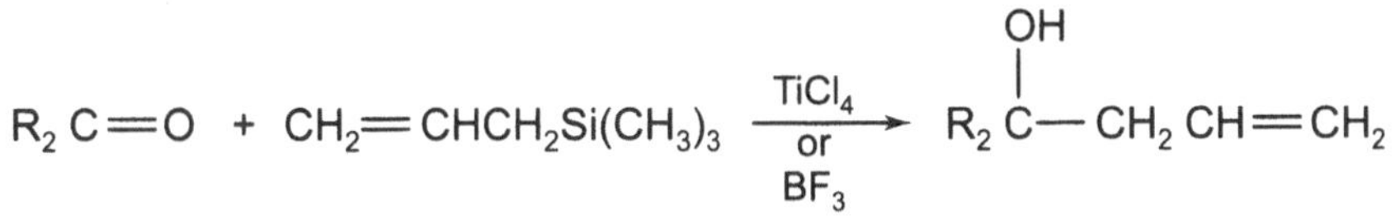

$$R_2\,C{=}O \;+\; CH_2{=}CHCH_2Si(CH_3)_3 \xrightarrow[\substack{\text{or}\\ BF_3}]{TiCl_4} R_2\,\overset{\displaystyle OH}{\overset{|}{C}}{-}CH_2\,CH{=}CH_2$$

(Scheme–4)

The above example exhibit the utility of organo silanes for carbon-carbon bond formation.

References

1. R.K. Boeckman, Jr. D.M. Blum and N. Haley, Org. Synth., 1978, **58**, 152.
2. R. Fleming and A. Pearce, J. Chem. Soc. Chem. Commun., 1975, 633; W.E. Fristad, D.S. Dime, T.R. Bailey and L.A. Paquette, Tetrahedron Lett., 1979, 1999.
3. K. Yamamoto, O. Nunokawa and J. Tsuji, Synthesis, 1977, 721.
4. J.P. Pillot, G. Déléris, J. Dnogués and R. Calas, J. Org. Chem., 1997, **44**, 3397; R. Calas, J. Dunogues, J.P. Pillot, C. Biran, F. Pisciotic and B. Arreguy, J. Organomet. Chem., 1975, **85**, 149.

2.1.29.4 Organostannanes

Organostannanes are obtained by the addition trialkylstannanes to carbon-carbon double[1] and triple[2] bonds (Scheme–1).

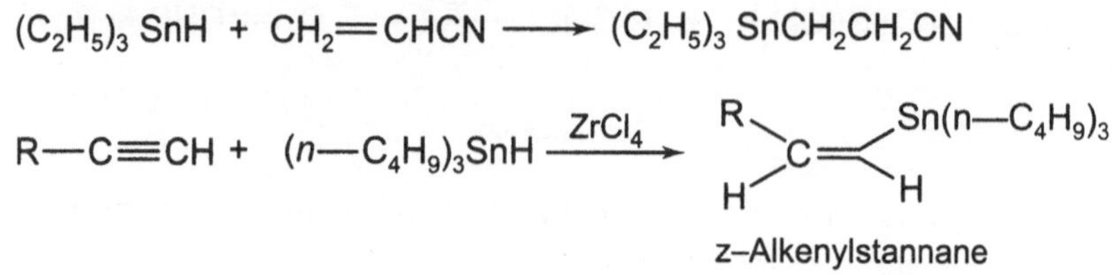

$$(C_2H_5)_3 \, SnH \; + \; CH_2{=}CHCN \longrightarrow (C_2H_5)_3 \, SnCH_2CH_2CN$$

$$R{-}C{\equiv}CH + \; (n{-}C_4H_9)_3SnH \xrightarrow{\;ZrCl_4\;} \begin{array}{c} R \\ \diagdown \\ H \diagup \end{array} C{=}C \begin{array}{c} Sn(n{-}C_4H_9)_3 \\ \diagup \\ \diagdown H \end{array}$$

z–Alkenylstannane

(Scheme–1)

The stannanes are considerably more reactive than the corresponding silanes; this is attributed to more ionic character of carbon in the C — Sn bond[3]. Organostannanes are useful in carbon-carbon bond formation. Thus, aldehydes react with allylic stannanes in presence of Lewis acid as shown below; in this case a double bond shift occurs in conjunction with destannylation[4] (Scheme–2).

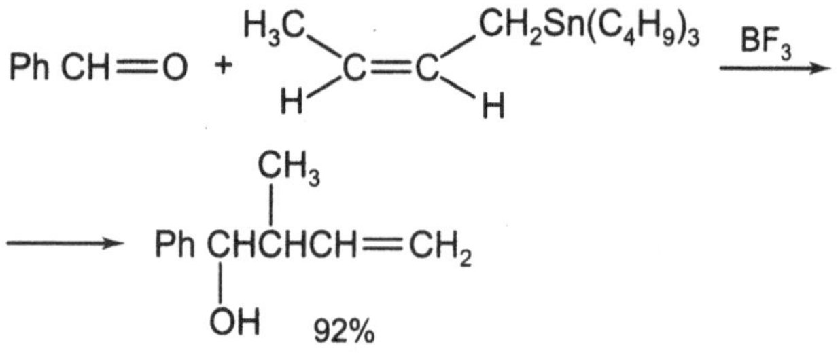

$$Ph\,CH{=}O \; + \; \begin{array}{c} H_3C \\ \diagdown \\ H \diagup \end{array} C{=}C \begin{array}{c} CH_2Sn(C_4H_9)_3 \\ \diagup \\ \diagdown H \end{array} \xrightarrow{\;BF_3\;}$$

$$\longrightarrow \; Ph\,CHCHCH{=}CH_2$$

$$\underset{OH}{\overset{CH_3}{|}} \quad 92\%$$

(Scheme–2)

Aldehydes also react with alkenyl stannane in presence of acyl chloride to give the keto compound; the reaction is catalysed by $SnCl_2$ (Scheme–3).

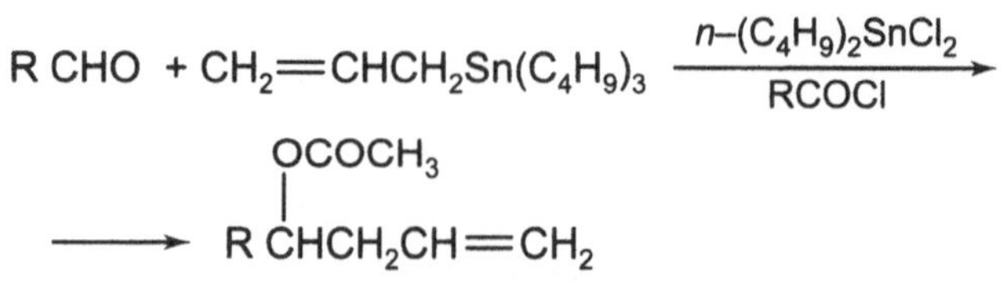

$$R\,CHO \; + \; CH_2{=}CHCH_2Sn(C_4H_9)_3 \xrightarrow[\;RCOCl\;]{n{-}(C_4H_9)_2SnCl_2}$$

$$\longrightarrow \; R\,\underset{OCOCH_3}{\overset{|}{CHCH_2CH}}{=}CH_2$$

(Scheme–3)

The addition of 2-butenylstannanes to benzaldehyde in presence of BF_3 gives the syn isomer irrespective of the stereochemistry of the butenyl group[5] (Scheme–4).

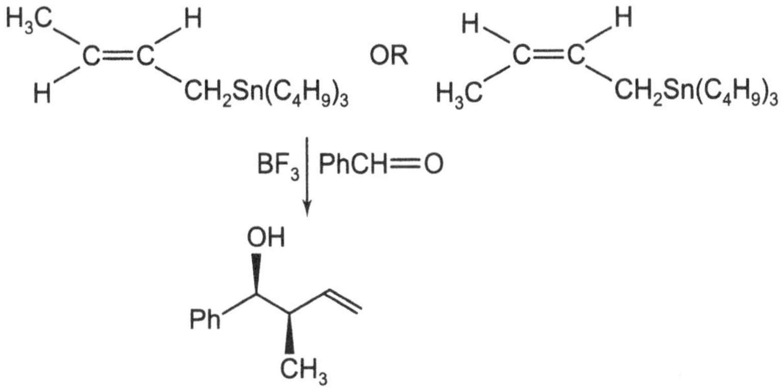

(Scheme–4)

In case TiCl$_4$ is used as catalyst, the stereochemistry depends on the order of the addition of reagents (Scheme–5).

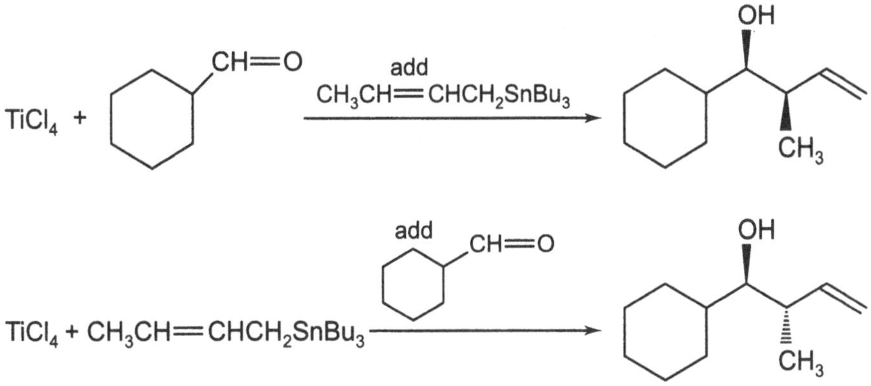

(Scheme–5)

See also stille coupling reaction (2.2.13).

References

1. A.J. Leusinsk and J.G. Noltes, Tetrahedron Lett., 1966, 335.
2. N. Asao, J. –X. Liu, T. Sudoh and Y. Yamamoto, J. Org. Chem., 1966, **61**, 4568.
3. J. Burfeindt, M. Patz, M. Müller and H. Mayr, J. Am. Chem. Soc., 1998, **120**, 3629.
4. H. Yatagai, Y. Yamamoto and K. Maruyama, J. Am. Chem., Soc., 1980, **102**, 4548; Y. Mamamoto, Acc. Chem., Res., 1987, 20, 243; Y. Yamamoto and N. Asao, Chem., Rev., 1993, **93**, 2207.
5. Y. Yamomoto, Acc. Chem., Res., 1987, **20**, 243.

2.1.29.5 Organocadmium Compounds

Organocadmium compounds are obtained by the action of cadmium chloride on a Grignard reagent or organo lithium compounds (Scheme–1).

$$2\ CH_3MgCl\ +\ CdCl_2\ \xrightarrow{\text{ether}}\ (CH_3)_2Cd\ +\ 2\ MgCl_2$$

Dimethyl
Cadmium

$$2\ RLi\ +\ CdCl_2\ \xrightarrow{\text{ether}}\ R_2Cd\ +\ 2\ LiCl$$

Dialkyl
Cadmium

(Scheme–1)

These are less reactive than the Grignard reagents or organolithium compounds. They are useful for carbon-carbon bond formation. Some examples are given below (Scheme–2).

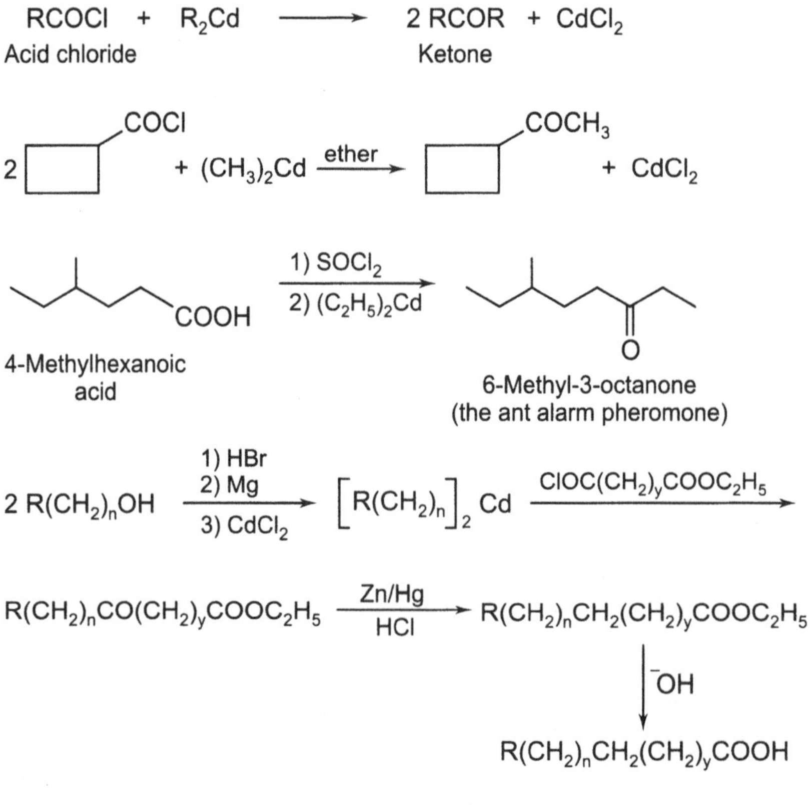

$$RCOCl\ +\ R_2Cd\ \longrightarrow\ 2\ RCOR\ +\ CdCl_2$$
Acid chloride   Ketone

4-Methylhexanoic acid

6-Methyl-3-octanone
(the ant alarm pheromone)

$$2\ R(CH_2)_nOH\ \xrightarrow[\substack{\text{2) Mg} \\ \text{3) CdCl}_2}]{\text{1) HBr}}\ \left[R(CH_2)_n\right]_2 Cd\ \xrightarrow{ClOC(CH_2)_yCOOC_2H_5}$$

$$R(CH_2)_nCO(CH_2)_yCOOC_2H_5\ \xrightarrow[\text{HCl}]{\text{Zn/Hg}}\ R(CH_2)_nCH_2(CH_2)_yCOOC_2H_5$$

$$\downarrow \overline{O}H$$

$$R(CH_2)_nCH_2(CH_2)_yCOOH$$

(Scheme–2)

2.1.29.6 Organocopper Reagents

The organocopper reagents are useful in organic synthesis. They are methyl copper and lithium dimethylcuprate. These are prepared as shown below (Scheme–1).

$$CH_3—X \xrightarrow{\text{Li}} CH_3—Li$$

Alkyl halide   Methyl lithium

$$2\ CH_3—Li\ +\ CuI \longrightarrow (CH_3)_2\ CuLi$$

Methyl   Cuprous   Lithium dimethyl
lithium   iodide   cuprate

(Scheme–1)

Much better yields are obtained by sonication. Lithium dimethyl cuprate (or in general lithium dialkyl cuprates) are known as **Gilman reagents**[1,2,3] and are prepared as follows (Scheme–2).

$$R—X\ +\ 2Li \xrightarrow{\text{Diethyl ether}} R\ Li\ +\ Li\ X$$

Alkyl halide   Alkyl lithium

$$2\ R\ Li\ +\ CuI \longrightarrow R_2\ CuLi$$

Alkyl lithium   Cuprous   Lithium dimethyl
              iodide   cuprate

(Scheme–2)

The lithium dialkyl cuprates are particularly useful for carbon-carbon bond formation.

> *(i)* **Coupling of two alkyl halides:** Following two examples illustrate the coupling of two alkyl halides.

$$(a)\ CH_3—I \xrightarrow[\text{Et}_2\text{O}]{\text{Li}} CH_3Li \xrightarrow{\text{CuI}} (CH_3)_2LiCu \xrightarrow{CH_3CH_2CH_2CH_2CH_2I}$$

Methyl   Methyl   Lithium   Pentyl iodide
iodide   lithium   dimethyl
                                   cuprate

$$\longrightarrow CH_3—CH_2CH_2CH_2CH_2CH_3$$

n–Hexane (98%)

(b) $CH_3CH_2CH_2CH_2Br$ $\xrightarrow[Et_2O]{Li}$ $CH_3CH_2CH_2CH_2Li$ $\xrightarrow{CuI}$

    n-Butyl bromide   Butyl lithium

$\longrightarrow$ $(CH_3CH_2CH_2CH_2)_2CuLi$ $\dfrac{CH_3CH_2CH_2CH_2Br}{\text{Pentyl bromide}}\longrightarrow$

        Lithium dibutyl
           cuprate

$\longrightarrow$ $CH_3CH_2CH_2CH_2{-}CH_2CH_2CH_2CH_2CH_3$

            n-Nonane (98%)

Thus, by this process even two different alkyl halides can be made to couple to give a single hydrocarbon. This synthesis is known as **Corey, Posner, Whitesides House synthesis.**

See also Ullmann reaction (2.1.43).

(ii) **Other coupling reactions:** The coupling reactions of lithium dialkyl cuprates with other substrates are shown below (Scheme–5).

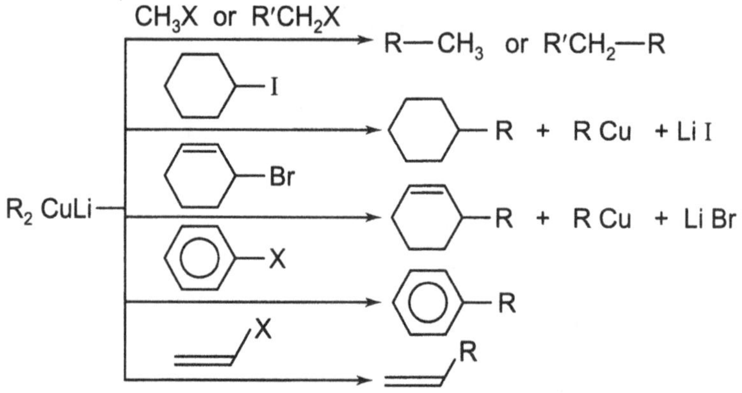

(Scheme–5)

In place of lithium dialkylcuprate, lithium divinylcuprate can be used to synthesise alkenes (Scheme–6).

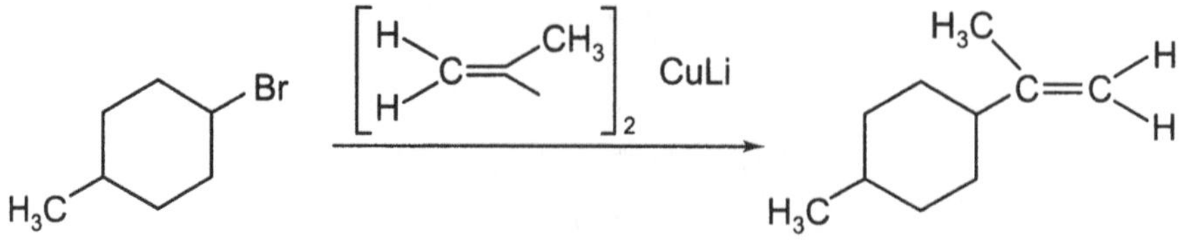

(Scheme–6)

(*iii*) **Reaction with epoxide:** The organocopper reagents attack the epoxide at the least substituted carbon to give the corresponding alcohol (Scheme–7).

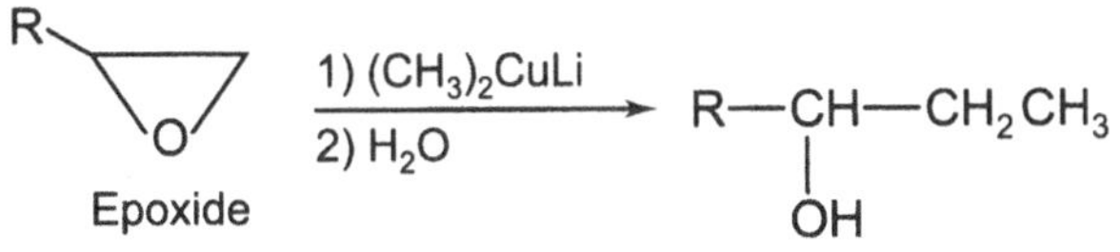

(Scheme–7)

(*iv*) **Reaction with acid chlorides** (Scheme–8)

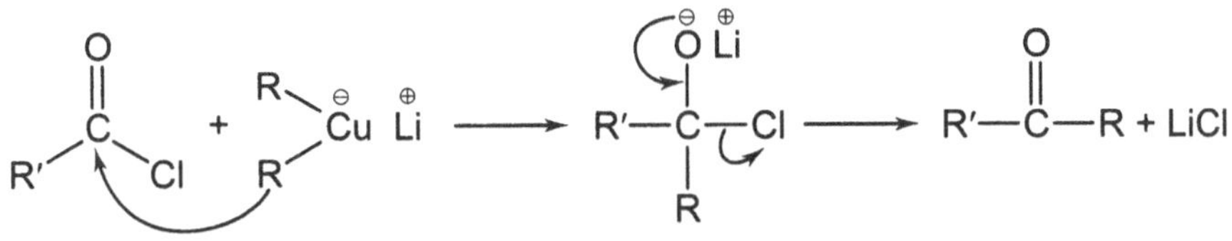

(Scheme–8)

Dialkyl cuprates, unlike Grignard reagents or organolithium reagents do not affect other functional groups (like —NO$_2$, CN, — CO —, —COOR); this makes it feasible to synthesise ketones having other reactive groups. These, reagents also do not react with the formed ketones.

(*v*) **Reaction with α, β-unsaturated compounds:** α, β-unsaturated compounds on reaction with lithium organocuprates give exclusive formation of 1, 4-addition product (Scheme–9).

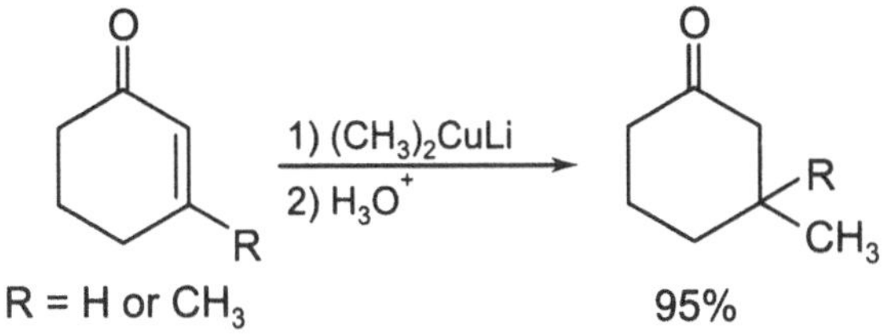

R = H or CH$_3$ 95%

(Scheme–9)

In case of enones, *viz.*, cyclohexenone, consecutive addition of lithium dimethyl cuprate and alkyl halide takes place; in this way, two different alkyl groups can be introduced in the same operation[4] (Scheme–10).

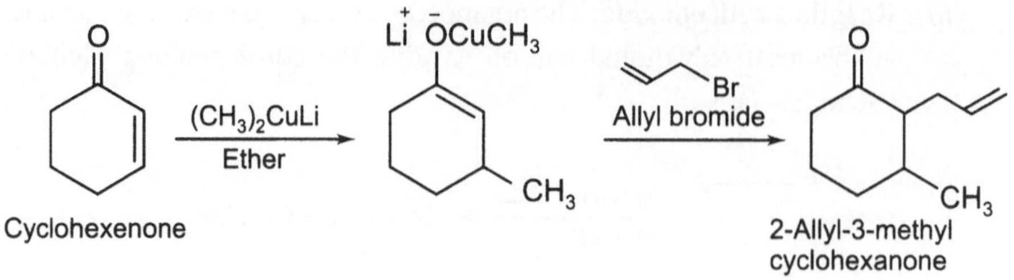
O
$(CH_3)_2CuLi$
Ether
Li OCuCH₃
CH₃
Br
Allyl bromide
O
CH₃
Cyclohexenone
2-Allyl-3-methyl cyclohexanone

(Scheme–10)

Two other organocopper reagents are high order cuprates and copper-isonitrile derivatives, the former, *viz.*, high order cuprates are prepared[5] by the reaction of organolithium reagent with cuprous cyanide (Scheme–11).

$$2\,R{-}Li \;+\; CuCN \longrightarrow R_2Cu(CN)Li_2$$

Alkyl lithium + Cuprous cyanide → High order cuprate

(Scheme–11)

Compared to Gilman reagents, the high order cuprates react much faster with 1° or 2° alkyl halides (Scheme–12).

$$R{-}X \;+\; 2n{-}Bu_2Cu(CN)Li_2 \longrightarrow R{-}nBu$$

(Scheme–12)

These reagents react with chiral halides to give chiral alkanes (Scheme–13).

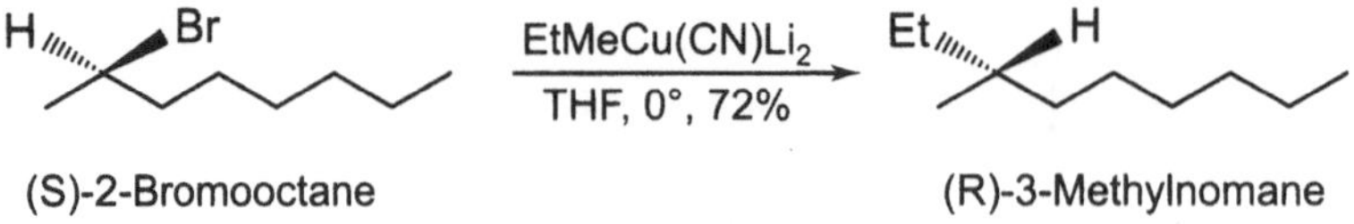
H
Br
EtMeCu(CN)Li₂
THF, 0°, 72%
Et
H
(S)-2-Bromooctane
(R)-3-Methylnomane

(Scheme–13)

Some more examples of carbon-carbon bond formation using high order cuprates are given as follows (Scheme–14).

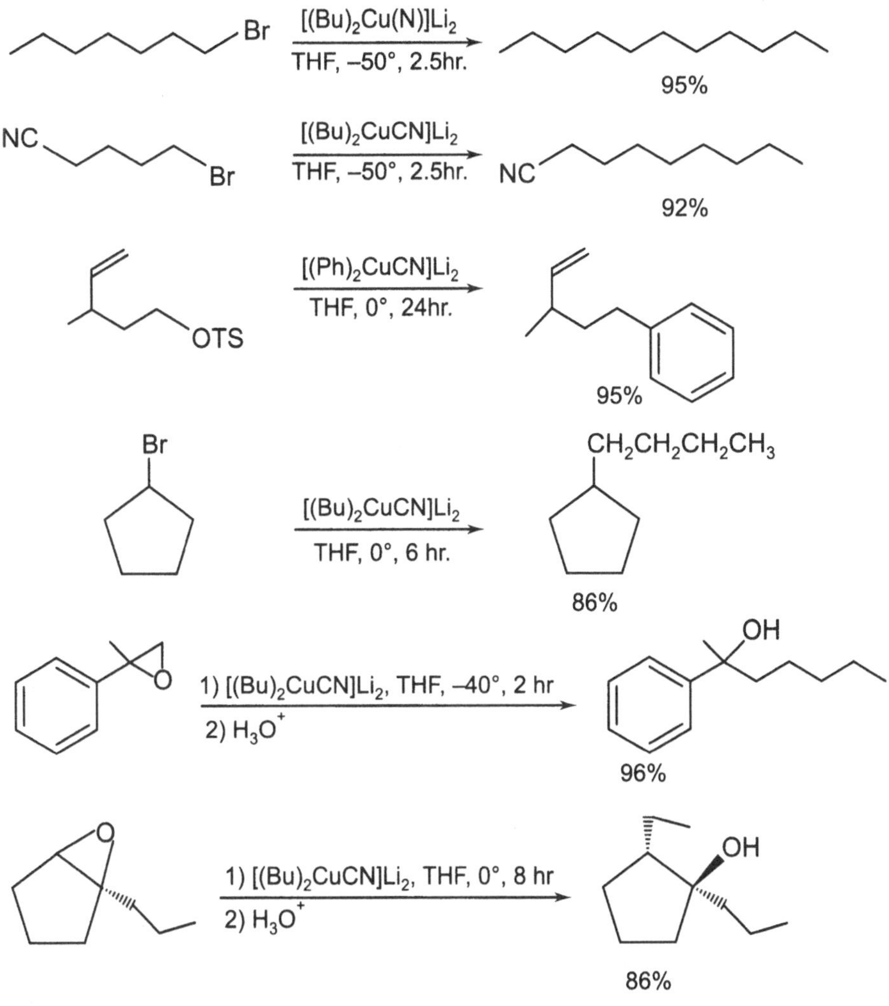

(Scheme–14)

Copper-isonitrile complex is obtained by the mixing of cuprous oxide and an alkyl isonitrile, and then reacting with an active methylene compound (Scheme–15).

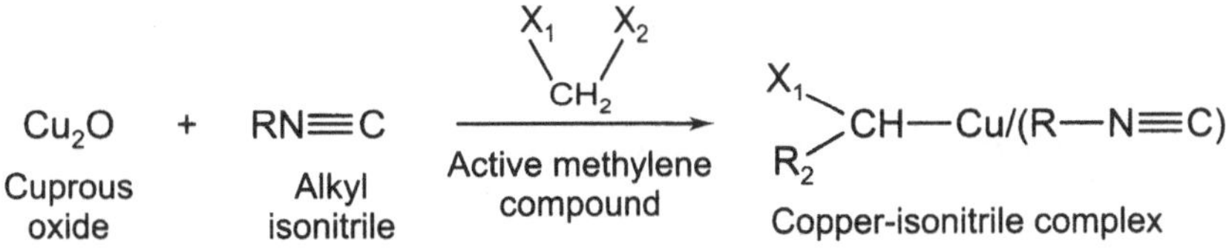

(Scheme–15)

These reagents on reaction with ethyl acrylate give 1, 4-addition products as in Michael addition (Scheme–16).

$$\underset{X_2}{\overset{X_1}{>}}CH-Cu/(R-N\equiv C) \; + \; CH_2\!\!=\!\!CH-COOEt \longrightarrow$$
$$\text{Ethyl acylate}$$

$$\longrightarrow \quad \underset{R_2}{\overset{X_1}{>}}CH-CH_2CH_2COOEt$$
$$\text{1,4-addition product}$$

(Scheme–16)

References

1. H. Gilman and J.M. Straley, Rect. Trav. Chem., Pays. Bas, 1936, **55**, 825.
2. H. Gilman, R.G. Jones and L.A. Woods, J. Org. Chem., 1952, **17**, 1630.
3. G.H. Posner, An introduction to Synthesis using Organocopper Reagents, Wiley, New York, 1980.
4. R.K. Boeckman, J. Org. Chem., 1973, **38**, 4450.
5. B.H. Lipshutz, R.S. Wilhelm and J.A. Kozolowski, Tetrahedron, 1984, **40**, 5005; B.H. Lipshutz and R.S. Wilhelm, J. Am. Chem. Soc., 1982, **104**, 4696.

2.1.29.7 Organozinc Compounds

Organozinc compounds, $R_2\,Zn$, are prepared by the treatment of alkyl iodides with zinc fillings. The formed alkylzinc iodides on distillation in CO_2 atmosphere give dialkylzinc derivatives. In place of zinc, zinc-copper alloy is also used. The reaction gives much better yield on sonication (Scheme–1).

$$RI \; + \; Zn \; \xrightarrow{\;))))\;} \; RZnI$$
$$\text{Alkyl} \quad \text{Zinc} \qquad\quad \text{Alkylzinc iodide}$$
$$\text{iodide}$$

$$RZnI \; \xrightarrow[\text{Distillation}]{CO_2} \; R_2Zn \; + \; ZnI_2$$
$$\qquad\qquad\qquad\qquad \text{Dialkyl}$$
$$\qquad\qquad\qquad\qquad \text{zinc}$$

(Scheme–1)

Organozinc compounds are also obtained as intermediates in the well known **Reformatsky reaction** (2.1.36).

Organozinc compounds are useful for carbon-carbon bond formation. Two such examples are given ahead (Scheme–2).

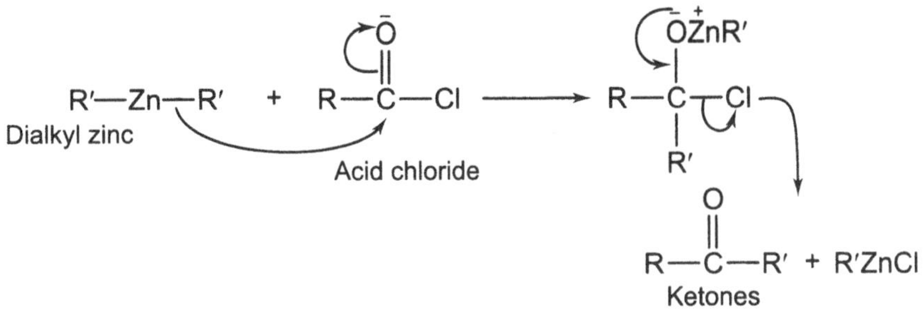

(Scheme–2)

This reagent reacts very slowly with ketones and this procedure is useful for the synthesis of ketones. Hydrocarbons can also be prepared (Scheme–3).

$$R_2Zn \quad + \quad (CH_3)_3\ CCl \longrightarrow (CH_3)_3CR \ + \ RZnCl$$

Dialkyl zinc   Tert. alkyl halide   Hydro carbons

(Scheme–3)

An extremely important and useful organozinc reagent is iodomethylzinc iodide (ICH_2ZnI). This reagent adds on to olefins to give cyclopropane derivatives (Scheme–4).

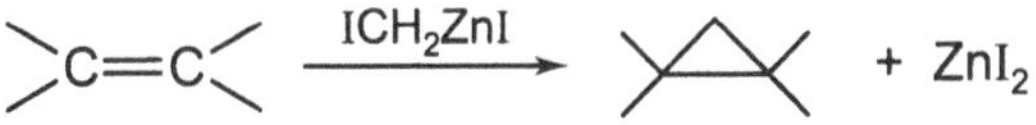

(Scheme–4)

The reaction is believe to take place in a concerted manner involving a cyclic transition state (Scheme–5).

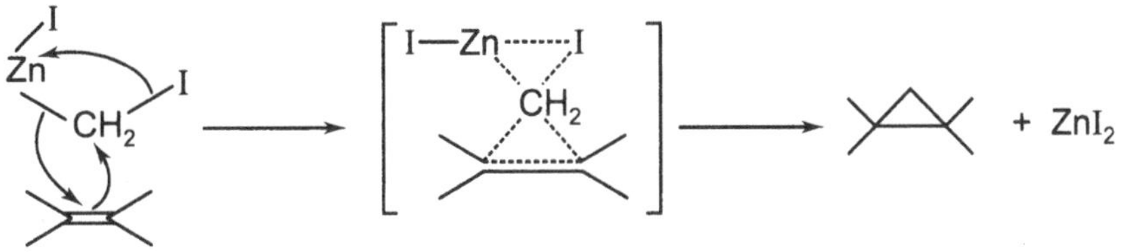

(Scheme–5)

The reaction is known as **Simmons-Smith reaction** and the organozinc reagent, **iodomethylzinc iodide** is called **Simmon-Smith reagent**. This reagent is obtained by the reaction of methylene diodide and zinc (Scheme–6).

$$CH_2I_2 \ + \ Zn \longrightarrow ICH_2ZnI$$

(Scheme–6)

This reagent generated *in situ* from methylene diiodide and zinc in THF on sonication. Using this reagent, methylenation of carbonyl group can be conveniently accomplished[1] (Scheme–7).

$$R^1\diagdown C=O \xrightarrow[\text{RT, }))))]{\text{CH}_2\text{I}_2/\text{Zn/THF}} R^1\diagdown C=CH_2$$
$$R^2\diagup \qquad\qquad R^2\diagup$$

(Scheme–7)

The reaction is generally applicable to aldehydes and not ketones. With benzaldehyde the yield is 70 per cent in 20 min.

This reagent can also be generated *in situ* by the action of methylene diiodide and zinc-copper couple. Thus, the reaction of cyclohexene with methylene diiodide and zinc-copper couple gives bicyclo [4.1.0] heptane (norcarene) (Scheme–8).

$$\xrightarrow[\text{Zn–Cu couple}]{\text{CH}_2\text{I}_2}$$

Norcarene

(Scheme–8)

Reference

1. J. Yamashila, Y. Inou, T. Kando and H. Hashimolo, Bull. Chem. Soc. Jpn., 1984, **57**, 2335.

2.1.30 Paterno-Büchi Reaction

Photochemical cycloadditions of carbonyl compounds with olefins give oxetanes (four membered ether ring). The reaction is known as Paterno-Büchi reaction[1] (Scheme–1).

(Scheme–1)

The Paterno-Büchi reaction usually occurs by the cycloaddition of the triplet state of the carbonyl compound with the ground state of an alkene. An interesting reaction is the photocyclisation of butyraldehyde with 2-methyl-

2-butene to yield a mixture of 2,3,3-trimethyl-4-propyloxetane and 2,2, 3-trimethyl-4-propyloxetane (Scheme–2).

$$CH_3CH_2CH_2CHO \quad + \quad CH_3—\underset{\underset{CH_3}{|}}{C}{=}CH—CH_3 \xrightarrow{h\nu}$$

Butyraldehyde

2-Methyl-2-butene

2,3,3-Trimethyl
-4-propyloxetane

2,2,3-Trimethyl
-4-propyloxetane

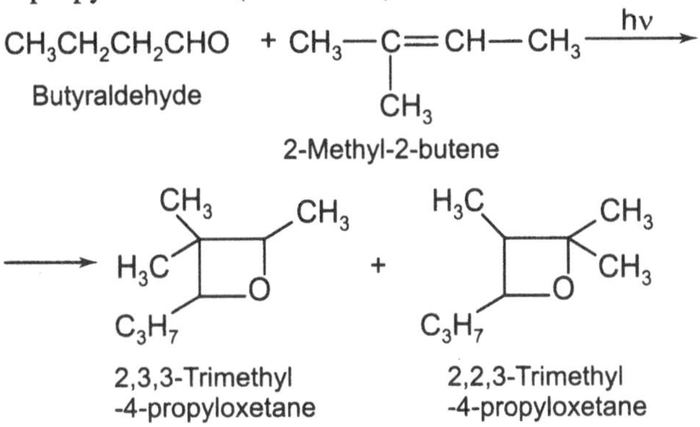

(Scheme–2)

Reference

1. E. Paterno and G. Cheffi, Gazz. Chem. Ital., 1909, **39**, 341; G. Büchi, C.G. Inmand and E.S. Lipinsky, J. Am. Chem., Soc., 1954, **76**, 4327; D.R. Arnold, Advan. Photochem., 1968, **6**, 301.

2.1.31 Pechmann Condensation

It involves condensation[1] of phenols with β-ketoesters in presence of conc. sulphuric acid or other condensing agents like P_2O_5, $POCl_3$, $AlCl_3$ *etc.*, to give 4-methyl cumarins. The reaction proceeds by the formation of β-hydroxy ester intermediate, which on cyclisation and subsequent dehydration gives coumarins (Scheme–1).

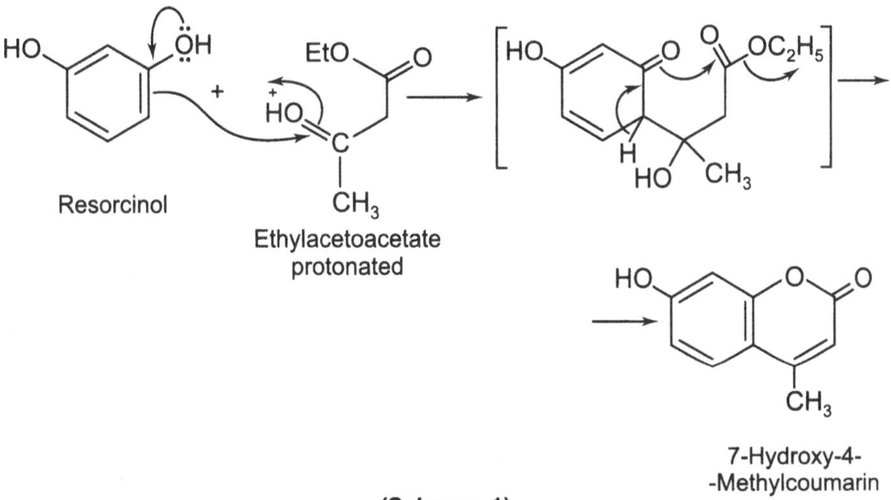

Resorcinol

Ethylacetoacetate
protonated

7-Hydroxy-4-
-Methylcoumarin

(Scheme–1)

2.1.31.1 Microwave Promoted Pechmann Condensation

Pechmann condensation of salicylaldehydes with alkyl substituted ethylacetates under basic conditions (piperidine) on microwave irradiation gave coumarins[2,3] (Scheme–2) No solvent is used.

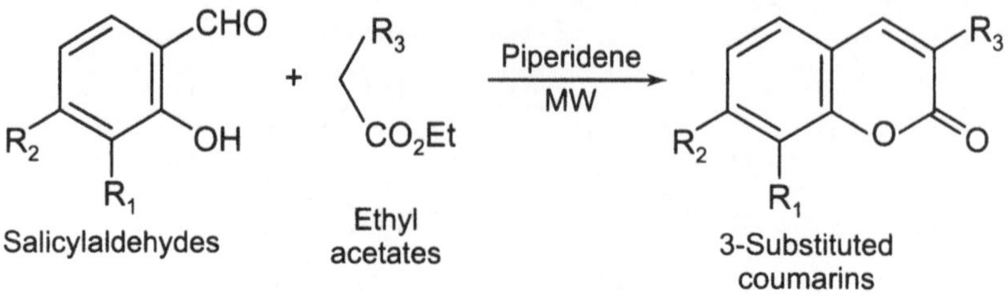

| Salicylaldehydes | Ethyl acetates | 3-Substituted coumarins |

(Scheme–2)

2.1.31.2 Pechmann Condensation in Presence of Ionic Liquids

Pechmann condensation of phenols with ethyl acetoacetate using a Bronsted-acidic ionic liquids[4], which acts both as catalyst and solvent gives the corresponding 4-methylcoumarins (Scheme–3).

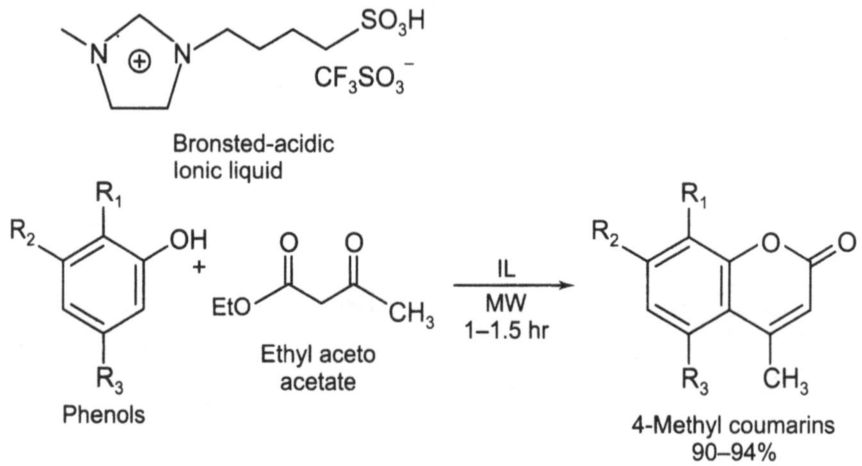

Bronsted-acidic
Ionic liquid

| Phenols | Ethyl aceto acetate | 4-Methyl coumarins 90–94% |

R₁	R₂	R₃
H	OH	H
H	OH	OH
H	OMe	H
OMe	OH	H
OH	OH	H
Me	H	H

(Scheme–3)

References

1. H.V. Pechmann and C. Duisberg, Ber., 1883, **16**, 2119; S. Sethna, Chem. Rev., 1945, **36**, 10; K.D. Kaufman, J. Org. Chem., 1967, **32**, 504.

2. V. Singh, P. Singh, P. Kaur and G.L. Kad, J. Chem., Res., (S), 1997, 58.

3. D. Bogdal, J. Chem. Res.,(s), 1998, 468.

4. Y. Gu, J. Zhang, Z. Duan and Y. Deng, Adv. Synth. Catal., 2005, 347, 512.

2.1.32 Photochemical Reactions

These are important carbon-carbon bond forming reactions. Compared to a thermal reaction, which is promoted by heat, a photochemical reaction derives its energy from the absorption of light radiation. It is possible that the products of a photochemical reactions are quite different than those obtained by thermal reactions. A typical reaction is the photoreductions of benzophenone by irradiation in propan-2-ol. Sunlight or a medium pressure mercury arc lamp can be used (Scheme–1).

$$(C_6H_5)_2C{=}O \xrightarrow[\text{isopropyl alcohol}]{hv} (C_6H_5)_2\underset{\underset{OH}{|}}{C}{-}\underset{\underset{OH}{|}}{C}(C_6H_5)_2$$

Benzophenone   Benzopinacol (98%)

(Scheme–1)

1,2- and 1, 4-cycloadditions occur photochemically with or without sensitizers. Examples of both types are given below (Scheme–2).

$$\begin{matrix} H_2C{=}CH_2 \\ H_2C{=}CH_2 \end{matrix} \xrightarrow{hv} \begin{matrix} CH_2{-}CH_2 \\ | \qquad | \\ CH_2{-}CH_2 \end{matrix}$$ [2 + 2 cycloaddition]

Ethylene   Cyclobutane

[4 + 2 cycloaddition]

Butadiene   Ethylene

Cyclohexene

A typical example is the photochemical dimerisation of 1,4-naphthoquinone.

1,4-Naphthoquinone   1,4-Naphthoquinone photodimer

(Scheme–2)

Photochemical cyclisation can also proceed in an intramolecular fashion. Some examples are given below (Scheme–3).

Hexatriene   hv   1,3-cyclohexadiene   ...(Ref. 1)

Stilbene   hv   ...(Ref. 2)

1,3-Cyclooctadiene   hv   Bicyclo [4.2.0] oct-3-ene   ...(Ref. 3)

1,4-Cyclooctadiene   hv   ...(Ref. 3)

(Scheme–3)

Following are given some examples of electrocyclic reactions and cycloadditions (Scheme–4).

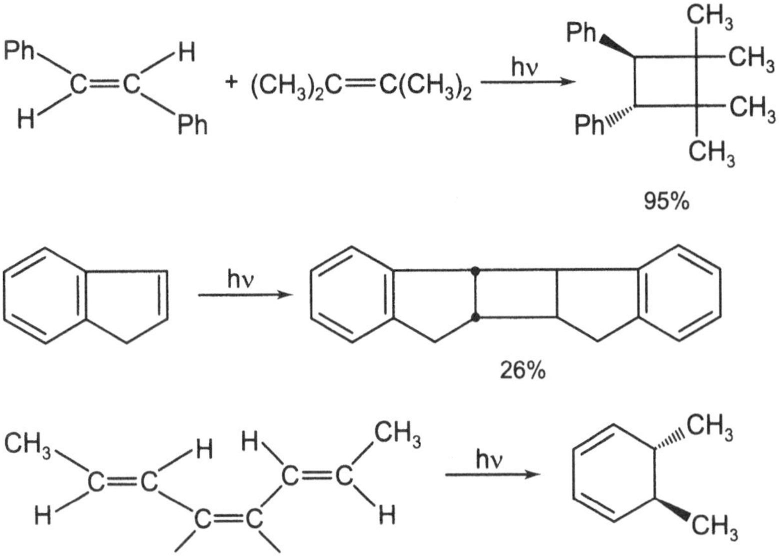

...(Ref. 4)

95%

...(Ref. 5)

26%

...(Ref. 6)

(Scheme–4)

See also Paterno-Büchi Reaction (2.1.30).

2.1.32.1 Photochemical Cycloadditions in Water

Photodimerisation of thymine, uracil and their derivatives increased considerably in water[7] in comparison to other organic solvents (Scheme–1).

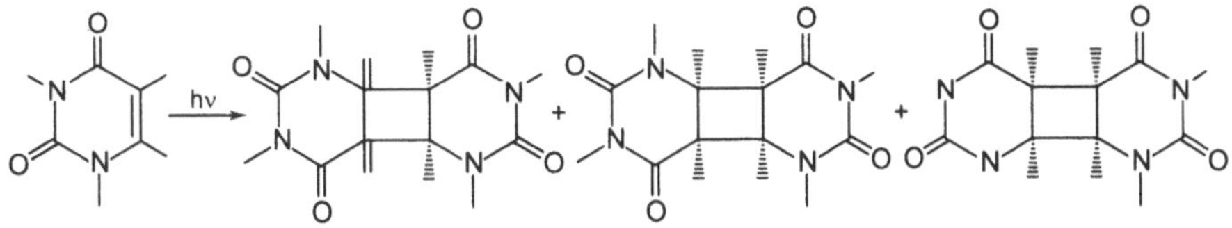

Water	27.8%	63.1%	9.9% θ = 0.015
Acetonitrile	24.9%	68.2%	6.7% θ = 0.0047
Methanol	31.4%	68.6%	— θ = 0.004

Data taken from reference 7

(Scheme–1)

A number of other photochemical cycloadditions like that of stilbene, coumarin and anthracene-2-sulfonate have also been performed in water[8].

2.1.32.2 Photochemical Reactions in Micellar Media

Addition of a surfactant aggregates, the formation of miscelles; this has a significant effect on the regio and stereoselectively of photochemical reactions.

Using this technique, photodimerisation of anthracene-2-sulfonate, cycloaddition of isobutylene to cyclohexanone, dimerisation of isophorone, photodimerisation of acenaphthylene and photocrossed cycloaddition of acenaphthylene has been reported. For details see reference[9].

2.1.32.3 Photoreactions in Solid State

A number of photoreactions like photodimerisation, photopolymerisation, photocyclisation, photorearrangements, photoisomerizations, photosolvolysis, Photodecarboxylation and photoaddition reactions between different molecules have been carried out in solid state. For details see reference[10].

References

1. D.H.R. Barton, Helv. Chem. Acta, 1959, **42**, 2604.
2. E.V. Blackburn and C.J. Timmons, Quat. Rev. Chem. Soc., London, 1969, **23**, 482; W.H. Laarhoven, Rev. Trav. Chim., 1983, **102**, 185.
3. S. Moon and C.R. Ganz, Tetrahedron Lett., 1968, 6275.
4. O.L. Chapman and W.R. Adams, J. Am. Chem. Soc., 1968, **99**, 2333.
5. A.G. Anastassiou, F.L. Setliff and W.G. Griffin, J. Org. Chem., 1966, **31**, 2705.
6. G. J. Fonken, Tetrahedron Lett., 1962, 549.
7. R. Ramamurty, Tetrahedron, 1986, **42**, 5753.
8. For details see organic reactions in aqueous media, C. –J. Li and T. –H. Chan., John Wiley, New York. Page 41–43.
9. V.K. Ahluwalia and R.S. Varma, Alternate Energy Processes in Chemical Synthesis, Narosa Publishing House, New Delhi, 2008, page 15.15 to 15.17 and the references cited therein.
10. K. Tanaka and F. Toda, Solvent-free Organic Synthesis, Chem., Rev., 2000, **100** page 1047–1071 and the references cited therein.

2.1.33 Pinacol Coupling

The reaction of ketones with magnesium in benzene give 1,2-diols. Thus, acetone under these conditions give pinacol (Scheme–1).

(Scheme–1)

This reaction is known as pinacol coupling. The use of zinc-copper couple for unsaturated aldehydes was known as early as 1892[1]. Subsequently, chromium and vanadium[2] and some ammonical $TiCl_3$ based reducing agents[3] were used. It has now been found that pinacol coupling takes place in aromatic aldehydes and ketones in presence of Ti(III), under alkaline conditions. However, in presence of acids, only the substrates (aromatic aldehydes and ketones) having electron withdrawing groups like CN, CHO, COMe, COOH, COOMe, pyridyl as activating groups under went pinacol coupling[5]. In case of non-activating carbonyl compounds, it was necessary to use excess of the substrate as solvent[6] (Scheme–2).

R_1 = Ph, 2–Py
R_2 = CN, CO_2Me

61–75%

(Scheme–2)

The coupling reaction of α, β-unsaturated ketones and acetone using a Zn-Cu couple and ultrasound in aqueous acetone suspension gave the corresponding pinacol[7] (Scheme–3).

(Scheme–3)

References

1. Grinder, Ann. Chem., Phys., 1892, **26**, 369.
2. J.B. Conant and H.B. Cutter, J. Am. Chem., Soc., 1926, **48**, 1016.
3. P. Karrer, Y. Yen and I. Reichstein, Helv. Chem., Acta; 1993, **13**, 1308.
4. A Clerici and O. Porta, Tetrahedron Lett., 1982, **23**, 3517.
5. A Clerici and O. Porta, J. Org. Chem., 1982, **47**, 2852; A Clerici, O. Porta and M. Riva, Tetrahedron Lett., 1981, **22**, 1043.
6. A Clerici and O. Porta, J. Org. Chem., 1983, **48**, 1690; Tetrahedron, 1983, 39, 1239; A Clerici, O. Porta and P. Zaga, Tetrahedron, 1986, **42**, 561; A Clerici and O. Porta, J. Org. Chem., 1989, **54**, 3872.
7. P. Deliar and J.L. Luchi, J. Chem., Soc., Chem., Commun., 1989, 398.

2.1.34 Pinacol-Pinacolone Rearrangement

Acid catalysed rearrangement of pinacols to pinacolone is known as pinacol Pinacolone rearrangement[1]. This is basically a rearrangement of carbon-carbon bonds. A typical example is given as follows (Scheme–1).

(Scheme–1)

2.1.34.1 Pinacol-Pinacolone Rearrangement in Water

The rearrangement can be performed in water[2,3] by heating in microwave oven at 270°; at this temperature water acts as an acid catalyst.

2.1.34.2 Pinacol-Pinacolone Rearrangement in Solid State

Irradiation of pinacols with Al^{3+} montmorillonite K10 clay in a microwave oven for 15 min. gives the rearranged product[4] in 98–99 per cent yield.

References

1. R. Fitting, Ann., 1859, **110**, 17; 1860, **114**, 54.
2. B. Kuhlman, E.M. Arnett and M. Siskin, J. Org. Chem., 1994, **59**, 5377.
3. J.M. Kremsner and C.O. Kappe, Eur. J. Org. Chem., 2005, 3672.
4. E. Gutirrez, A. Loupy, G. Bram and E. Ruiz-Hitzky, Tetrahedron Lett., 1989, **30**, 945.

2.1.35 Prins Reaction

It is a useful carbon-carbon bond forming reaction and is notable in the formation of tetrahydropyran derivatives. It consists in the condensation of olefins with aldehydes under strongly acedic conditions and high reaction temperature. This limits its potential as an useful synthetic methodology.

Using simple homoallylic alcohol and an aldehyde in presence of a catalytic amount of cerium triflate, the direct stereoselective formation of tetrahydropyranol derivative in ionic liquid has been achieved[1] (Scheme–1).

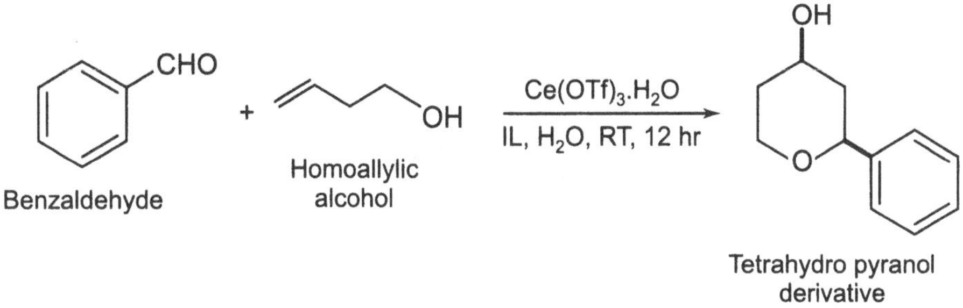

(Scheme–1)

Tetrahydropyranols are present in a number of natural products including carbohydrates, polyether antibiotics and marine toxins.

Reference

1. C.C.K. Keh, V.V. Namboodri and R.S. Varma, Tetrahedron Lett., 2002, **43**, 4993.

2.1.36 Reformatsky Reaction

The reaction of an α-haloester (usually an α-bromoester) with carbonyl compounds (aldehyde or ketone) in presence of zinc metal in an inert solvent give β-hydroxyester[1]. This reaction extends the carbon skeleton of an carbonyl compound (Scheme–1).

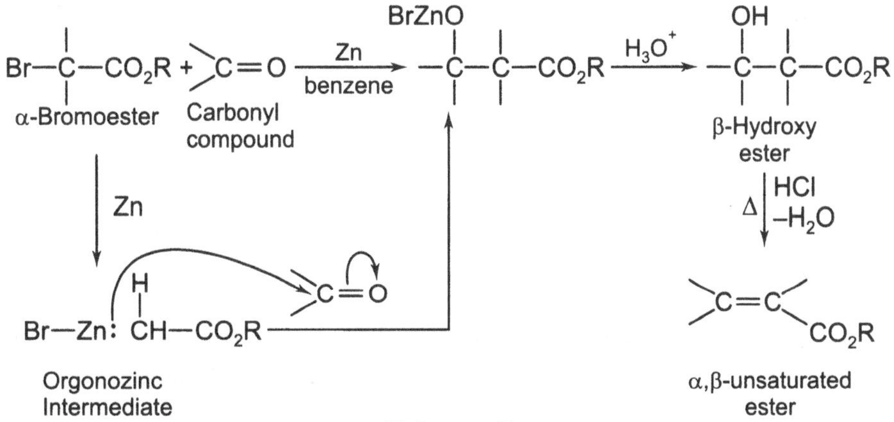

(Scheme–1)

The formed β-hydroxyesters on heating in presence of acid gives α,β-unsaturated ester.

In a modified Reformatsky reaction, using nitriles in place of aldehydes or ketones, the product obtained is an imine, which undergoes readily hydrolysis to give the ketones (Scheme–2). This reaction is called **Blaise Reaction**.

$$Br-Zn-\overset{\underset{|}{R}}{C}H-CO_2Et \;+\; R'-C\equiv N \xrightarrow{\;\;H_2O\;\;} R'-\overset{\underset{\|}{O}}{C}-\overset{\underset{|}{R}}{C}H-CO_2Et$$

Organo zinc intermediate   Nitrile

(Scheme–2)

2.1.36.1 Reformatsky Reaction using Sonication

Excellent yields are obtained on sonication[2] compared to more traditional methods, such as those employing activated zinc or trimethylborane as cosolvent[3] (Scheme–3).

$$\overset{R}{\underset{R'}{>}}C=O \;+\; BrCH_2CO_2Et \xrightarrow[\substack{\text{Dioxane, RT,}\\ \text{)))), 5–30 min}}]{Zn/I_2} \overset{R}{\underset{R'}{>}}C\overset{\diagdown}{\underset{OH}{}}CO_2Et$$

R = H; R' = Ph or $(CH_2)_2CH_3$

R—R' = $-(CH_2)_4-$

(Scheme–3)

In the sonication procedure, it is necessary to activate zinc with iodine to carry out the reaction in dioxane.

Schiffs bases on subjecting to Reformatsky reaction under sonication gives much better results to give β-lactams[2] (Scheme–4).

$$Ar\diagdown\diagup^{N}\diagdown_{Ar} \;+\; BrCH_2CO_2Et \xrightarrow[\text{Dioxane}]{Zn/I_2}$$

Schiffs base
Ar = Ph, 4-MeC$_6$H$_4$,
    4-ClC$_6$H$_4$

(Scheme–4)

The Reformatsky reaction with nitriles lead to the formation of imines, which are readily hydrolysed to the ketone[4] (Scheme–5).

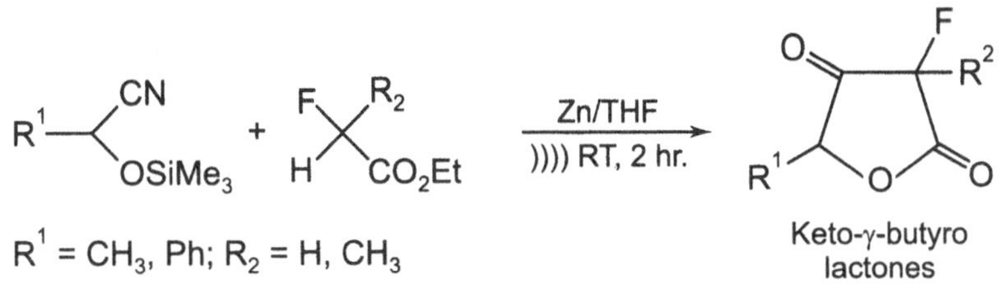

R¹ = CH₃, Ph; R₂ = H, CH₃

$R^1 = CH_3, Ph; R_2 = H, CH_3$

Keto-γ-butyro lactones

(Scheme–5)

2.1.36.2 Reformatsky Reaction in Solid State

Treatment of aromatic aldehydes with ethyl bromoacetate and Zn — NH₄Cl in the solid state gives the corresponding reformatsky product[4] (Scheme–6).

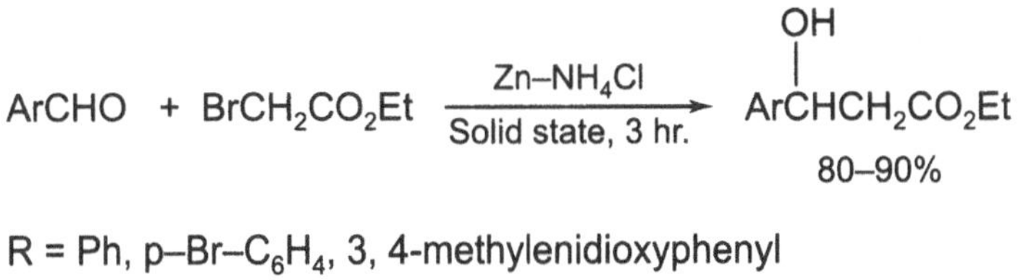

$$ArCHO + BrCH_2CO_2Et \xrightarrow[\text{Solid state, 3 hr.}]{Zn-NH_4Cl} Ar\overset{OH}{\underset{|}{C}}HCH_2CO_2Et$$

80–90%

R = Ph, p–Br–C₆H₄, 3, 4-methylenidioxyphenyl

(Scheme–6)

References

1. S. Reformatsky, Ber., 1887, **20**, 1210; J. Russ. Chem., Soc., 1890, **22**, 44.
2. A.K. Bose, K. Gupta and M.S. Manhas, J. Chem., Soc., Chem. Commun., 1984, 86; C. Petrier, L. Gemai and J.L. Luchi, Tetrahedron Lett., 1982, 3361.
3. B. Hans and P. Boudjouk, J. Org. Chem., 1982, **47**, 5030.
4. T. Kitazume, Synthesis, 1986, 855.

2.1.37 Reimer-Tiemann Reaction

Phenols on reaction with chloroform in presence of sodium or potassium hydroxide give *o*-hydroxy aldehydes as a major product[1] (Scheme–1).

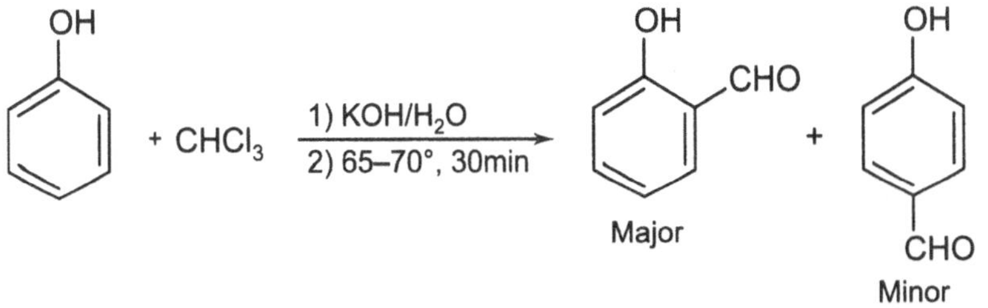

$$+ CHCl_3 \xrightarrow[\text{2) 65–70°, 30min}]{\text{1) KOH/H}_2\text{O}}$$

Major

Minor

(Scheme–1)

The reaction takes place *via* the formation of dichloro carbene (Scheme–2).

$$CHCl_3 \xrightarrow{\bar{O}H} \bar{C}Cl_3 + H_2O$$

$$\xrightarrow[-Cl]{} :CCl_2$$

Dichloro carbene

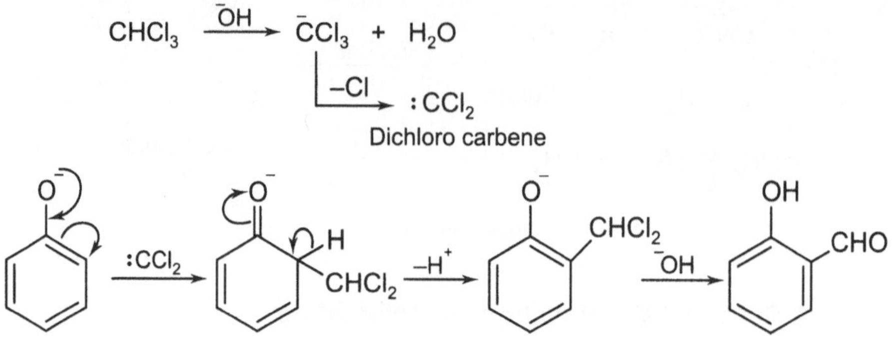

(Scheme–2)

The aldehyde is obtained in much better yield when the reaction is carried out is presence of a PTC (Tetrabutyl ammonium bromide)[2] (Scheme–3).

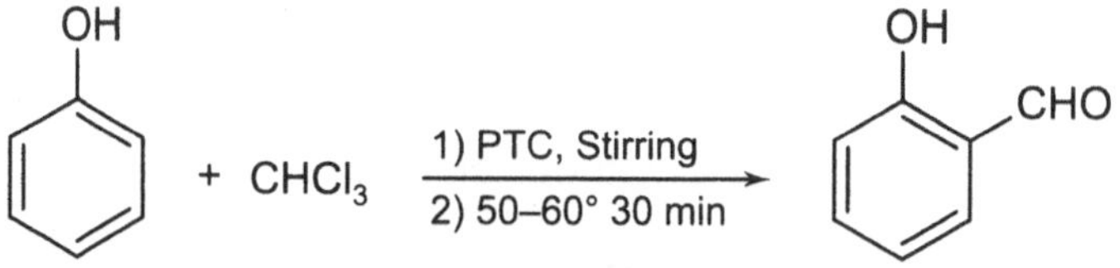

(Scheme–3)

References

1. K. Reimer and F. Tiemann, Ber., 1876, **9**, 824, 1285, H. Wynberg, Chem., Rev., 1960, **60**, 169.

2. S.L. Regan and A. Singh, J. Org. Chem., 1982, **47**, 1587.

2.1.38 Simmons-Smith Reaction

This is a versatile reaction[1] for the conversion of alkenes into cyclopropane derivatives. It involves reacting an alkene with methylene iodide and zinc-copper or better zinc-silver couple (Scheme–1).

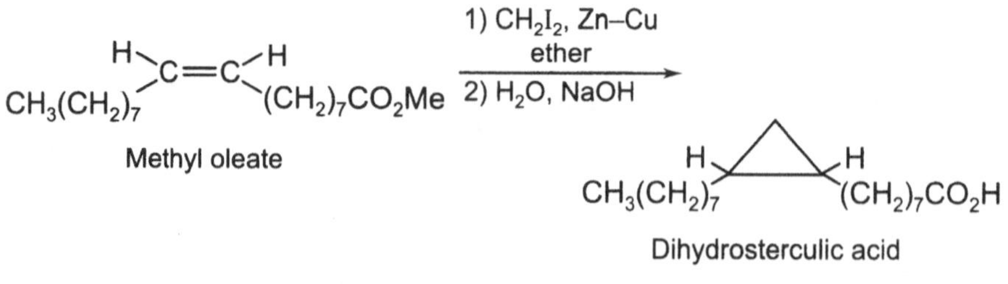

Methyl oleate

Dihydrosterculic acid

(Scheme–1)

In this case, the reactive intermediate is iodomethylene zinc-iodide complex, which is generated *in situ* (Scheme–2).

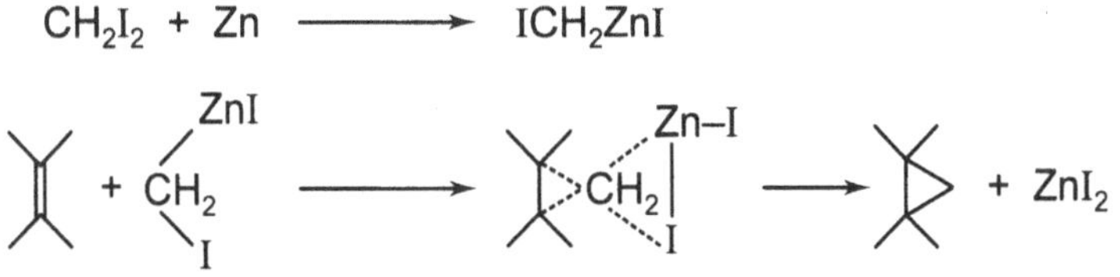
CH_2I_2 + Zn
ICH_2ZnI
ZnI
+ CH_2
I
Zn–I
CH_2
I
+ ZnI_2

(Scheme–2)

Simmons-Smith reaction gives much better yield if sonochemically activated zinc and methylene iodide are used[2] (Scheme–3).

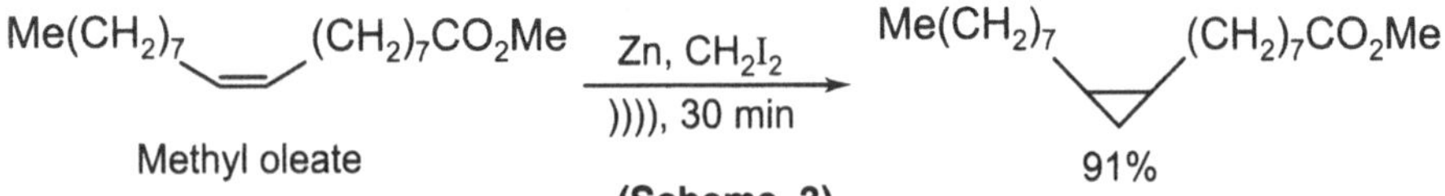
$Me(CH_2)_7$
$(CH_2)_7CO_2Me$
Methyl oleate
Zn, CH_2I_2
)))), 30 min
$Me(CH_2)_7$
$(CH_2)_7CO_2Me$
91%

(Scheme–3)

The above process can be scaled up[3].

References

1. H.E. Simmons and R.D. Smith, J. Am. Chem., Soc., 1958, **80**, 5323.
2. O. Repic and S. Vogt, Tetrahedron Lett., 1982, **23**, 2729.
3. H. Tso, T. Chou and H. Hung, J. Chem., Soc. Chem., Commun., 1887, 1552.

2.1.39 Stetter Reaction

The reaction of aldehydes with olefins to give 1, 4-dicarbonyls is known as stetter reaction[1]. It is possible to carry out the reaction in ionic liquid using triethylamine as catalyst[2] (Scheme–1).

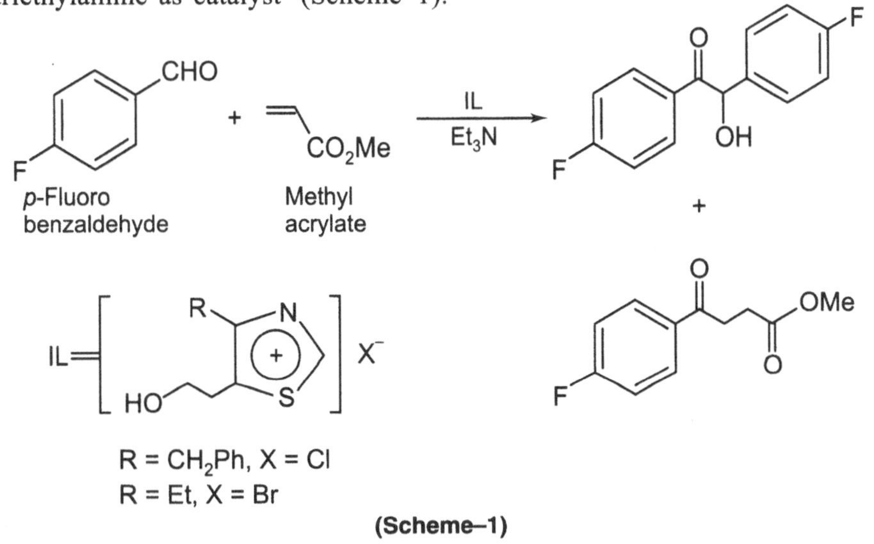
CHO
F
p-Fluoro
benzaldehyde
+
CO_2Me
Methyl
acrylate
IL
Et_3N
O
F
F
OH
+
O
OMe
F
O
IL=
R
N
+
S
HO
X^-
R = CH_2Ph, X = Cl
R = Et, X = Br

(Scheme–1)

References

1. H. Stetter, Angew. Chem., 1973, **85**, 89; Chem., Ber., 1974, **107**, 210; Synthesis, 1975, 379.

2. A. Anjaiah, S. Chandrasekhar and R. Gree, Adv. Synth. Catal., 2004, **346**, 1329.

2.1.40 Strecker Synthesis

Aldehydes on treatment with ammonia and hydrogen cyanide give α-amino nitriles, which on hydrolysis give α-amino acid. This synthesis is called strecker synthesis[1] (Scheme–1).

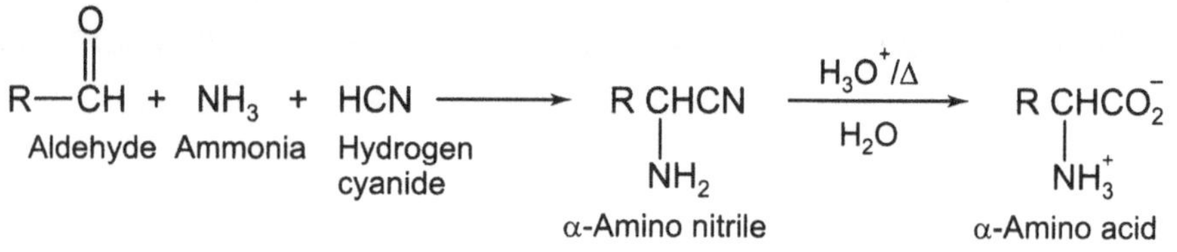

(Scheme–1)

In an alternative procedure (Erlenmeyer modification[2] of the strecker synthesis), the aldehyde is treated with HCN and the formed cyanohydrin is reacted with ammonia; the final step is the same as given above (Scheme–2).

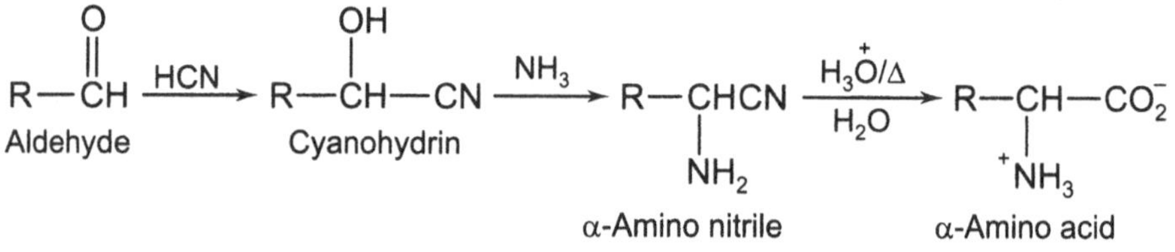

(Scheme–2)

A one step procedure for strecker synthesis involves treatment of the aldehyde with ammonium chloride and sodium cyanide (this mixture is equivalent to ammonium cyanide, which in turn decomposes into ammonia and HCN). This procedure is known as **Zelinsky-Stadnikoff modification**[3]. The final step involves the hydrolysis of the formed α-amino nitrile under acidic or basic conditions to give the corresponding α-amino acid (Scheme–3).

$$\text{Ph CH}_2\text{CHO} \xrightarrow{\text{NH}_3,\text{HCN}} \text{Ph CH}_2\overset{\overset{\displaystyle NH_2}{|}}{\text{CHCN}} \xrightarrow{\text{H}_3\text{O}^+} \text{Ph CH}_2\overset{\overset{\displaystyle +NH_3}{|}}{\text{C}} \text{COO}^-$$

Phenyl acetaldehyde   Phenylalanine

$\downarrow \text{NH}_3$ $\uparrow \text{H}_3\text{O}^+ \ \Delta$

$$\left[\text{PHCH}_2\overset{\overset{\displaystyle OH^-}{|}}{\underset{\underset{\displaystyle NH_2}{|}}{\text{CH}}} \right] \xrightarrow{-\text{H}_2\text{O}} \text{PhCH}_2\text{CH}=\text{NH} \xrightarrow{\text{CN}^-} \text{PhCH}_2\overset{\overset{\displaystyle}{}}{\underset{\underset{\displaystyle CN}{|}}{\text{CH}}}-\overset{}{\text{NH}} \xrightarrow{\text{H}_3\text{O}^+} \text{PhCH}_2\overset{}{\underset{\underset{\displaystyle CN}{|}}{\text{CH}}}-\text{NH}_2$$

α-Amino
nitrile

(Scheme–3)

2.1.40.1 Strecker Synthesis using Sonication

Strecker synthesis of α-aminonitriles in much better yield is possible using ultrasonic acceleration[4] (Scheme–4).

$$\text{R}_2\text{CO} \xrightarrow[\text{))))}]{\text{R}'\text{NH}_2,\text{KCN},\text{AcOH}} \text{R}_2\text{C} \overset{\diagup\text{CN}}{\underset{\diagdown\text{NHR}'}{}}$$

Aldehyde
or
Ketone   α-Aminonitriles

(Scheme–4)

A modified strecker synthesis for the preparation of α-amino nitriles in excellent yield involves the adsorption of the reagent on the surface of a catalyst before reaction. In this procedure, side reactions are suppressed[5] (Scheme–5).

$$\text{RCHO} \xrightarrow[\text{50}°, \text{5–48 hr.))))}]{\text{KCN/Al}_2\text{O}_3\text{/CH}_3\text{CN/NH}_4\text{Cl}} \text{R}-\text{CH} \overset{\diagup\text{CN}}{\underset{\diagdown\text{NH}_2}{}}$$

82–100%

(Scheme–5)

References

1. A. Strecker, Ann., 1850, **75**, 25; 1954, **91**, 349; D.T. Mowry, Chem., Rev., 1948, **42**, 236.

2. D.T. Mowry, Chem., Rev., 1948, **42**, 189.

3. K. Weinges, Chem., Ber., 1971, **104**, 3594.

4. J. Menedez, G.G. Trigo and M.M. Solthuber, Tetrahedron Lett., 1986, **27**, 3285.

5. T. Hanafusa, J. Ichichara and T. Ashida, Chem., Lett., 1987, 687.

2.1.41 Suzuki Coupling Reaction

Suzuki coupling reaction[1] is a versatile method for making carbon-carbon bonds. It is a palladium catalysed coupling of an organohalide and an organoborane. It is a convenient method for the synthesis of biaryls (Scheme–1).

$$R{-}X \; + \; R'{-}B{\Big<} \quad \xrightarrow[\text{NaOH}]{\text{Pd (Ph}_3\text{P)}_4 \text{ catalyst}} \quad R{-}R' \; + \; HO{-}B{\Big<} \; + \; NaX$$

R = Alkenyl or alkyl group
X = I or Br
R' can be practically anything

(Scheme–1)

It is advantageous to use the organoborane that has been prepared using catecholborane or 9-BBN, since these reagents have a single B — H bond, and the stoichiometry of the hydroboration reaction is easy to control (Scheme–2).

9–BBN   Catecholborane

(Scheme–2)

Also, with these reagents, the group that derives from the alkene is the only one that is transferred in the coupling reaction (Scheme–3).

Octene–1

Only this group is
transferred in the
suzuki reaction

(Scheme–3)

As an illustration, alkenyl boranes couple with alkenyl or aryl halides as shown below (Scheme–4).

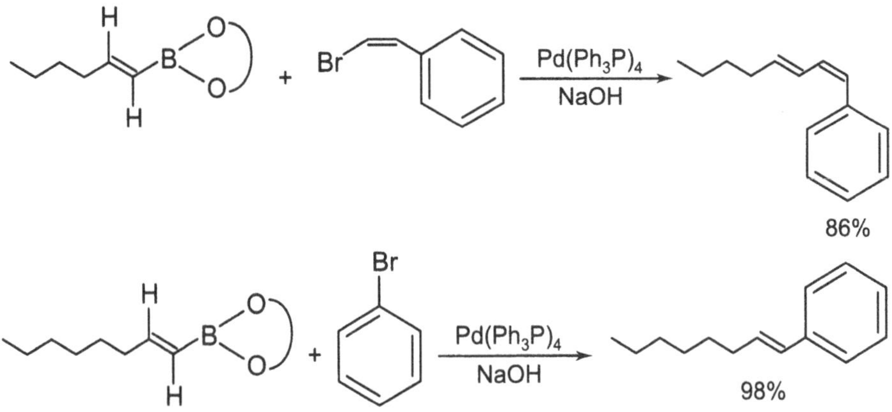

(Scheme–4)

2.1.41.1 Suzuki Coupling in Aqueous Medium

A number of biaryl derivatives were synthesised[3,4] by the reaction of various aryl halides with phenyl boranic acid in aqueous medium using microwave irradiation (Scheme–5).

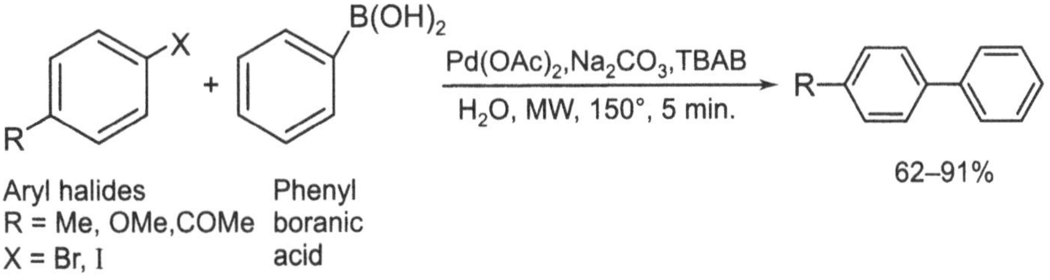

Aryl halides   Phenyl
R = Me, OMe,COMe   boranic
X = Br, I   acid

62–91%

(Scheme–5)

The above reaction gave poor yields if chloride (p-ClC$_6$H$_4$R) is used as the substrate. This problem was overcome by carrying out the reaction of aryl chloride and phenyl boronic acid catalysed by Pd/C in aqueous medium using simultaneous cooling technique in conjugation with M.W. heating[5].

Suzuki coupling reaction has also been used for the synthesis of heterocyclic system. Thus, the reaction imidazo [1,2-*a*] pyridines with aryl boronic acid in aqueous medium under microwave irradiation gave a convenient synthesis of heterocyclic compounds. This method is more efficient than the under conditions[6] (Scheme–6).

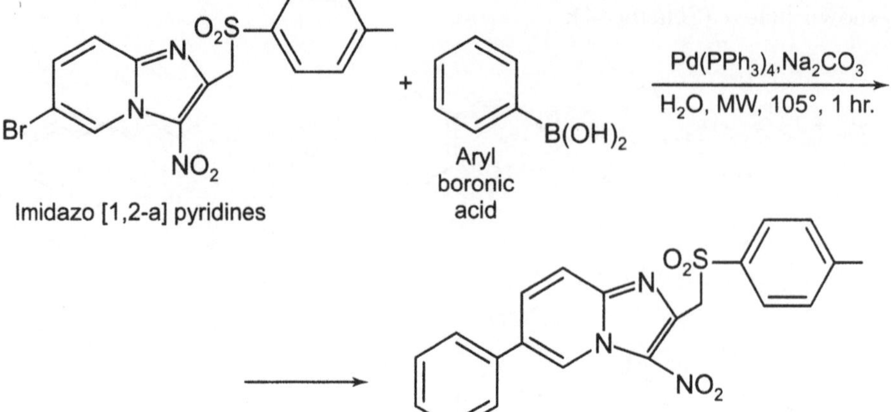

(Scheme–6)

Suzuki coupling reaction was also used for the synthesis of 5-arylthiazole acylnucleosides[7], potential candidates for combating various viruses. The process is an efficient one step procedure in aqueous solution for the synthesis of 5-aryltriazole acylnucleosides (Scheme–7).

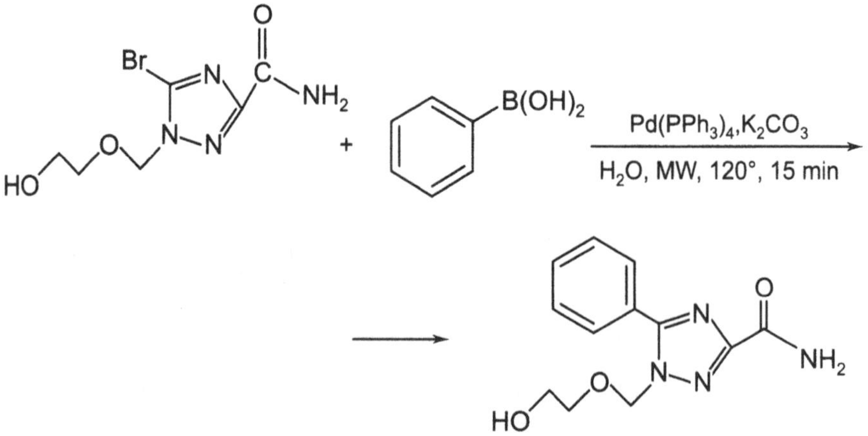

(Scheme–7)

In the suzuki coupling reaction (cited above) the benzothiazole-based Pd(III) complexes (A) and (B) were found to be very active and efficient catalyst (the immobilized catalyst B was found to have high durability compared to mobilized catalyst A). These catalysts can be reused in a number of reactions and this aspect is very important for mass production[8] (Scheme–8).

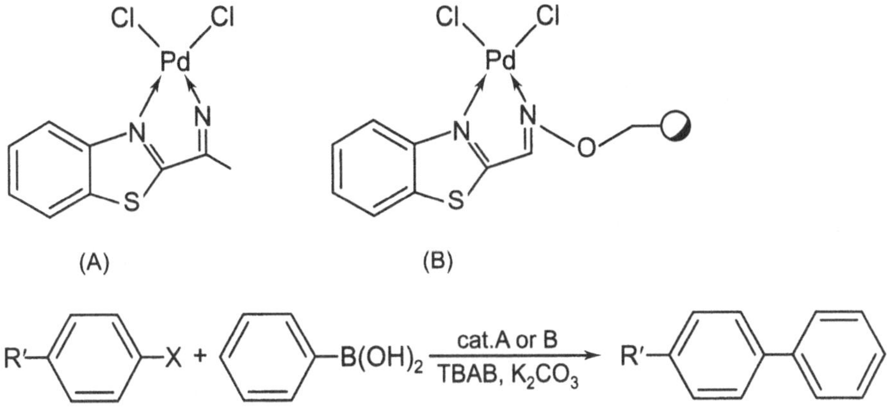

(A)   (B)

X = Cl, Br
R' = Me, OMe, MeCO, NO$_2$

(Scheme–8)

A good substitute for boronic acid in suzuki coupling reaction is sodium tetraphenyl borate (Ph$_4$BNa)[8a] (Scheme–9).

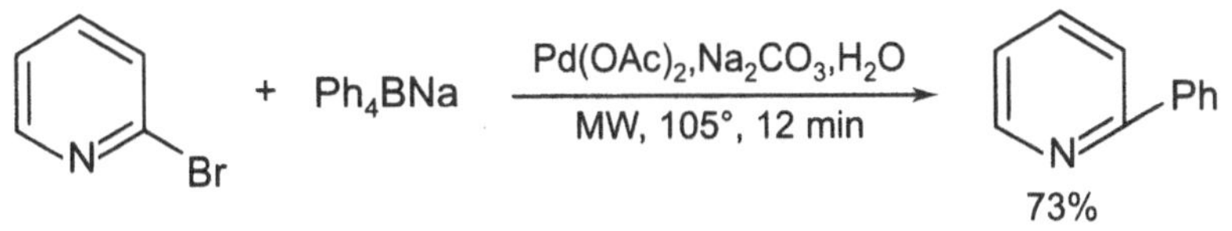

73%

(Scheme–9)

2.41.2 Suzuki Coupling Reaction in Ionic Liquids

The suzuki coupling using a Pd catalyst in an ionic liquid as the solvent gave[9,10] excellent yield and turnover numbers at room temperature. An example in given below (Scheme–10).

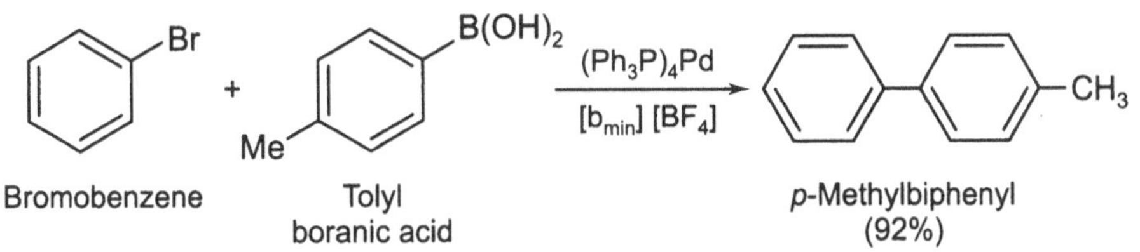

Bromobenzene   Tolyl   *p*-Methylbiphenyl
                  boranic acid   (92%)

(Scheme–10)

The suzuki coupling reaction has also been carried out under mild conditions in an ionic liquid with methanol as a cosolvent (necessary to solubilize the phenyl boranic acid) using ultrasound[11] (Scheme–11).

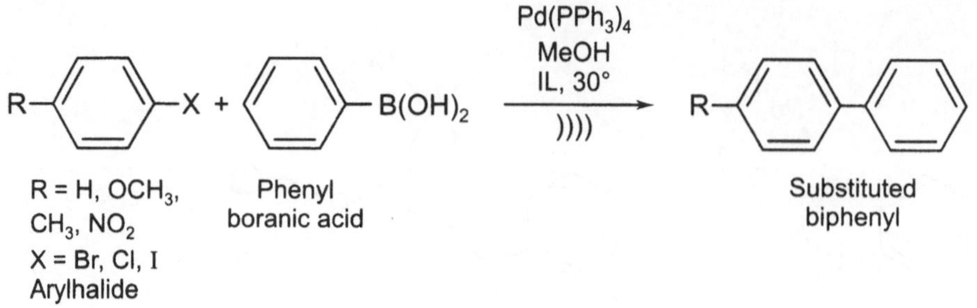

$$R-\langle\!\!\!\rangle-X \ + \ \langle\!\!\!\rangle-B(OH)_2 \quad \xrightarrow[\text{))))}]{\substack{Pd(PPh_3)_4 \\ MeOH \\ IL, 30°}} \quad R-\langle\!\!\!\rangle\!\!-\!\langle\!\!\!\rangle$$

R = H, OCH₃,   Phenyl   Substituted
CH₃, NO₂   boranic acid   biphenyl
X = Br, Cl, I
Arylhalide

R = H, OCH$_3$, CH$_3$, NO$_2$ Phenyl boranic acid   Substituted biphenyl
X = Br, Cl, I
Arylhalide

(Scheme–11)

In the above reaction, due to the formation of inactive Pd black, the recycling of the catalyst was not possible. This problem was overcome by using Pd biscarbene complex as a catalyst and using only methanol under sonochemical conditions (Scheme–12).

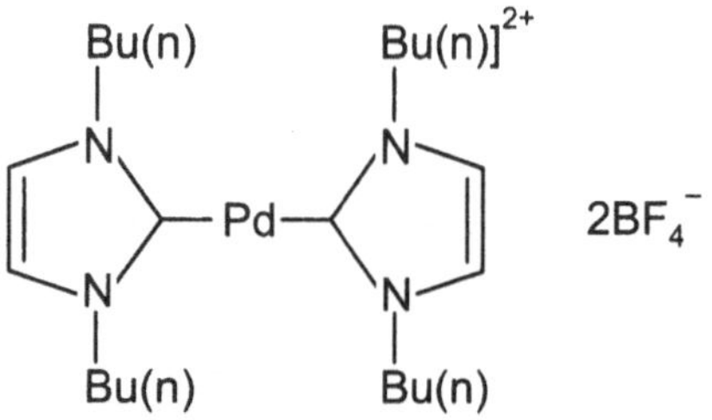

(Scheme–12)

2.1.41.3 Suzuki Coupling Reaction in Polyethylene Glycol (PEG)

In this procedure, substituted aromatic bromides/iodides and aromatic boronic acids in PEG-400 as reaction medium are made to react in presence of a base (potassium fluoride). The reaction in conducted in presence of PdCl₂ using microwave irradiation[12] (Scheme–13).

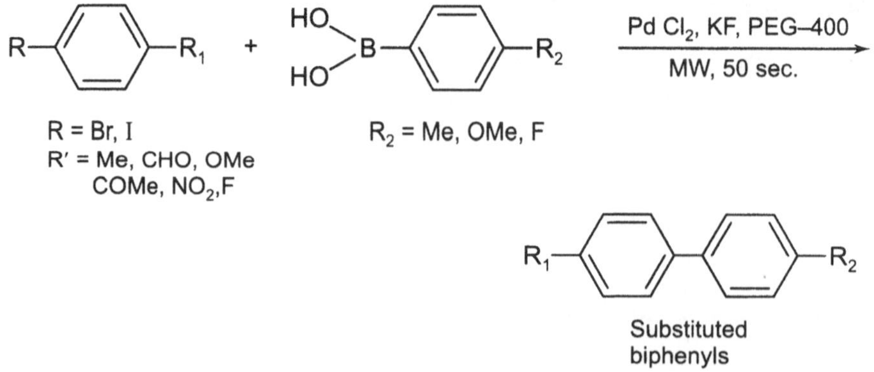

R = Br, I   R₂ = Me, OMe, F
R′ = Me, CHO, OMe
COMe, NO₂,F

Substituted
biphenyls

(Scheme–13)

Suzuki coupling reactions have also been performed using Fluorous Phase Technique[13].

References

1. A. Suzuki, Tetrahedron Lett., 1979, **20**, 3437; 1981, **22**, 127; Tetrahedron, 1983, **39**, 8271; J. Am. Chem., Soc., 1987, **109**, 4756.
2. N.E. Leadbeater, Chem. Commun., 2005, 2881; N. Miyaura and A. Suzuki, Chem., Rev., 1995, **95**, 2457.
3. N.E. Leadbeater and M. Marco, J. Org. Chem., 2003, **68**, 888.
4. L. Bai, J. –X. Wang and Y. Zhang, Green Chem., 2003, **5**, 615.
5. R.K. Arvela and N.L. Leadbeater, Org. Lett., 2005, **7**, 2101.
6. M.D. Crozet, C. Castera-Ducros and P. Vanelle, Tetrahedron Lett., 2006, **47**, 7061.
7. R. Zhu, F. Qu, G. Queleverb and L. Peng, Tetrahedron Lett., 2007, **48**, 2389.
8. K.M. Dawood, Tetrahedron, 2007, **63**, 9642.
8a. R.K. Arvela, N.E.Leadbeater, T.L. Mack and C.M. Kormos, Tetrahedron Lett., 2006, **47**, 217.
9. T. Welton, P.J. Smith, C.J. Mathew, 221st American Chemical Society National Meeting, 1 EC –**311**, 2001.
10. C.J. Methew, P.J. Smith and T. Welton, Chem. Commun., 2000, 1249.
11. R. Rajgopal, D.V. Jarikote and K.V. Srinivasan, Chem., Commun., 2006, 616.
12. V.V. Namboodri and R.S. Varma, Green Chemistry, 2001, **3**, 46.
13. D. Clarke, M.A. Ali, A.A. Clifford, A. Parratt, P. Rose, D. Schwinn, W. Bannwarth and C.M. Rayner, Current Topics in Medicinal Chemistry, 2004, **4**, 740 and the references cited therein.

2.1.42 Trost-Tsuji Coupling Reaction

The coupling reaction[1,2] involving nucleophilic allylic substiution catalysed by Pd(O) complexes is a convenient method for the formation carbon-carbon bonds (Scheme–1).

Methyl geranyl carbonate

Methyl aceto acetate

74% (92% ee)

(Scheme–1)

This reaction has been performed in ionic liquid, [b$_{min}$] [BF$_4$]^3 using Pd(OAc)$_2$ — PPh$_3$/K$_2$CO$_3$ and PdCl$_2$ — TPPTS (TPPTS = triphenyl trisulphonate sodium salt) in [b$_{min}$] [Cl]/cyclohexene^4 (Scheme–2).

(Scheme–2)

References

1. J. Tsuji, Tetrahedron Lett., 1965, 4387; J. Org. Chem., 1985, **50**, 1523; Acc. Chem., Res., 1969, **2**, 144.

2. B. Trost, J. Am. Chem., Soc., 1973, **95**, 292; 1980, **102**, 5699; Acc. Chem., Res., 1980, **13**, 385.

3. W. Chen., L. Xu, C. Chatterton and J. Xiao, Chem., Commun, 1999, 1247.

4. C-de Bellefon, E. Pollet and P. Grenouillet, J. Mol. Catal A. Chem., 1999, **145**, 121.

2.1.43 Ullmann Reaction

It is an useful carbon-carbon bond forming reaction and involves^1 the synthesis of biaryls by copper induced coupling of aryl halides (Scheme–1).

(Scheme–1)

The order of reactivity of aryl halides is ArI > ArBr > ArCl. Aromatic rings having electron withdrawing groups, such as nitro group or carbomethoxy at ortho position, activate the ring towards coupling reaction. Unsymmetrical biaryls can be prepared by taking two different aryl halides. Formation of side products is avoided when one aryl halide is very active and the other is relatively unreactive. (See also Corey, Postner, White codes House Synthesis which is useful for the coupling of the alkyl halides in organocopper reagents, section 5.2.5.3).

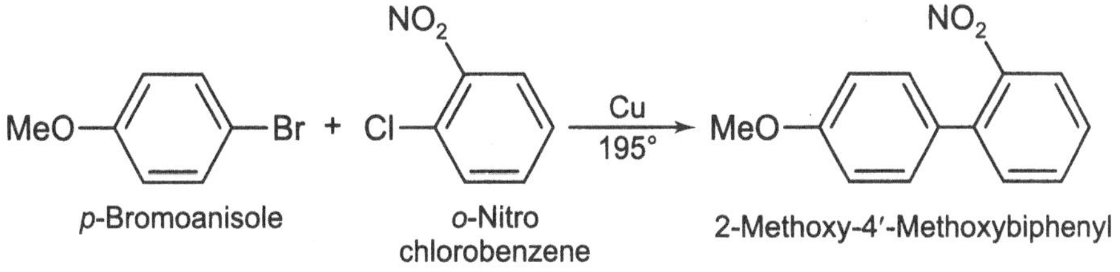
NO2
NO2
MeO
Br
+
Cl
Cu
195°
MeO
p-Bromoanisole
o-Nitro chlorobenzene
2-Methoxy-4′-Methoxybiphenyl

(Scheme–2)

Ullmann coupling takes place via a radical mechanism.

$$\mathrm{Ar\,X} + \mathrm{Cu} \longrightarrow \dot{\mathrm{Ar}} \xrightarrow{\mathrm{Cu}} \mathrm{ArCu} \xrightarrow{\mathrm{ArX}} \mathrm{Ar{-}Ar}$$

(Scheme–3)

2.1.43.1 Ullmann Coupling under Sonication

Under sonication, the size of the copper powder in Ullmann coupling is considerably reduced[2] and this helps in enhanced yields in the coupling. The reactivity is not even hindered by the usual oxide layer on copper powder. The coupling is carried out in dimethylformamide. Much lower yields are obtained in decalin (20%) and toluene (5%) (Scheme–4).

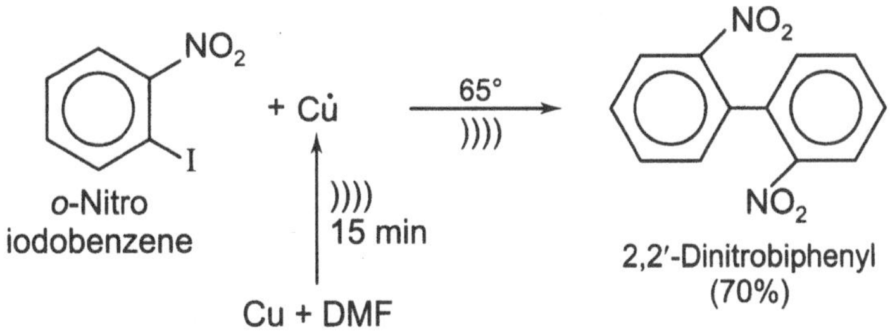
NO2
I
o-Nitro iodobenzene
+ Cu
65°
))))
))))
15 min
Cu + DMF
NO2
NO2
2,2′-Dinitrobiphenyl
(70%)

(Scheme–4)

Another ullmann type coupling is the cross coupling reactions[3,4,5] of perfluoroalkyl zinc reagents with vinyl, alkyl or aryl halides, which can be achieved by using a cleaning bath (35–45 KHz) (Scheme–5).

$$R_f X + R' \overset{X}{=\!\!=} \quad \xrightarrow[\text{))))}]{\text{Zn, Pd}^\circ} \quad R' \overset{R_f}{=\!\!=}$$

$$R_f X + R' \diagdown\!\!\diagup\!\!\diagdown\, Br \quad \xrightarrow[\text{))))}]{\text{Zn, Pd(OAc)}_2} \quad R' \diagdown\!\!\diagup\!\!\diagdown\, R_f$$

(Scheme–5)

2.1.43.2 Ullmann Type Coupling in Water

The homocoupling of arylsulfinic acids in presence of Pd(II) is aqueous solvents gave the biaryls[6] (Scheme–6).

$$2\,\text{ArSO}_2\text{Na} + \text{Na}_2\text{PdCl}_4 \xrightarrow{\text{H}_2\text{O}} \text{Ar—Ar} + 2\text{SO}_2 + \text{Pd} + 4\text{NaCl}$$

(Scheme–6)

The coupling requires stoichiometric amount of palladium. In the presence of hydrogen gas, aryl halides homocoupled to give biaryl compounds in moderate yield[7] (Scheme–7).

$$2\,\text{Ar X} \xrightarrow[\text{H}_2\text{O–BuOH}]{\text{H}_2,\text{Cat. PdCl}_2,\text{K}_2\text{CO}_3} \text{Ar—Ar} \atop 30\text{–}50\%$$

(Scheme–7)

A facile coupling of aryl halides via a palladium-catalysed reductive coupling using zinc in air and aqueous acetone at room temperature using Pd/C as a catalyst was reported[8] (Scheme–8).

$$2\,\text{Ar X} \xrightarrow[\substack{\text{Zn, H}_2\text{O/18–crown 6}\\ \text{air atmosphere, rt,}}]{\text{Pd/C cat.}} \text{Ar—Ar}$$

(Scheme–8)

Using crown ether in the above procedure gave better isolated yields of the products[9] in water alone.

Subsequently, polyethylene glycol (PEG) was used as an additive at elevated temperature. In this case, aryl chloride also worked effectively[10]. Reductive coupling of chlorobenzene to biphenyls give high yields (93–95%) in the presence of catalytic PEG-400 and 0.4 mol % of recyclable heterogeneous trimetallic catalyst (4% Pd, 1% Pt and 5% Bi as carbon)[11].

Carbon dioxide also promotes the palladium-catalysed zinc mediated reductive Ullmann coupling of aryl halides. It is believed that in presence of CO_2, Pd/C, and Zn, various aromatic halides including less reactive aromatic chlorides coupled to give the corresponding homocoupled products in good yields[12].

References

1. F. Ullmann, Ann., 1904, **332**, 38; F. Ullmann and P. Sponagel, Ber., 1905, **38**, 2211; P.E. Fana, Chem., Rev., 1946, **38**, 139.

2. J. Lindley, T.J. Manson and J.P. Lorimer, Ultrasonics, 1987, **25**, 45.

3. K. Kitazumi and N. Ishikawa, Chemistry Lett., 1981, 1679.

4. T. Kitazumi and N. Ishikawa, J. Am. Chem. Soc., 1981, 1679.

5. N. Ishikawa and T. Kitazumi, European Patent, 0082, 252 A1, 1982.

6. K. Garves, J. Org. Chem., 1970, **35**, 3273.

7. D.V. Davydov and I.P. Beletskaya, Russ, Chem., Bull, 1955, **44**, 1139.

8. S. Vankataraman and C. –J. Li, Org. Lett., 1999, **1**, 1133.

9. S. Vankataraman and C. –J. Li, Tetrahedron Lett., 2000, **41**, 4831; S. Venkataraman, T. Huang and C. –J. Li, Adv. Synth. Catal., 2002, **344**, 399; S. Mukhopadhyay, A. Yaghmur, M. Baidossi, B. Kundi and Y. Sasson, Org. Process Res. Dev., 2003, **7**, 641.

10. S. Mukhopadhyay, G. Rothenberg, D. Gitcs and Y. Sasson Org. Lett., 2000, **2**, 211.

11. S. Mukhopadhyay, G. Rothenberg and Y. Sasson, Adv. Synth. Catal., 2001, **343**, 274.

12. J. –H. Li, Y. –X. Xie and D. –L. Yin, J. Org. Chem., 2003, **68**, 9867; J. –H. Li, Y. –X. Xie, Chin. J. Chem., 2004, **22**, 966; J. –H. Li, Y. –X. Xie, J. Jiang, and M. Chen, Green Chem., 2002, 4, 424.

2.1.44 Vilsmeier Reaction

Also known as **Vilsmeier-Haack Reaction**, it consists[1] in the reaction of activated aromatic or heterocyclic compounds with disubstituted formamides and phosphorus oxychloride to give aldehydes (Scheme–1).

(Scheme–1)

Reference

1. A. Vilsmeier, A. Haack, Ber., 1927, **60**, 119; M.R. Demaheas, Bull Soc., Chim. France, 1962, 1989, L.N. Ferguson, Chem., Rev., 1946, **38**, 230.

2.1.45 Weiss-Cook Reaction

The reaction of dimethyl-3-oxyglutarate with glyoxal in aqueous acidic solution gives [3.3.0] octane-3,7-dione-2,4,6,8-tetracarboxylate, which on acid hydrolysis followed by decarboxylation gives cis-bicyclo [3.3.0] octane-3,7-dione. The reaction is believed to involve double Knoevengal reaction that gives[2] an α, β-unsaturated-γ-hydroxy cyclopentenone, which reacts with another molecule of dimethyl-3-oxyglutarate by Michael addition.

(Scheme–2)

References

1. W. Weiss and J.M. Edwards, Tetrahedron Lett., 1968, 4885.
2. C. Mannich and W. Krosche, Arch. Pharm., 1912, **250**, 647; F.E. Blike, Organic Reactions, 1942, **1**, 303.

2.1.46 Wurtz Reaction

It involves[1] the coupling of alkyl halides with sodium in dry ether to give hydrocarbons (Scheme–2).

$$CH_3CH_2CH_2Br \xrightarrow[\text{ether}]{\text{Na}} CH_3(CH_2)_4CH_3$$

n-Propylbromide   *n*-Hexane

(Scheme–1)

Using two different alkyl halides results in the formation of three different alkanes (RX + R'X $\longrightarrow$ R — R + R' — R' + R — R').

Coupling of alkyl halides with grignard reagents or RLi in presence of Cobalt chloride also give hydrocarbons. This is a modified form of Wurtz reaction (Scheme–2) (See organometallic compound, section 2.1.29).

$$C_6H_5MgBr \quad + \quad C_6H_5CH_2Br \xrightarrow{CoCl_2} C_6H_5CH_2C_6H_5$$

Phenyl   Benzyl   Diphenylmethane
magnesium   bromide
bromide

(Scheme–2)

The reaction is known as **Kharasch reaction**. Another procedure for carbon-carbon bond formation is the reaction of alkyl halide with alkyl lithium (Scheme–3).

$$R—X \quad + \quad R'—Li \quad \longrightarrow \quad R—R' \quad + \quad LiX$$

Alkyl halide   Alkyl
               lithium

(Scheme–3)

This procedure gives hydrocarbons of the type R—R', which are difficult to prepare by other methods (See also organometallic compound, lithium reagents section 2.1.29.2. Also see Corey. Posner-Whitesides-House synthesis, Interconversion of Functional Groups, page 251).

Another variation of Wurtz reaction is known as **Wurtz-Fitting reaction**[2]. In this procedure, an aryl halide and an alkyl halide couple to form alkylated aromatic compounds (Scheme–5).

$$C_6H_5Br \quad + \quad n—C_4H_9Be \xrightarrow{Na} C_6H_5C_4H_9 \quad + \quad 2NaBr$$

Bromobenzene   *n*-Butylbenzone   *n*-Butylbenzene

(Scheme–5)

The biproducts in this reaction are R — R and Ar — Ar, which can be easily separated.

2.1.46.1 Wurtz Reaction under Sonication

Under sonication, Wurtz reaction gives much better yields see (Ullmann coupling under sonication, section 2.1.43.1).

2.1.46.2 Wurtz Reaction in Water

Wurtz coupling can also be carried out[3] in water using Zn/H_2O (Scheme–6).

(Scheme–6)

References

1. A. Wurtz, Ann. Chem., Phys., 1885, **3(44)**, 275; Ann., 1855, **96**, 364; R.E. Buntrock, Chem., Rev., 1968, **68**, 209.
2. B. Tollens and R. Fittig, Ann., 1864, 131, 303; R. Fittig and J. Koneg., Ann., 1867, **144**, 277.
3. C.J. Li and T.H. Chan, Organometallics, 1991, **10**, 2548.

2.1.47 Miscellaneous Procedures for Carbon-Carbon Bond Formation

2.1.47.1 Electrochemical Synthesis

Electrolysis of sodium or potassium salt of carboxylic acid to give hydrocarbons is known as **Kolbe electrolytic synthesis**[1] (Scheme–1).

(Scheme–1)

Using this procedure adiponitrile (a raw material for preparation of hexamethylene diamine and adipic acid, which are used for the manufacture of Nylon-66) are obtained[2] commercially by electrolysis of acrylonitrile in presence of a concentrated solution of quaternary ammonium salts (QASs), such as tetraethyl ammonium-p-toluene sulfonate using lead or mercury cathode (Scheme–2).

(Scheme–2)

Another important intermediate sebacic acid (for the manufacture of polymide resins) is obtained by electrochemical process by electrolysis of monomethyl ester of adipic acid, followed by the hydrolysis of the formed dimethyl ester of sebacic acid (Scheme–3).

$$HOOC-(CH_2)_4-COOH \xrightarrow[\text{esterification}]{CH_3OH} HOOC-(CH_2)_4-COOCH_3 \longrightarrow$$

Adipic acid   Monomethylester of adipic acid

$$\xrightarrow[55°]{\text{Electrolysis}} CH_3OOC(CH_2)_8 COOCH_3 \xrightarrow{\text{Hydrolysis}} HOOC(CH_2)_8COOH$$

Dimethylester of sebacic acid   Sebacic acid

(Scheme–3)

In the above process, anodic coupling of the monomethyl ester of adipic acid takes place. The electrolyte is a 20 per cent aqueous solution of monomethyl adipate, neutralised with sodium hydroxide. The anode is a platinium plated with titanium and cathode is of steel[3].

Some other electrochemical synthesis are:

- Coupling of acetone to yield pinacol[4].
- Epoxidation of alkenes[5]
- Conversion of alkenes to ketones[6]
- Oxidation of primary alcohols to carboxylic acids[7]

An interesting electrochemical synthesis in the conversion of halogen into COOH. This procedure is used for the synthesis of Ibuprofin[8] (Scheme–4).

(Scheme–4)

2.1.47.2 Carbon-Carbon Bond Formation Involving Carbocations

In aromatic compounds, the formation of carbon-carbon bond is well known (see Friedel-Crafts alkylation and acylation, section 2.1.17.1 and 2.1.17.6). The carbocation can also react with alkenes to form a carbon-carbon bond. However, in this case a new carbocation is obtained, which in turn attacks the alkene and the reaction continues till a polymer is obtained as in the case of polyethylene.

(Scheme–5)

For utility of the above procedure, it is necessary to have a termination step. This has been successfully achieved by using alkenyl and allylic silanes and stannanes with electrophilic carbon (See organosilanes and organo stannanes, section 2.1.29.3 and 2.1.29.4). In these reactions, the silyl or stannyl substituent is elimated, a stable alkene with a new carbon-carbon bond is formed (Scheme–6).

(Scheme–6)

Using this procedure, the synthetic utility of carbocation-alkene reaction is achieved. As an example, trimethylsilyl ether of cyclopentanone on reaction with the carbocation generated from tertiary chlorides and a Lewis acid, such as $TiCl_4$ to give the carbon-carbon bond formed product[9].

Trimethylsilyl
ether of cyclopentanone

$TiCl_4$,–50°

$(CH_3)_2C$—CH_2CH_3
|
Cl

62%

(Scheme–7)

The above is a one step process involving the reactions of trimethylsilyl ether of cyclopentanone with the tertiary alkyl halide with $TiCl_4$ at $-50°$.

Secondary benzylic bromides and allylic bromides also undergo analogous reaction using $ZnBr_2$ as catalyst[10]. However, primary iodides react with silyl enol ethers in presence of Ag_2CCF_3[11].

In some cases, carbocations undergo molecular rearrangement and may lead to ring expansion. A typical example is given below (Scheme–8).

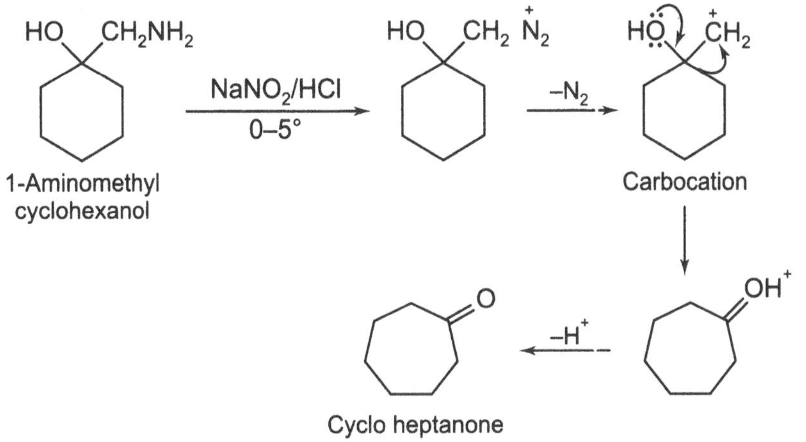

(Scheme–8)

2.1.47.3 Carbon-Carbon Bond Formation Involving Carbanions

A number of reaction leading to carbon-carbon bond formation include carbanions as intermediates. These include Claisen-Schmidt condensation, Dieckmann condensation, Perkin reaction *etc.*, (Sections, 2.2.5., 2.1.13, 2.2.12).

In certain cases, the carbanion, can be oxidised with one-electron oxidizing agents (*e.g.*, iodine) to give dimerised product (Scheme–9).

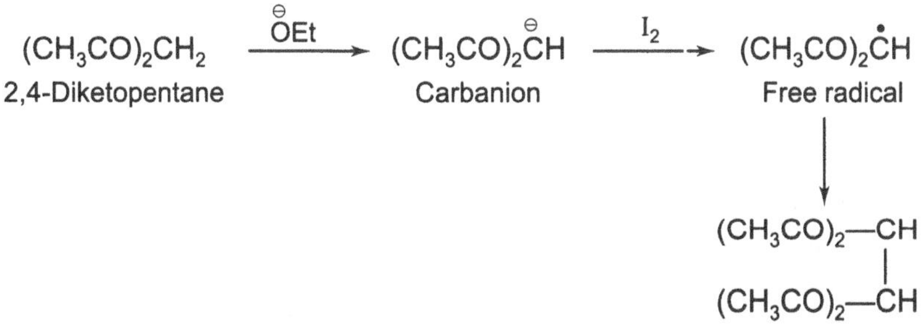

(Scheme–9)

This is a useful reaction forming a C — C bond via dimerisation.

2.1.47.4 Carbon-Carbon Bond Formation Involving Free Radicals

The polymerisation of ethylene to polyethylene in presence of a free radical initiator, such as benzoyl peroxide is well known. The polymerisation involve C — C bond formations via free radical mechanism. Besides this, a number of reactions leading to carbon-carbon bond formation include free radicals as intermediates. These include Sandmeyer reaction, Gomberg reaction, Wurtz-Fittig reaction, Hunsdiecker reaction, Kolbe's electrolytic method *etc.*

2.1.47.5 Carbon-Carbon Bond Formation Involving Carbenes

The well known **Simmon-Smith reaction** involving the reaction of an alkene with chloroform and alkali to form the corresponding cyclopropane proceed via the formation of carbene. Some other reactions involving carbene intermediate are Reimer-Tiemann reaction, carbylamine reaction, Wolff rearrangement *etc.*

A typical reaction is the conversion of indole into 3-chloroquinoline by reaction with chloroform and alkali. The reaction involves ring expansion (Scheme–10).

(Scheme–10)

2.1.47.6 Carbon-Carbon Bond Formation via Electrophilic Substitution of Aromatic Hydrogens

The well-known example is the Friedel-Crafts type reactions. A typical example is the reaction of indole derivatives with equimolar amounts of 3 per cent aqueous CH_2O and 33 per cent Me_2NH at 70–75° in 96 per cent ethanol gave Mannih type products[12]. A lanthanide catalysed reaction of indole with benzadehyde in ethanol water system gave the condensation product[13] (Scheme–11).

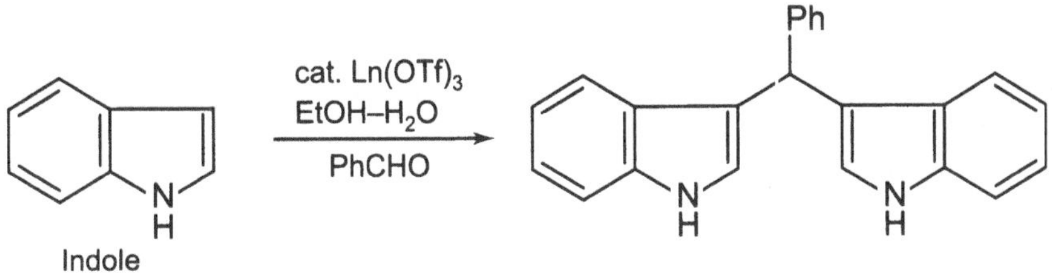

(Scheme–11)

The reaction of N-methyl indole and N-methylpyrrole via Friedel-Crafts reaction with $OCHCO_2Et$ in aqueous medium yielded substituted indoles and pyrroles[14] (Scheme–12).

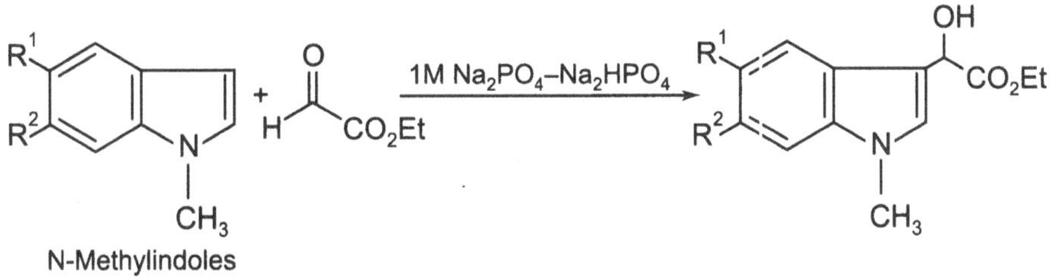

(Scheme–12)

2.1.47.7 Carbon-Carbon Bond Formation using Enamines

For details see Enamine reaction (section 2.1.15).

2.1.47.8 Carbon-Carbon Bond Formation using Sonogashira Reaction

Though it is a powerful method for the creation of carbon-carbon bonds, but is also a general method for the preparation of unsymmetrical alkynes. For details see sonogashira reactions (section 2.3.2).

2.1.47.9 Carbon-Carbon Bond Formation using Stille Coupling Reaction

Besides $C = C$ double bond formation, still coupling reaction has also been used for the formation of carbon-carbon single bonds.

For details see stille coupling reaction section 2.2.13.

References

1. H. Kolbe, Ann., 1849, **69**, 257; B.C.L. Weldon, Quart. Rev., 1952, **6**, 380.
2. D.E. Danly and C.J.H. King in Organic Electrochemistry, 3rd ed., H. Lund and M.M.

Baizer, Eds., Marcel Dekker, New York, 1991; M.M. Baizer and D.E. Danly, Chemtech, 1980, **10(3)**, 161.

3. T. Isoya, R. Kakuka and C. Kawamura, U.S. Patent, 3, 896, 611 (1975)-Asahi Kasu.

4. T.T. Sugano, B.A. Schenber, J.A. Walburg and N. Shuster, U.S. Patent 3, 992, 269 (1976) to Diamrock Corpn.

5. J. Yoshida, J. Hasmoto and N. Kawabatu, J. Org. Chem., 1982, **47**, 3575.

6. J. Tsuji and M. Minato, Tetrahedron Lett., 1987, **28**, 3683.

7. J. Kaulen and H.J. Schafer, Synthesis, 1979, 513.

8. A. Sasaki, H. Kudoh, H. Senboku and M. Kokuda, in Novel Trends in Electrochemical Synthesis, S. Torii Ed., Springer-Verlag, Tokyo, 1998.

9. M.T. Reetz, I. Chatziiosifidis, U. Löwe and W.F. Maier, Tetrahedron Lett., 1979, 1427; M.T. Reetz, I. Chatziiosifidis, F. Hübner and H. Heimbach, Org. Synth., 1984, **62**, 95.

10. I. Paterson, Tetrahedron Lett., 1979, 1519.

11. C.W. Jefford, A.W. Sledeski, P. Lelandais and J. Bolukouvalas, Tetrahedron Lett., 1992, **33**, 1855.

12. J. Thesing and P. Binger, Chem., Ber., 1957, 90, 1419.

13. D. Chen, L. Yu and P.G. Wang, Tetradron Lett., 1996, **37**, 4467.

14. W. Zhuang and K.A. Jorgensen, Chem., Commun., 2002, 1336.

2.2 CARBON-CARBON DOUBLE BOND (C = C) FORMATION

2.2.1 Introduction

Next to carbon-carbon single bond (C — C), the carbon-carbon double bond (C = C) formation is equally important for organic synthesis. Following are given some of the common strategies for C = C formation.

- Reduction of Alkynes (for details, see section 2.3.1).

- The reaction of Grignard reagents (RMgX) with unsaturated alkyl halides. For details see section 2.1.29.1.

- Coupling reaction of lithium dialkyl cuprate with vinyl halide. For details see section 5.2.5.3.

- Dehydration of alochol. For details see section 5.2.6.3.

- Alkenes are known to be obtained by the elimination reactions like dehydrohalogenation of alkyl halides.

Besides the above procedure, $C = C$ bond formation can also be affected by other well known reactions. These are given below.

2.2.2 Bamford-Stevens Reaction

It involves the formation of alkenes by base catalysed decomposition of p-toluenesulfonyl hydrazones of aldehydes and ketones[1] (Scheme–1).

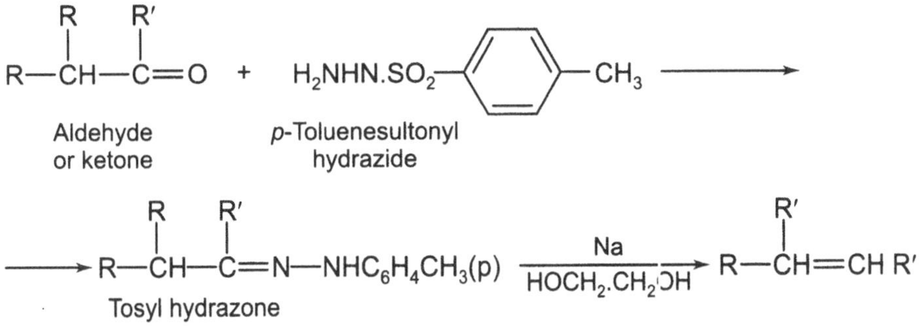

(Scheme–1)

The reaction is believed to take place as shown below (Scheme–2).

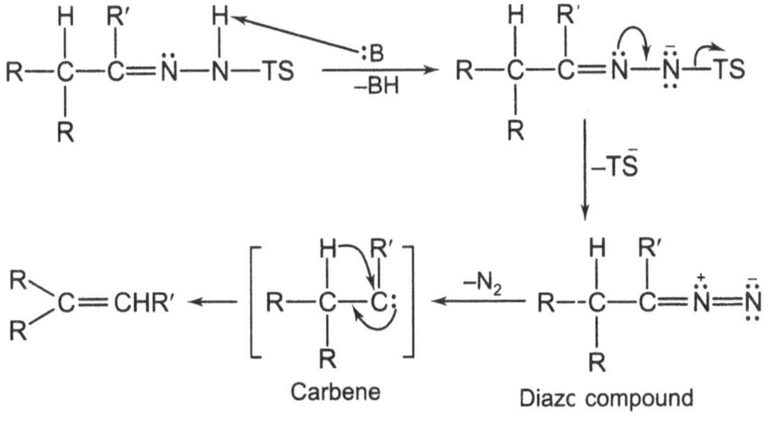

Carbene   Diazo compound

(Scheme–2)

Reference

1. W.R. Bamford and T.S. Stevens, J. Chem., Soc., 1952, 4735; A.G.M. Barrett., Acc. Chem., Res., 1983, **16**, 55.

2.2.3 Baylis-Hillman Reaction

It is also known as **Baylis-Hillman Vinyl Alkylation reaction**[1] and is a very useful C — C and $C = C$ bond forming reaction. It consists in the reaction of aldehydes with acrylates to give α-(hydroxyalkyl) acrylates or of vinyl ketones to α-hydroxyalkyl) vinyl ketones[2]. A typical example is given below (Scheme–1).

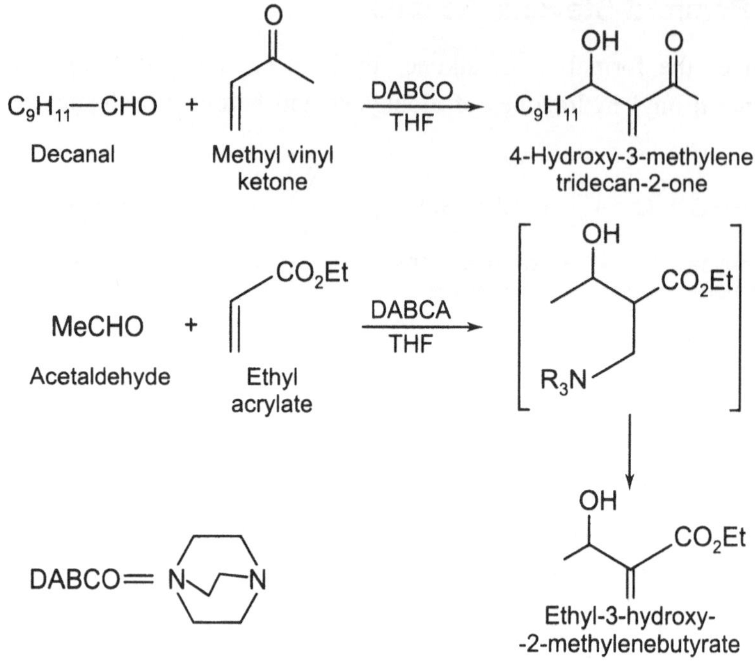
O
OH O
C_9H_{11}—CHO +
DABCO
THF
C_9H_{11}
Decanal
Methyl vinyl ketone
4-Hydroxy-3-methylene tridecan-2-one
OH
CO_2Et
CO_2Et
MeCHO +
DABCA
THF
Acetaldehyde
Ethyl acrylate
R_3N
OH
CO_2Et
DABCO= N N
Ethyl-3-hydroxy- -2-methylenebutyrate

(Scheme–1)

2.2.3.1 Baylis-Hillman Reaction using Microwaves

The use of microwave improved[3] the yield of the product as well as reduction in reaction time. Thus, the reaction of benzaldehyde with methyl crotonate in presence of DABCO gave the product in good yield in 10 min (Scheme–2).

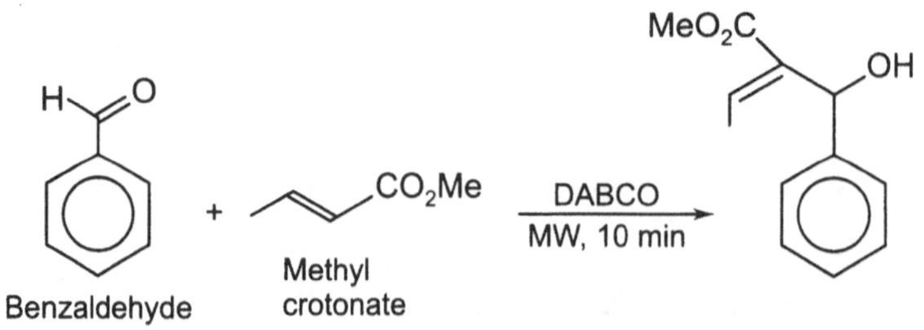
MeO_2C
OH
H O
CO_2Me
+
DABCO
MW, 10 min
Benzaldehyde
Methyl crotonate

(Scheme–2)

2.2.3.2 Baylis-Hillman Reaction in Supercritical Carbon Dioxide

The Baylis-Hillman reaction has been carried out in SC-CO$_2$ and gives better conversion and reaction rates[4] as compared to solution phase reactions (Scheme–3).

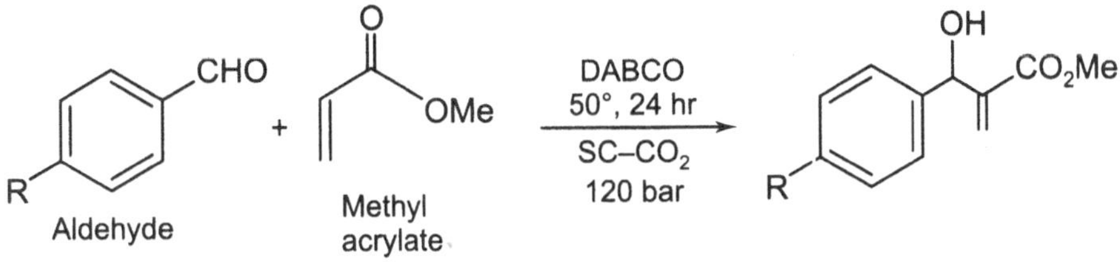

(Scheme–3)

In case, Baylis-Hillman reaction is carried out in presence of an alcohol, the major product is an ether from a three component coupling reaction, which occurs only in the presence of SC–CO$_2$ (Scheme–4).

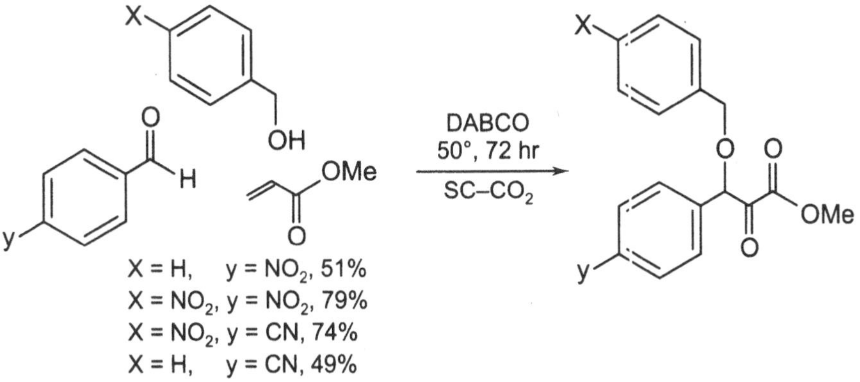

X = H,   y = NO$_2$, 51%
X = NO$_2$, y = NO$_2$, 79%
X = NO$_2$, y = CN, 74%
X = H,   y = CN, 49%

(Scheme–4)

2.2.3.3 Baylis-Hillman Reaction in Ionic Liquids

The Baylis-Hillman reaction has been conducted in ionic liquids[5]. It is believed[5] that the reaction proceeds via an addition-elimination mechanism. The formed zwitterionic species(A) attacks, the aldehyde to give the product (Scheme–5).

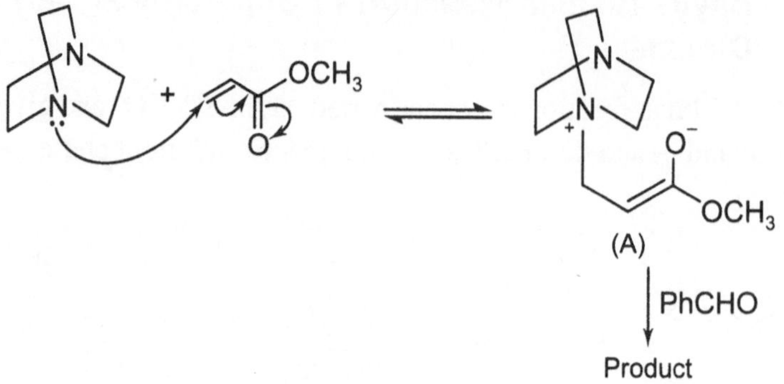

(Scheme–5)

Though a number of tertiary amines have been used, as the base. However the choice is diazabicyclo [2.2.2] octane (DABCO).

The Baylis-Hillman reaction between benzaldehyde and methyl acrylate in the ionic liquid $[b_{min}]$ $[PF_6]$ was formed to be 33 times faster than the reaction is CH_3CN, although only moderate yield of the desired product was obtained[6] (Scheme–6).

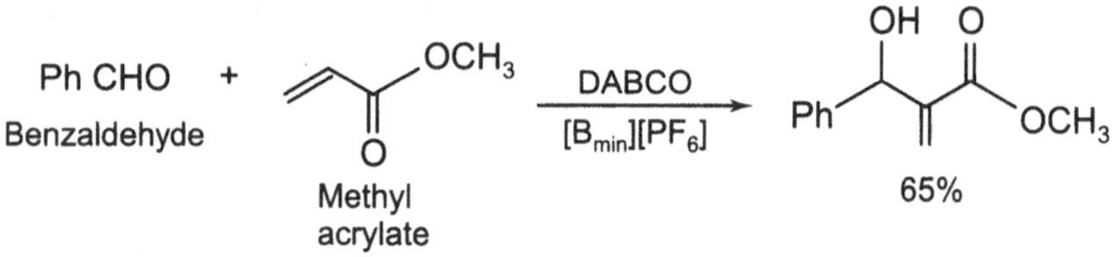

(Scheme–6)

It was found[6] that under the basic reaction conditions, the aldehyde was being consumed with imidazolium cation. This accounts for the low yields and also demonstrated that ionic liquids are not always inert solvents[6]. Thus, it was shown that the acidic nature of the C(2) hydrogen of the imidazolium cation was responsible for the side reaction (Scheme–7).

(Scheme–7)

In each successive recycle, more and more of the addition product is formed between the deprotinated imidazolium salt and the aldehyde was consumed. Thus, less of ionic liquid was left for reacting with the aldehyde to give the side reaction product and so more of aldehyde was available for the normal Baylis-Hillman reaction. Thus, for subsequent recycling the yield improved.

On the basis of results obtained, it was concluded that caution must be exercised when using ionic liquids from one reaction for another; in such case, a mixture of products were obtained.

In order to overcome the problem of low yields due to the formation of side reaction products (because of the acidity of C(2) imidazolium cation), ionic liquids substituted at the position 2 were used[6a]. It was found[6a] that the Baylis-Hillman reaction between a variety of aldehydes and methyl acrylate proceeded smoothly in the ionic liquid [b_{min}] [PF_6], in contrast to the results obtained with [b_{min}] [PF_6].

The Baylis-Hillman reaction has been succesfully carried out in presence of imidazolinium-based ionic liquids having a phenyl group at C(2) position [cation = mPh$_{min}$(A),] which are stable to a variety of strong basic conditions[6b].

(X = Br, TfN, PF$_6$)

The Baylis-Hillman reaction has been performed in presence of chloroaluminate ionic liquids. It has been found that 1-ethyl-3-methylimidazolium chloride [E_{MIC} + $AlCl_3$] is a more efficient chloroaluminate ionic liquid.

2.2.3.4 Baylis-Hillman Reaction in Polyethylene Glycol (PEG)

Polyethylene glycol (PEG-400) has been used[8] for the Baylis-Hillman reaction using the conventional basic catalyst, DABCO, between unreactive aldehydes and olefins. Thus, the reaction of benzaldehyde, ethyl acrylate and DABCO in PEG-400 at room temperature for 2 hr. gave the Baylis-Hillman product in 92 per cent yield. Similar products were obtained by the reaction of benzaldehyde with acrylonitrile or methyl vinyl ketone (Scheme-8).

(Seheme-8)

4-Nitrobenzaldehyle reacted with ethyl acrylate and acrylonitrile giving the expected products in 90 and 93 per cent yield, respectively after 2 hr. 4-Fluorobenzaldehyde reacted much faster forming the expected product in 4 hr; the same reaction took over 60 hrs. using Et_3N (Scheme-9).

4-Nitrobenzaldehyde, R = NO_2
4-Fluorobenzaldehyde, R = F

(Scheme–9)

Some other aldehydes like 2-chloro-5-nitrobenzaldehyde showed similar result. Aliphatic aldehydes like 3-phenylproponal ($C_6H_5CH_2CH_2CHO$), isobutyraldehyde [$(CH_3)_2CHCHO$], and hexanal ($\sim\!\sim\!\sim\!CHO$) also underwent Baylis-Hillman reaction with activated olefins like acrylo nitrile and ethyl acrylate (75, 86 and 80 per cent yields, respectively). Also, formaldehyde and trans cinnamaldehyde reacted with acrylates in PEG to provide good yields of the expected products.

Polyethyene glycol is a rapid and recyclable medium for the Baylis Hillman reaction, which is one of very few reactions where is there in 100 per cent atom economy.

The Baylis-Hillman reaction is also known as **Morita-Baylis-Hillman reaction** (MBH reaction)[4,9] and has also been carried out in supercritical carbon dioxide (see section 2.2.3.2).

References

1. A.B. Baylis and M.E.D. Hillman, Ger. Pat. 215, 5113, C.A. 1972, **77**, 3417.
2. D. Basavaiah, Tetrahetron Lett., 1986, **27**, 2031; 1987, **28**, 4591 and 4351; 1990, **31**, 1621.
3. M.K. Kundu, S.B. Mukerjee and N. Balu, R. Padmakuma and S.V. Bhat, Synlett., 1994, 444.
4. P.M. Rose, R.A. Clifford and C.M. Rayner, Chem. Commun., 2002, 968.
5. D. Basavaiah, P.D. Rao and R.S. Huma, Tetrahedron, 1996, **52**, 8001.
6. V.K. Aggarwal, A. Mereu, G.J. Tarver and R. McCague, J. Org. Chem., 1998, **63**, 7183.
6a J.–C. Hsu, Y.–H. Yen and Y.–H. Chu, Tetrahedron Lett., 2004, 45, 4673.
6b W. Jucik and R. Wilheim, Green Chem., 2002, **7**, 844.

7. A. Kumar and S.S. Pawar, J. Mol. Catal. A: Chem, 2004, **211**, 43.

8. S. Chandrasekher, Ch. Narsihmulu, B. Saritha and S.S. Sultana, Tetrahedron Lett., 2004, **45**, 5865.

9. D. Basavaiah, A.J. Rao and T. Satyanarayana, Chem. Rev., 2003, **103**, 811.

2.2.4 Claisen Reaction

It is a base catalysed reaction between an aldehyde or a ketone with no α-hydrogen atom and an ester, which has an active hydrogen to give on α, β-unsaturated ester (Scheme-1)

$$C_6H_5CHO + CH_3COOEt \xrightarrow{\ ^-OEt\ } C_6H_5CH=CHCOOEt$$

Benzaldehyde   Ethyl   Ethyl cinnamate
               acetate

(Scheme-1)

Reference

1. L. Claisen, Ber., 1912, **45**, 3157; L. Claisen and E. Tietz., Ber., 1925, **58**, 275; 1926, **59**, 2344; D.S. Tarbell, Chem. Rev., 1940, **27**, 495.

2.2.5 Claisen-Schmidt Reaction

The condensation of aromatic aldehydes (without α-hydrogen) with an aliphatic aldehyde or ketone (having α-hydrogen) in presence of strong base (hydroxide or alkoxide ion) to form α, β-unsaturated aldehydes or ketones is known as Claisen-Schmidt condensation (Scheme-1)

$$C_6H_5CHO + CH_3CHO \xrightarrow{\ 10\%\ NaOH\ } C_6H_5CH=CHCHO$$

Cinnamaldehyde

$$C_6H_5CHO + CH_3COCH_3 \xrightarrow[\text{2–3 days 70\%}]{\ 10\%\ NaOH,\ 30°\ } C_6H_5CH=CHCOCH_3$$

Benzylidine acetone
(Benzalacetone)

$$C_6H_5CHO + CH_3COC_6H_5 \xrightarrow[\text{<30°, 85\%}]{\ 10\%\ NaOH\ } C_6H_5CH=CHCOC_6H_5$$

Benzylidene acetophenone
(Benzalactophenone)

(Scheme-1)

In above synthesis, the concentration of alkali is crucial, otherwise **Cannizzaro Reaction** may give major products.

2.2.5.1 Claisen-Schmidt Reaction using PTC

Phase transfer catalysts, such as cetyl trimethyl ammonium compounds (like CTACl, CTABr, (CTA)$_2$SO$_4$ and CTAOH have been successfully used for Claisen-Smith reaction of acetophenones with benzaldehydes (Scheme-2) under weak alkaline conditions. This permits the synthesis of biologically interesting compounds, such as chalcones and flavanols in water only[2,3].

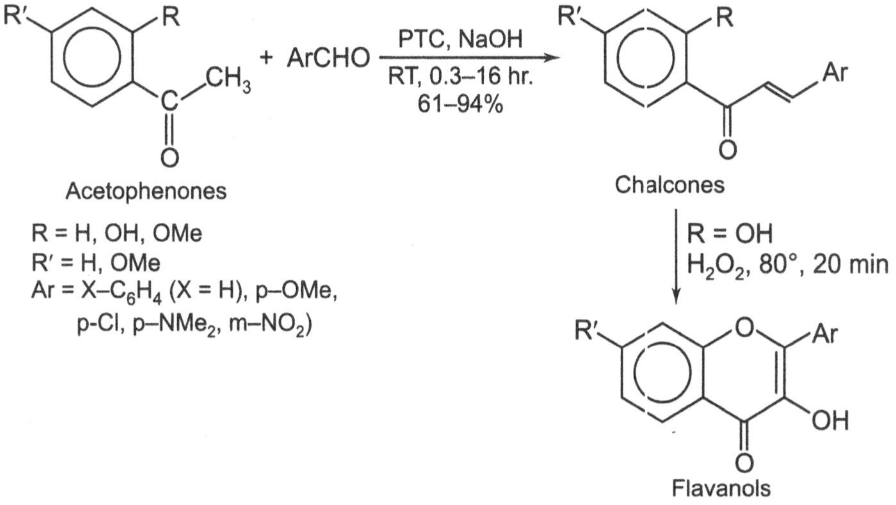

Acetophenones

R = H, OH, OMe
R' = H, OMe
Ar = X–C$_6$H$_4$ (X = H), p–OMe,
p-Cl, p–NMe$_2$, m–NO$_2$)

Chalcones

R = OH
H$_2$O$_2$, 80°, 20 min

Flavanols

(Scheme–2)

2.2.5.2 Claisen-Smith Reaction in Ionic Liquids

The Claisen-Smith reaction between acetophenone and benzaldehyde (Scheme-3) in presence of sodium hydroxide using Ionic liquid, [bmin] [PF$_6$] gave low yields of the condensation product (Scheme–3). The reason for low yields by using [bmin] [PF$_6$] has already been explained (see Baylis-Hillman reaction, section 2.2.3).

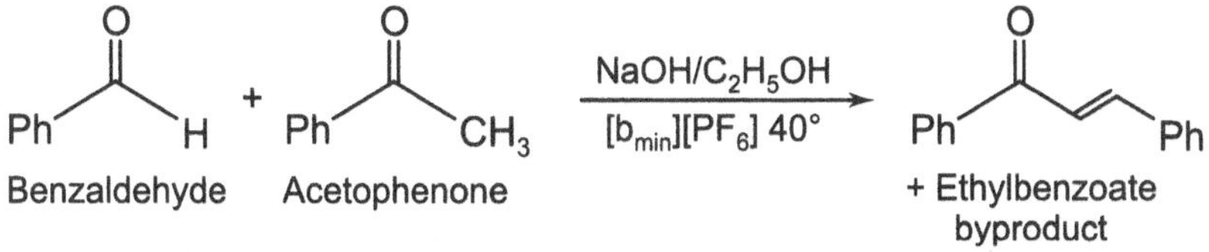

Benzaldehyde   Acetophenone

+ Ethylbenzoate
byproduct

(Scheme–3)

It is best to carry out the above Claisen-Schmidt reaction in presence of imidazolium based ionic liquid having a phenyl group at (2) position.[5]

$(X = Br, Tf_2N, PF_6)$

A reaction related to Claisen-schmidt reaction is **Mukaiyama reaction**[6]. It involves the reaction of silyl enol ether of the ketone with an aldehyde in an organic solvent in presence of $TiCl_4$ (Scheme-4).

Manicone
(an alarm pheromone)

(Scheme–4)

References

1. L. Claisen and A. Claparede, Ber., 1981, **14**, 2460; J. G. Schmidt, Ber., 1881, **14**, 1459.
2. K. R. Nivalkar, C.D. Mudaliar and S.H. Mashraqui. J. Chem. Res (s), 1992, 98.
3. F. Fringulli, G. Pani, O. Piermatti and F. Pizzo, Tetrahedron, 1994, **50**, 11499; F. Fringulli, G. Pani, O. Piermath and F. Pizzo, Life. Chem. Rep., 1995, **13**, 133.
4. P. Formentin, H. Garcia and A.J. Mol., Catal A. Chem., 2004, 214, 137.
5. V. Jurcik and R. Wiehelm, Green Chem., 2002, **7**, 844.
6. T. Mukaiyama, K. Banno and K. Narasaka, J. Am. Chem. Soc., 1974, **96**, 7503; T. Mukaiyama, Chem. Lett., 1982, 353; J. Am. Chem, Soc., 1973, **95**, 967; Chem.. Lett., 1986, 187; K. Banno and T. Mukaiyama, Chem. Lett., 1976, 279; Org. Reactions, 1982, **82**, 187.

2.2.6 Heck Reaction

It is a very useful carbon-carbon single and double bond forming reaction for organic synthesis. It involves coupling of an alkene with a halide or triflate in presence of Pd(0) catalyst to form a new alkene (Scheme-1)

R = aryl, vinyl or alkyl group without β-hydrogen on a Sp^3 carbon atom.
X = halide or triflate (OSO_2CF_3).

(Scheme–1)

The base used in Heck reaction is a mild base like Et_3N or anion like ^-OH, $^-OCOCH_3$, CO_3^{2-} etc. and the reaction is carried out in anhydrous polar solvents. It is mostly used for the alkylation or arylation of alkene[5].

2.2.6.1 Heck Reaction in Aqueous Phase

Normally Heck reaction uses anhydrous solvents (*e.g.*, DMF, MeCN). It has been found that the reaction can proceed very well in water. The role of water in the Heck reaction, as well as other reactions catalyses by Pd(O) in presence of phosphine legend is the transformation of catalyst precursor into Pd(O) species and the generation of zero-valent palladium species capable of oxidative addition by oxidation of phosphine ligand by Pd(II) precussor.

An interesting application of Heck reaction is the synthesis of substituted cinnamic acids by the reaction of aryl halides and acrylic acid in presence of a base ($NaHCO_3$ or K_2CO_3) in water[2] (Scheme–2).

Aryl halides
X = I, Br
R = H, p–Cl, p–OMe, p–Me,
p–NO$_2$, p–CHO, p–OH,
m–COOH etc.

Substituted
cinnamic acids
87–97%

(Scheme–2)

Use of acrylonitrile in place of acrylic and give the corresponding cinnamonitriles.

Heck reactions have also been carried out using a water soluble phosphine legend[3] (Triphenylphosphine trisulfonate sod. salt, TPPTS). Thus, iodobenzoic acid can be directly used for coupling with acrylic acid (Scheme–3)

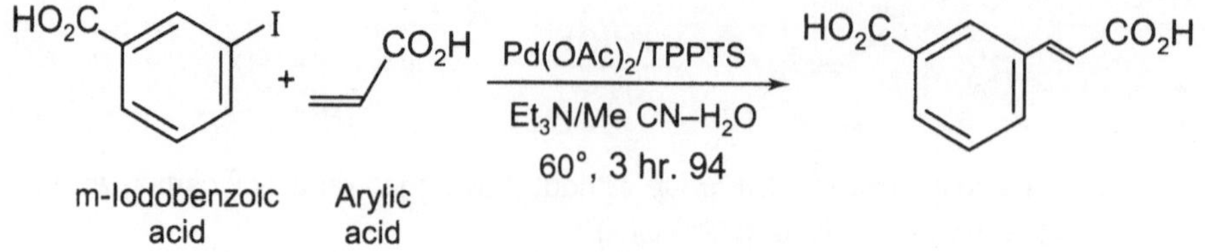

(Scheme–3)

Diaryliodonium salts react in a similar way[4] (Scheme-4).

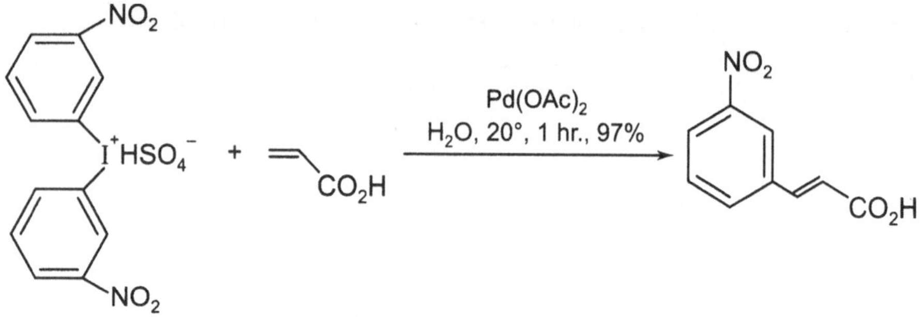

(Scheme–4)

It has been found that the Heck reaction can be successfully carried out under PTC conditions[5] with inorganic carbonates as base under very mild conditions at room temperature. Using this procedure, even substrates like methyl ketone, which is unstable under the conventional conditions of Heck arylation also react.

The Heck reaction has also been performed under mild conditions by the addition of acetate ion (Scheme–5).

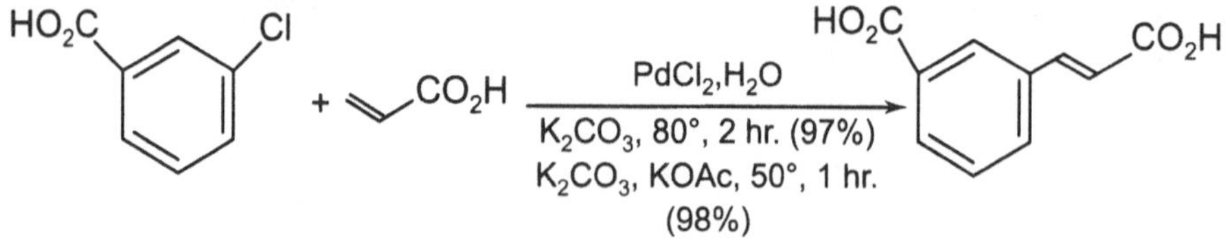

(Scheme–5)

A large number of other applications of Heck reaction have been described in literature[2].

Heck coupling reaction in water using microwave heating (Scheme-6) also gives the coupled products[6] (Scheme-6).

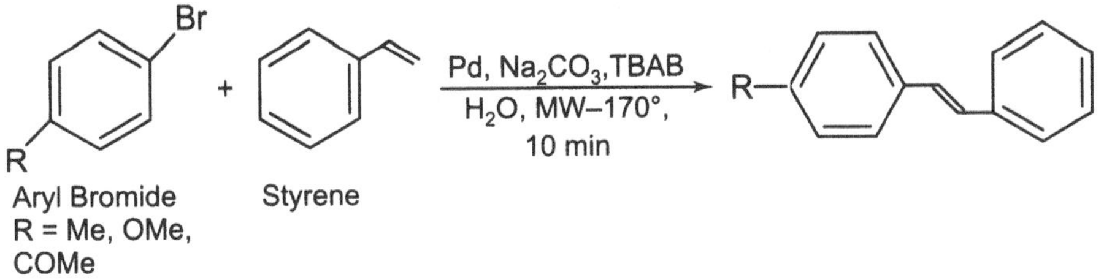

(Scheme–6)

It was found that in Heck reaction Pd-catalyst concentration as low as 500 ppb was sufficient for these reactions with good yield.

2.2.6.2 Heck Reaction in Supercritical Carbon Dioxide

The Heck reaction between iodobenzene and methyl acrylate in SC–CO_2 is catalysed by Pb $(OAc)_2$ in the presence of ligand (A) to give much better yield of methyl cinnamate (92%) than reported in conventional solvents (Scheme–7).

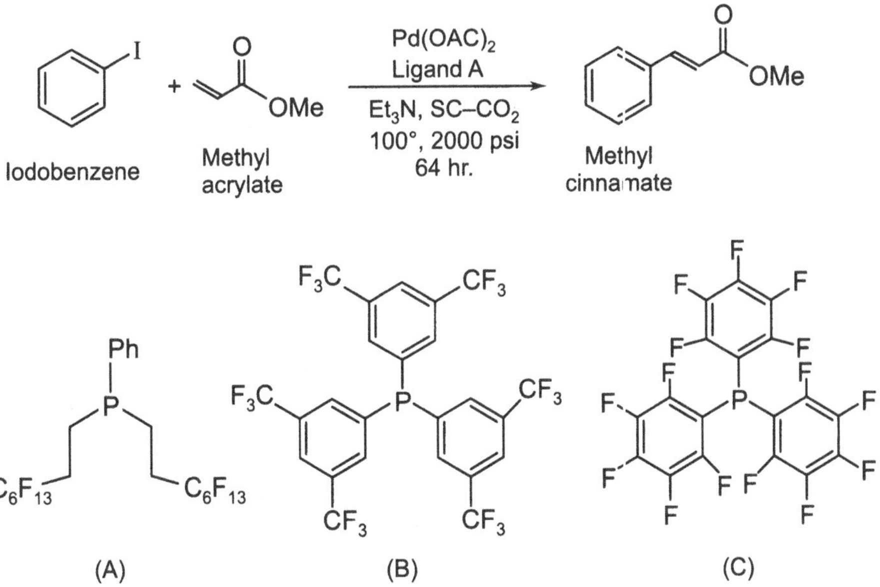

(Scheme–7)

Heck coupling was also reported in SC–CO_2 using ligands B and C.

Heck reaction have also been carried out using water soluble catalyst in SC–CO_2 (water-biphasic systems)[8]. Thus, the coupling of butyl acrylate in

SC–CO$_2$ using Pd(OAc)$_2$ and triphenylphosphine trisulfonate sodium Salt (TPPTS) (Scheme-8) gave the coupled product. Use of ethylene glycol gave better conversion. The importance of this reaction is that there is no catalyst leaching (Scheme–8).

(Scheme–8)

A related reaction, **Mizaoroki-Heck arylation** has also been conducted in SC–CO$_2$. Thus, the reaction of ethylene with arylhalides give styrene[9] (Scheme–9).

(Scheme–9)

An interesting Heck coupling of iodobenzene and methyl acrylate in presence of Dendrimer-encapsulated (DEC) palladium nanoparticles gave exclusive formation[10] of methyl 2-phenylacrylate instead of the usual product (methyl cinnamate) (Scheme-10).

(Scheme–10)

2.2.6.3 Heck Reaction in Ionic Liquids

The Heck reaction has been performed in ionic liquids, which are excellent solvents. Use of ionic liquids allows recycling of the catalyst[11], which is not possible under usual conditions. Thus, the Heck reaction of iodobenzene with ethyl acrylate in both N-hexylpyridinium, [C$_6$Py] and N,N-dialkyl imidazolium based ionic liquid (Scheme-11), higher yields were obtained in the former ionic liquid than the corresponding reaction in the imidazolium salts.

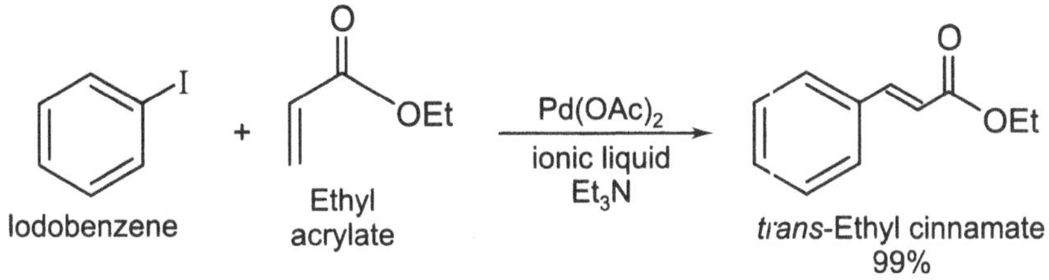

(Scheme–11)

The low yield in the imidazolium ionic liquid is due to the formation of carbene, which reacts with palladium to form a mixture of palladium carbene complex (for more details see Baylis-Hillman reaction section 2.2.3).

The Heck reaction has also been carried out by the reaction of aryl halides with acrylates as well as with styrene in the presence of Pd (OAc)$_2$ in ionic liquid [b$_{min}$] [BF$_4$] and [b$_{min}$] [Br] (Scheme–12)[12].

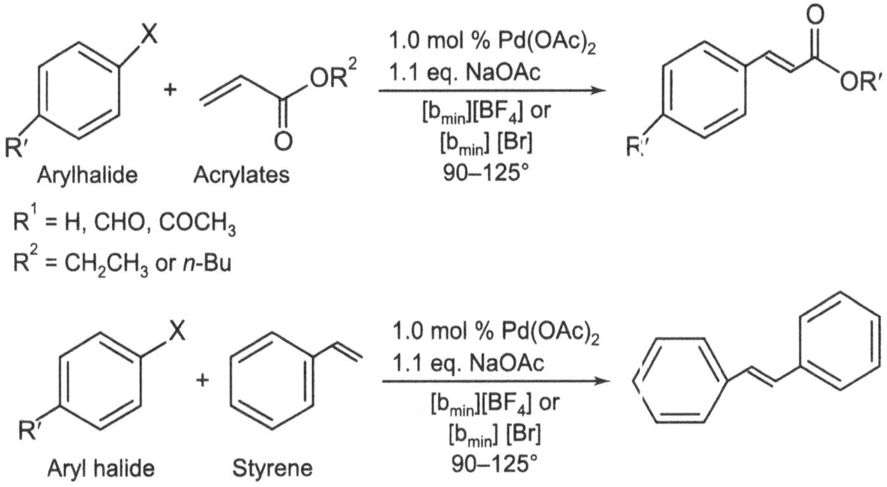

R^1 = H, CHO, COCH$_3$

R^2 = CH$_2$CH$_3$ or *n*-Bu

(Scheme–12)

Better selectivity and conversion was observed in reaction carried out in [b$_{min}$] [Br]. The low yield in case of [b$_{min}$] [BF$_4$], as already stated, is due to the formation of carbene intermediate, which complexes with palladium to give a mixture of palladium carben complex.

It has been suggested[12] that the active catalyst in the Heck reaction is a palladium nanoparticle generated *in situ* from palladium-carbene species. In fact, solution of ammonium stabilized Pd clusters are useful catalyst[13] for the Heck reaction. Thus, the Heck reaction between iodobenzene and *n*-butylacrylate in presence of [N$_{8888}$] [Br] stabilized 73 nm Pd clusters to afford *n*-butyl cinnamate (Scheme-13).

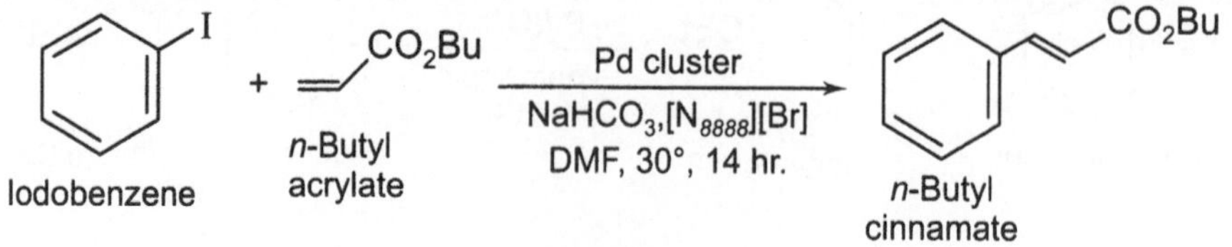

(Scheme–13)

2.2.6.4 Heck Reaction in Polyethylene Glycol (PEG)

Polyethylene glycol having molecular weight 2000 (or lower) has been used[14] as an efficient reaction medium for the Heck reaction. The stereo and regioselectivities are also different from those with conventional ionic liquids[15].

Thus, the reaction of bromobenzene with ethyl acrylate, Pd (OAc)$_2$ and TEA in PEG-2000 at 80° for 8 hr. gave exclusive formation of ethyl cinnamate (90% yield and 90% purity). However, the reaction of bromobenzene with styrene gave exclusively trans stilbene (93% yield). Finally, the reaction of bromobenzene with *n*-butyl vinyl ether gave butyl styrylether (Scheme-14). The results obtained are different than when the reaction was performed in ionic liquids[15]. However, in conventional solvents (DMF, DMSO, CH$_3$CN), a mixture of products was obtained in varying ratios[16]. Thus, PEG is unique in obtaining a single regioisomer with good diasteroselection (80/20 E/Z).

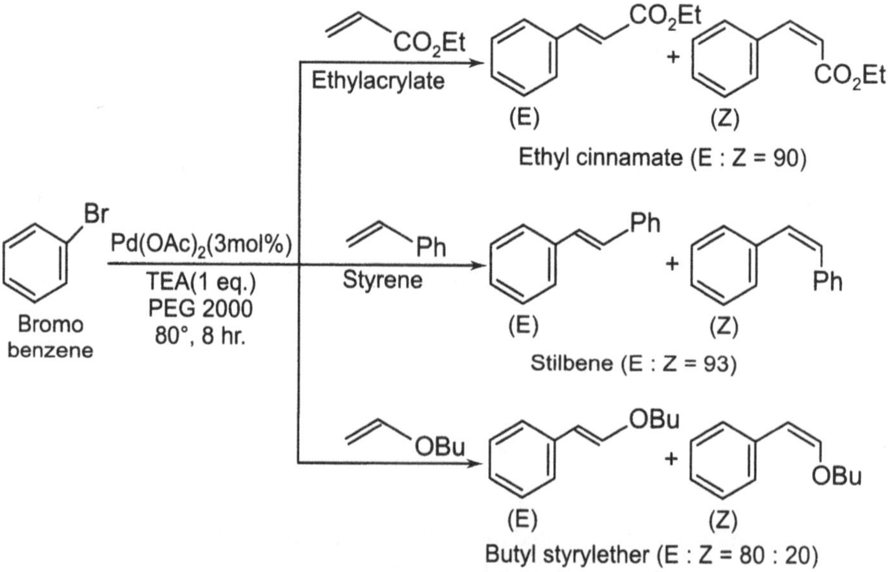

(Scheme–14)

2.2.6.5 Heck Reaction using Fluorous Phase Technique

Using Fluorous Phase Technique[17], palladium catalysed Heck reaction can be carried out. In this procedure, there is advantage of product isolation and catalyst recovery.

2.2.6.6 Heck Reaction using Polymer Supported Reagents in SC–CO₂

Supercritical CO_2 has the ability to plasticize polymers[18] and this has been utilized in using polymer supported reagents in SC–CO_2 for the Heck reaction[19]. Thus, the reaction of iodobenzene with polymer supported acrylic acid in SC–CO_2 using using tri-tert-butylphosphine gives the polymer supported Heck condensation product, from which the polymer support is cleaved by treatment with HF (Scheme–15).

(Scheme–15)

For details about polymer supported reagents and synthesis see Ref .20.

References

1. R.F. Heck, Acc. Chem. Res., 1979, **12**, 146; Organic Reactions, 1982, **27**, 346; H.A. Dieck J. Organometallic Chem.; 1975, **93**, 259.

2. N.A. Bumagin, N.P. Andryuchova, and I.P. Beletskaya, Izv. Akad. Nauk. SSSR, 1988, 6, 1449; N.A. Bumagin, P.G. More and I.P. Beletskaya, J. Organomet. Chem. 1989, **371**, 397; N.A. Bumagin. V.V. Bykov, L.I. Sukhomlinova, T.P. Tolstaya and I.P. Beletskaya, J. Organomet. Chem., 1995, **486**, 259.

3. J.P. Genet, E. Blart and M. Savignac, Synlett., 1992, 715.

4. N.A. Bumagin, L.I. Sukhomlinova, A.N. Banchikov, T.P. Tolstaya and J.P. Beletskaya, Bull. Russ. Acad. Sci., 1992, **41**, 2130.

5. T. Jeffery, Tetrahedron Lett., 1994, 35, 3051.

6. R.K. Arvela and N.E. Leadbeater, J. Org. Chem., 2005, **70**, 1786.

7. D.K. Morita, D.R. Pesiri, S.A. David, W.H. Glaze and W. Tumas, J. Chem., Soc. Chem. Commun, 1998, 1397–1398.

8. B.M. Bhanage, Y. Ikushima, M. Shirai and M. Arai, Tetrahedron Lett., 1990, **40**, 6427.

9. Y. Kayaki, Y. Noguchi and T. Karya, Tetrahedron Lett., 1990, **40**, 6427.

10. L.K. Yeng, Jr., K.P. Johnstone and R.M. Cooks, Chem. Commun., 2001 2290.

11. A. Carmichacl, M.J. Earle, I.D. Holbrey., P.B. McCormac and K.R. Seddon, Org. Lett., 1991, **1**, 997.

12. L. Xu, W. Chen and J. Xiao, Organometallics, 2000, **19**, 1123.

13. M.T. Rutz, R. Breinbauer and K. Wanninger, Tetrahedron Lett., 1996, **37**, 4499.

14. S. Chandrasekhar, Ch. Narsihmulu, S.S. Sultana and N.R. Reddy, Org. Lett., 2002, 4390.

15. L. Xu, W. Chen, 1. Ross and J. Xiao, Org. Lett., 2001, **3**, 293.

16. W. Cabri, I. Candiani, A. Bedeschi and S. Penco, J. Org. Chem., 1992, **57**, 1481; W. Cabri, I. Candiani and A. Bedeschi, J. Org. Chem. 1993, 58, 7421; W. Cabri and I. Candiani, Acc. Chem. Res., 1995, **28**, 2; M. Larhed and H. Hallberg, J. Org. Chem., 1997, **62**, 7858.

17. For details about Fluorous Phase Technique See V.K. Ahluwalia, Green Solvents for Organic Synthesis, Narosa Publishing House, New Delhi, 2009, Page 9.1 to 9.10 and the references cited therein.

18. T.R. Early, R.S. Gordon, M.A. Carroll, A.B. Holmes, R.E. Shute. I.F. McConvey, Tessa R. Early, Richard S. Gordon, Michael A. Carroll, Andrew B. Holmes, Richard E. Shute, Ian P. McConvey, Palladium catalysed cross-coupling reactions in SC–SO$_2$, Chem. Commun., 2001, 1966–1967.

19. A.I. Cooper, Polymer synthesis and processing using SC–CO$_2$, J. Mater. Chem., 2000, **10**, 207–234.

20. V.K. Ahluwalia and Renu Aggarwal, Organic Synthesis, Special Techniques, Narosa Publishing House, 2006, page 150–194 and the references cited therein.

2.2.7 Hiyama Reaction

It is a carbon-carbon double bond forming reaction and involves palladium-catalysed cross coupling of organo silicon compounds with organic halides. Organo silanes are environmentally benign and stable to many reaction conditions. It is a good alternative to other coupling reactions which employ different organo metallic reagents[1].

The reaction consists in coupling of aryl or vinyl halides with vinyl alkoxysilanes under thermal or microwave heating using either Pd (OAc)$_2$ or oxime-derived from palladacycle (A) as catalyst to give the corresponding styrene derivatives (Scheme-1).

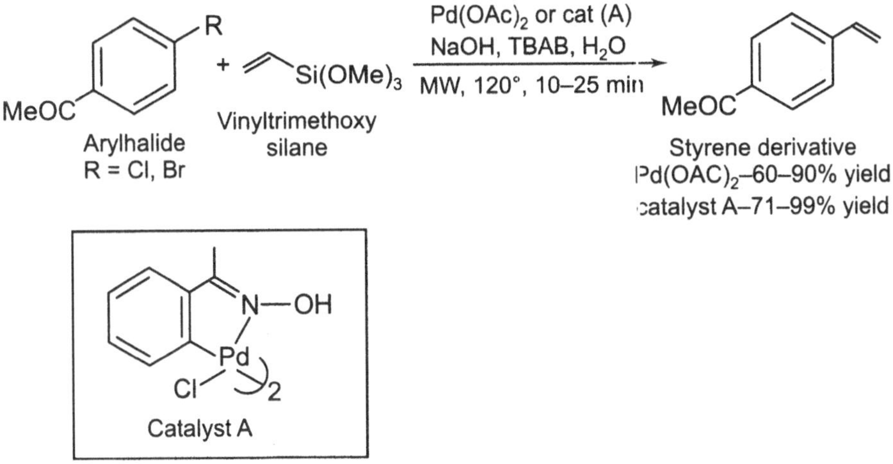

(Scheme–1)

It is found[2] that use of TBAB as additive lowers Pd loadings, in the range of 0.5–1 mol per cent of arylbromides and 2 mole per cent for arylchlorides. Also, the reaction time can be reduced from about 24 hr. under thermal heating in a pressure tube under the same conditions to 10–25 min. by using microwave heating. The reaction of activated aryl chlorides with vinyltrimethyoxsilane to the corresponding styrene proceeded only under microwave conditions. It is believed[2] that the active catalytic species could be Pd nanoparticles.

References

1. T.H. Hiyama in Metal-catalysed cross coupling reactions: F. Diederich and P.J. Stand Eds., Wiley-VCH, New York, 1998, Chapter 10, p. 421; S.E. Denmark and M.H. Ober, Aldrichim. Actd; 2003, **36**, 75; C.J. Handy, A.S. Manoso, W.T. McEloy; W.M. Seganish and P. Deshony, Tetrahedron, 2005, **61**, 12201.

2. D. Dallinger and C.O. Kappe, Chem. Rev., 2007, **107**, 2263–2591 and the references cited therein.

2.2.8 Hofmann Elimination

It involves the formation of olefins by heating quaternary ammonium hydroxides (Scheme-1).

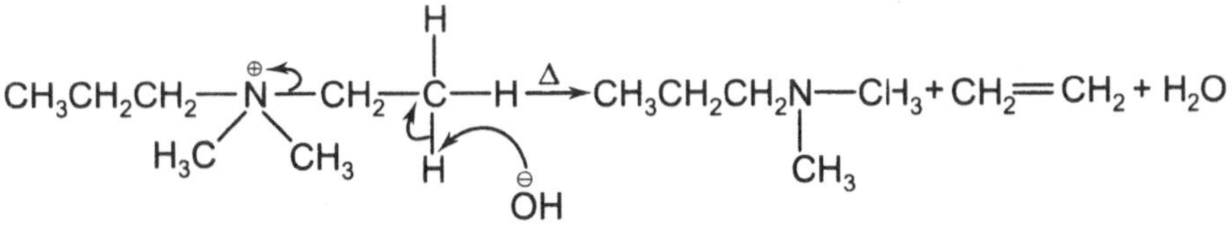

(Scheme–1)

These eliminations are governed by the well-known **Hofmann rule**, which states that the charged substrates yield the least substituted olefins.

2.2.8.1 Hofmann Elimination under Microwave Irradiation

On microwave irradiation in water-chloroform, the quaternary ammonium salts yield a thermally unstable Hofmann elimination product (Scheme–2).

(Scheme–2)

97%

The near quantitative yield are twice than those obtained by traditional methods[2].

References

1. A.W. Hofmann, Chem. Ber., 1881, **14**, 659.

2. C.R. Strauss, K.D. Raner, R.W. Trainor and J.S. Thorn, Aust. Pat. 677876 (1977).

2.2.9 Knoevenagel Condensation

It is the base catalysed condensation[1] between an aldehyde or ketone with any compound containing reactive methylene group (*e.g.*, malonic acid or ester) to give unsaturated carboxylic acids (Scheme–1).

$$CH_3CHO \ + \ CH_2(COOH)_2 \xrightarrow{\text{base}} CH_3CH{=}C(COOH)_2$$

Acetaldehyde   Malonic
                    acid

$$\Delta \Big\downarrow {-}CO_2$$

$$CH_3CH{=}CHCOOH$$

Crotonic acid

(Scheme–1)

The base used in the above is a weak base like ammonia or amine (primary or secondary). However, when condensation is carried out in presence of pyridine as a base, decarboxylation usually occurs during the condensation. This is known as **Doebner Modification**[2] (Scheme–2).

$$C_6H_5CHO \ + \ CH_2(CO_2Et)_2 \xrightarrow[\text{benzene}]{\text{Pyridine}} C_6H_5CH{=}C(CO_2Et)_2$$

Benzaldehyde   Ethyl malonate

$$\downarrow \begin{array}{l}\text{hydrolysis}\\ H_3O^+\end{array}$$

$$C_6H_5CH{=}CHCOOH \xleftarrow[-CO_2]{\Delta} C_6H_5CH{=}C(COOH)_2$$

Cinnamic acid

(Scheme–2)

Knoevenagel condensation is more useful for aromatic aldehydes, since with aliphatic aldehydes, the product obtained undergoes Michael condensation (Scheme–3).

$$O{=}CH_2 + \bar{C}H(CO_2Et) \xrightarrow{Et_3N} HO{-}CH_2{-}CH(CO_2Et)_2$$

Formaldehyde

$$\downarrow -H_2O$$

$$(EtO_2C)_2CHCH_2CH(CO_2Et)_2 \xleftarrow[\substack{\text{Michael}\\ \text{addn.}}]{\bar{C}H(CO_2Et)_2} CH_2{=}C(CO_2Et)_2$$

Tetraethylpropane
1,1,3,3-tetracarboxylate

(Scheme–3)

Ketones do not undergo knoevenagel condensation with malonic ester, but can react with more active cyanoacetic or its ester (Scheme–4).

$$(CH_3)_2C{=}O + \underset{\substack{CO_2Et}}{CH_2}{\overset{CN}{\diagup}} \longrightarrow (CH_3)_2C{=}C{\overset{CN}{\diagdown}}_{CO_2Et}$$

Acetone   Ethyl cyanoacetate   Isopropylidene cyanoacetic ester

(Scheme–4)

Ketones or aldehydes also react with a succinic ester in presence of sodium hydride to give the corresponding condensation product (Scheme–5).

$$\begin{array}{l}Ph\\ \diagdown\\ \diagup C{=}O\\ Ph\end{array} + \begin{array}{l}CH_2CO_2Et\\ |\\ CH_2CO_2Et\end{array} \xrightarrow[20°]{NaH} \begin{array}{l}Ph\\ \diagdown\\ \diagup C{=}C{-}CO_2Et\\ Ph \qquad |\\ \qquad CH_2CO_2Et\end{array}$$

Benzophenone   Diethyl succinate   97%

(Scheme–5)

This condensation in known as **Knoevenagel-Strobe condensation**[3].

2.2.9.1 Knoevenagel Condensation in Water

The knoevenagel reaction between aldehydes and acrylonitrile has been carried out in water. Thus, salicyaldehyde react with malononitrile in heterogeneous aqueous alkaline medium at room temperature to give *o*-hydroxy benzylidenemalonitrile[4], which is converted by acidification and heating into 3-cyanocoumarins in good yield (Scheme–6).

Salicylaldehydes   Malono
R = H, OH, OMe   nitrile

3-Cyanocoumarins
75–95%

(Scheme–6)

The condensation of substituted acetonitriles with salicyaldehydes requires the presence of catalytic amount of PTC (CTABr). The aqueous phase reaction gives better yields (Scheme–7).

Salicylaldehyde   Acetonitrile   3-Substituted
(substituted)   coumarins

R = CN, CO$_2$Et, NO$_2$
2-Py

(Scheme–7)

The yields of coumarins are much better in water (66–98 per cent) than in ethanol (35–80%).

The condensation of benzaldehyde with aryl acetonitrile does not take place in water, but required the presence of CATACl or CTABr to give high yields of aryl cinnamonitriles[5] (Scheme–8).

ArCH$_2$CN   +   PhCHO
Arylacetonitrile
Ar = Ph, p–NO$_2$C$_6$H$_4$,
PhSO$_2$

$\xrightarrow[\text{r.t., 0.5–9 hr.}]{\text{CTACl/NaOH}}$

85–90%

Aryl cinnamonitriles

(Scheme–8)

Knoevenagel-type addition products can be obtained[6] by the reaction of acrylic derivatives in presence of 1,4-dibenzocyclo (2.2.2) octane (DABCO) (Scheme 9).

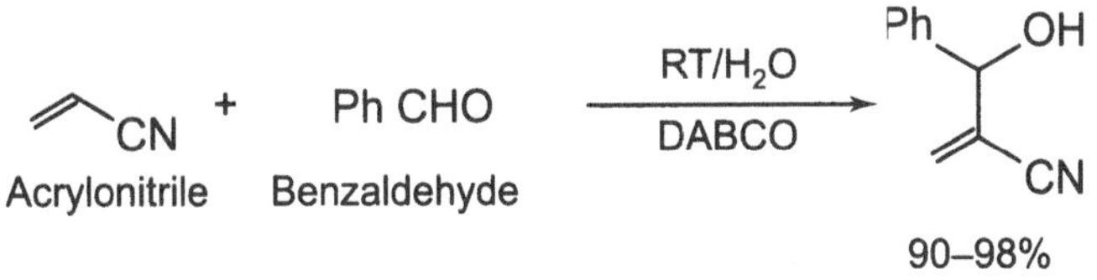

(Scheme–9)

2.2.9.2 Knoevenagel Condensation under Microwave Irradiation

The knoevenagel condensation reaction involving active methylene compounds and carbonyl compounds has been reported[7] using microwave irradiation. The reactions are conducted in open vessels that lead to efficient removal of water, thus, circumventing the use of dean-stark apparatus.

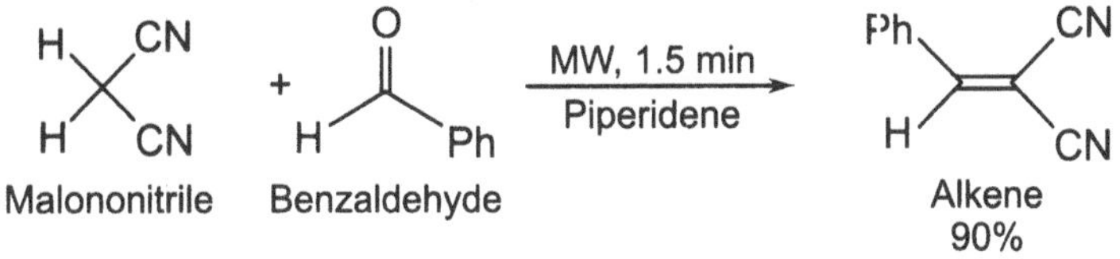

(Scheme–10)

Salicylaldehydes undergo knoevenagel condensation with alkyl substituted ethyl acetates under basic conditions (piperidine) on MW irradiation to yield coumarins[8] (Scheme–11).

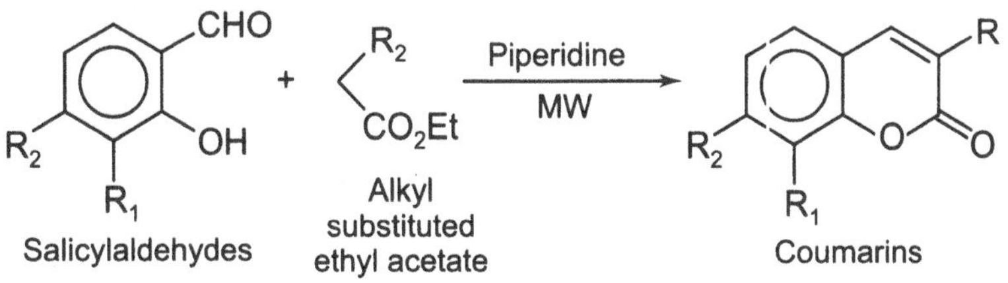

(Scheme–11)

2.2.9.3 Knoevenagel Condensation in Solid State

The knoevenagel condensation has been carried out in dry media[9] without the use of any solvent. The method consists in adding a solid inorganic support to a solution of aromatic aldehyde and diethylmalonate in acetone. The adsorbed material is mixed properly, dried in air (beaker) and placed in an alumina bath[10] in the MW oven for 2-3 min at medium power level (600 W) intermittently at 0.5 min intervals at 102°. The product is obtained by extraction of the reaction mixture with alcohol.

An expedious knoevenagel condensation of creatinine with aldehydes has been achieved using focused MW irradiation (40–60 W) under solvent-free reaction conditions at 160–170° (Scheme–12)[11].

Creatinine

(Scheme–12)

2.2.9.4 Knoevenagel Condensation in Ionic Liquids

The knoevenagel condensation of benzaldehyde with malononitrile in the presence of KOH dissolved in ionic liquid, $[b_{min}]$ $[PF_6]$ gave only low yield of the styrene derivative (Scheme–13).

Benzaldehyde   Malono
                    nitrile

(Scheme–13)

The low yield of the product was due to the formation of anion of the ionic liquid, which reacted with benzaldehyde (For details see Baylis-Hillman reaction in ionic liquid see. 2.2.3). The yield, could be increased as the substrate concentration was increased and the ionic liquid was reused. The ionic liquid could be reused up to 5 times without the additional base.

Kneovenagel condensation can also be carried out using chloroaluminate ionic liquids[13], which have a variable lewis activity, such as 1-butyl-3-methyl imidazoliun chloroaluminate, $[b_{min}]$ Cl. $AlCl_3$, $X(AlCl_3) = 0.67$, where X is the mole fraction and 1-butylpyridiniumchloroaluminate, $[b_{py}]$ Cl. $AlCl_3$,

X(AlCl$_3$) = 0.67. Ionic liquids work as lewis acid catalyst and solvent in the Kneovenagel condensation. A typical example is given below (Scheme–14).

(Scheme–14)

References

1. F. Knoevenagel, Ber., 1896, **29**, 1720; J.R. Johnson, Org. Reactions, 1942, **1**, 210.

2. Doebner, Ber., 1900, **33**, 2140.

2a. F. Texter-Boullet and A. Foucand, Tetrahedron Lett., 1982, 4927.

3. H. Stobbe, Ber., 1893, **26**, 2312.

4. K. Takai, C.H. Heathcock, J. Org. Chem., 1985, **50**, 3247; A.E. Vougioakas and H.B. Kagan, Tetrahedron Lett., 1987, **28**, 5513; K. Mikami, M. Terada and N. Nakai, J. Chem. Soc. Chem. Commun, 1935, 343 and the references cited therein.

5. Paul A. Grieco, Ed. In Organic synthesis in water, Blackie Academic and Professional, London, 1998, P 225.

6. J. Aug, M. Lubin and A. Lubineau, Tetrahedron Lett., 1994, **35**, 7947.

7. S.A. Ayoubi, F. Texier-Boullet and J. Hamelin, Synthesis, 1994, 258.

8. V. Singh, J. Singh, P. Kaur and Y.L. Kad, J. Chem. Res (S), 1997, 58; D. Bogdal, J. Chem.. Res (S), 1998, 468.

9. M. Kidwai, P. Sapra and K.R. Bhutani, J. Ind. Chem. Soc., 2002, **79**, 591.

10. B. Gram, A. Loupy and M. Majoub, Tetrahedron Lett., 1990, **46**, 5167.

11. D. Villemin and B. Martin, Synth. Commun., 1995, **25**, 3135.

12. P. Formentin, H. Garcia and A. Leyva, J. Mol. Catal A : Chem., 2004, **214**, 137.

13. J.R. Harjani, S.J. Nara and M.M. Salunkhe, Tetrahedron Lett., 2002, **43**, 1127.

2.2.10 Mannich Reaction

Compounds containing at least one active hydrogen (ketones, nitroalkanes, β-ketoesters, β-cyanoacids *etc.*) condense with formaldehyde and primary or secondary amine or ammonia (in the form of hydrochloride) to give β-aminocarbonyl compounds, known as mannich base[1] (Scheme–1).

$$\underset{\text{Acetophenone}}{C_6H_5\overset{O}{\overset{\|}{C}}CH_3} \; + \; \underset{\text{Formaldehyde}}{H{-}\overset{O}{\overset{\|}{C}}{-}H} \; + \; \underset{\substack{\text{Dimethylamine}\\\text{hydrochloride}}}{(CH_3)_2\,\overset{+}{N}H_2\overset{-}{Cl}}$$

$$\longrightarrow \; C_6H_5\overset{O}{\overset{\|}{C}}{-}CH_2CH_2\,\overset{\oplus}{N}(CH_3)_2\overset{-}{Cl}$$

Mannich base

(Scheme–1)

The Mannich reaction also proceeds with other activated compounds like indole, furan, pyrrole and phenols. The mannic bases are of great synthetic compounds. On heating, they yield α, β-unsaturated ketones (Scheme–2).

$$Ar{-}\overset{O}{\overset{\|}{C}}{-}\underset{\underset{H}{|}}{C}H{-}CH_2\overset{\oplus}{-}NH(CH_3)_2\overset{-}{Cl} \longrightarrow H_2\overset{+}{N}(CH_3)_2\overset{-}{Cl} \; + \; Ar\overset{O}{\overset{\|}{C}}{-}CH{=}CH_2$$

(Scheme–2)

In this way, mannich reaction is useful for C—C double bond formation.

Reference

1. C. Mannich, W. Krosche, Arch. Pharm, 1912, **250**, 647; F.F. Blicke, Organic Reactions, 1942, **1**, 303.

2.2.11 Mukaiyama Reaction

It is stereoselective aldol condensation of silylenol ethers of ketones with aldehydes in presence of titanium tetrachloride. Thus, the reaction of silylenol ether of 3-pentanone with 2-methylbutyraldehyde in presence of $TiCl_4$ gives the aldol product, which on hydrolyrisis yield, manicone, an alarm pheromone (Scheme–1).

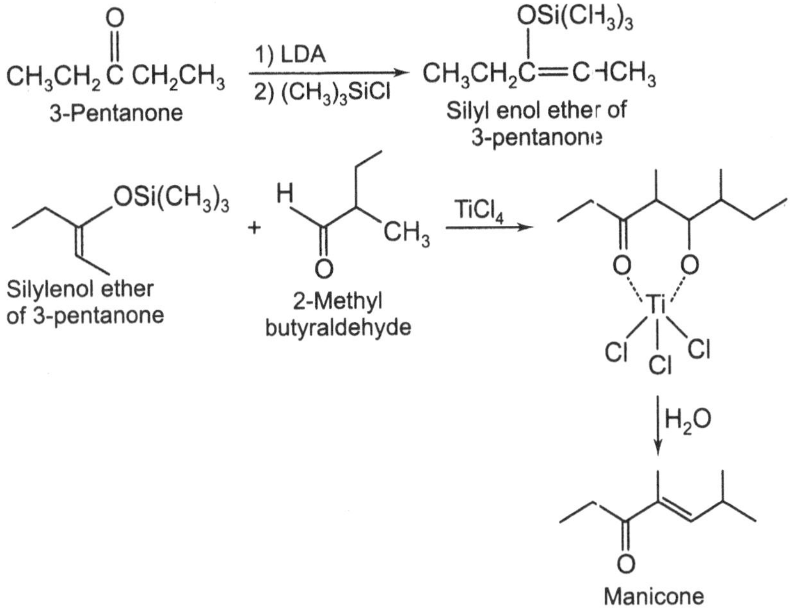

(Scheme–1)

Reference

1. T. Mukaiyama, Chem. Lett. 1982, 353; J. Am. Chem. Soc., 1973, **95**, 967; Chem. Lett., 1986, 175, T. Mukaiyana, Org. Reactions, 1982, **28**, 186.

2.2.11a Meyer-Schuster Rearrangement

Acid catalysed rearrangement of acetylenic alcohols to α, β-unsaturated carbonyl derivatives is known as Meyer-Schuster Rearrangement[1] (Scheme–1).

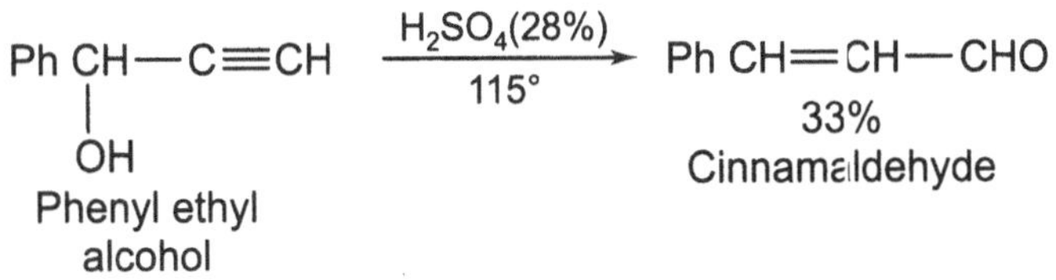

(Scheme–1)

The Meyer-Schuster rearrangement has also been reported[2] in SC–H_2O at elevated temperature.

References

1. K.M. Meyer and K. Schuster, Chem. Ber., 1922, 55, 819.
2. J. An. L. Bagnell, T. Cablewski C.R. Strauss and R.W. Trainer, J. Org. Chem., 1997, **62(8)** 2505.

2.2.12 Perkin Reaction

Aromatic aldehydes on heating with acetic anhydride in presence of anhydrous sodium acetate give[1] the corresponding cinnamic acid (Scheme–1).

$$Ar\,CHO \ + \ (CH_3CO)_2O \ \xrightarrow[\text{120°, 4 hr.}]{CH_3COONa} \ Ar\,CH{=}CH{-}COOH + CH_3COOH$$

Aromatic   Acetic   Cinnamic acid
aldehyde   anhydride   40%

(Scheme–1)

In the above use of potassium acetate gives much better yields (60–70%).

Reference

1. W.H. Perkin, J. Chem. Soc., 1868, **21**, 58, 181; 1877; **31**, 388; J.R. Johnson, Org. Reactions 1942, **1**, 210.

2.2.13 Stille Coupling Reaction

Coupling of organotin reagents with aryl or vinyl halides or triflates in presence of palladium give the corresponding coupled product[1] (Scheme–1).

4-tert. butyl   tributyl   1-Vinyl-4-tert. butyl
1-cyclohexene   vinyl tin   cyclohexene
triflate

(Scheme–1)

This reaction is of general applicability with respect to both, the halides and the types of stannanes that can be used. Benzylic, aryl, alkenyl, and allylic halides all can be used[2]. The group that can be transferred from tin include of alkyl, alkenyl, aryl and alkynyl.

Some examples of carbon-carbon double bond formation are given below: (Scheme–2).

96%

(Contd.)

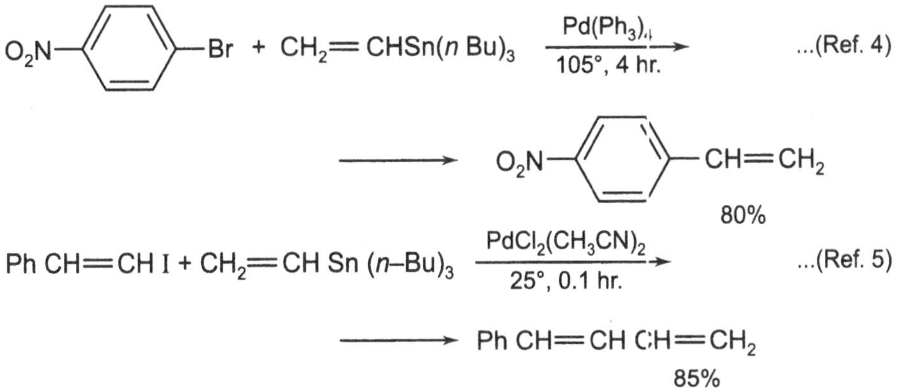
O2N
Br
+ CH2=CHSn(n Bu)3
Pd(Ph3)4
105°, 4 hr.
...(Ref. 4)
O2N
CH=CH2
80%
Ph CH=CH I + CH2=CH Sn (n–Bu)3
PdCl2(CH3CN)2
25°, 0.1 hr.
...(Ref. 5)
Ph CH=CH CH=CH2
85%

(Scheme–2)

Besides carbon-carbon double bond formation, stille coupling reaction has also been used for carbon-carbon single bonds. Sorne exmples are given below (Scheme–3)

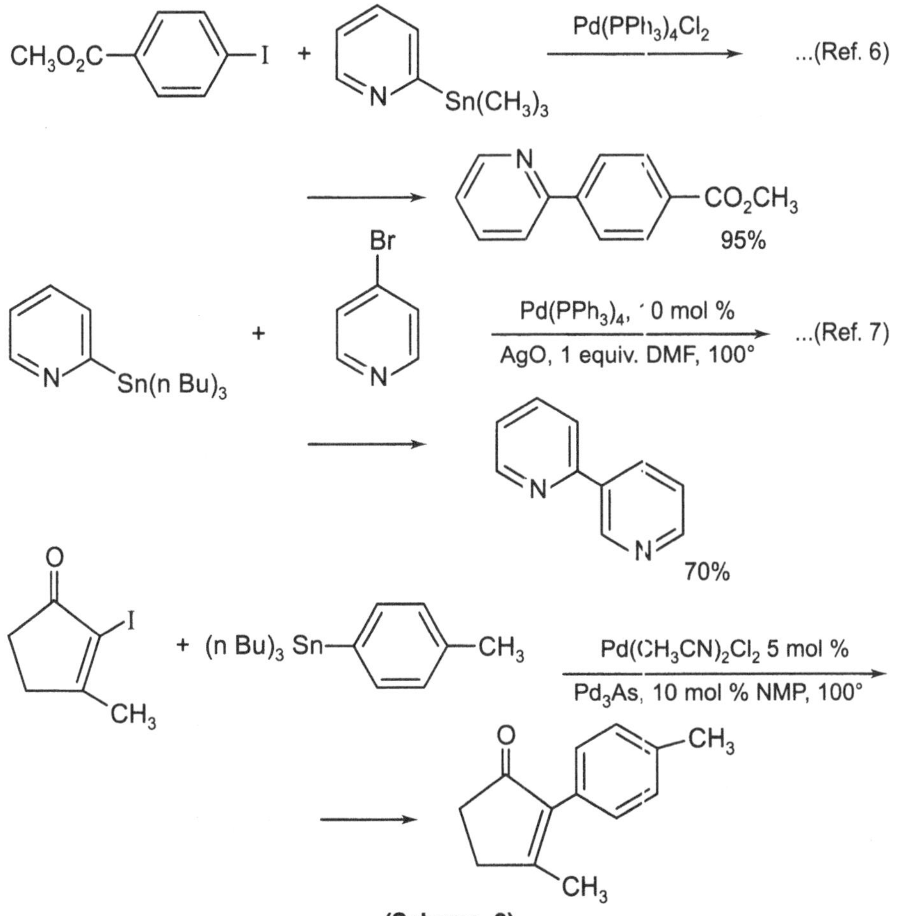
CH3O2C
I
+
N
Sn(CH3)3
Pd(PPh3)4Cl2
...(Ref. 6)
N
CO2CH3
95%
N
Sn(n Bu)3
+
Br
N
Pd(PPh3)4, 0 mol %
AgO, 1 equiv. DMF, 100°
...(Ref. 7)
N
N
70%
O
I
CH3
+ (n Bu)3 Sn
CH3
Pd(CH3CN)2Cl2 5 mol %
Pd3As, 10 mol % NMP, 100°
O
CH3
CH3

(Scheme–3)

2.2.13.1 Stille Coupling Reaction in Water

An aqueous microwave assisted still reaction has been reported[9] in the 2(H) pyrazinone derivatives at the C–3 position (Scheme–5).

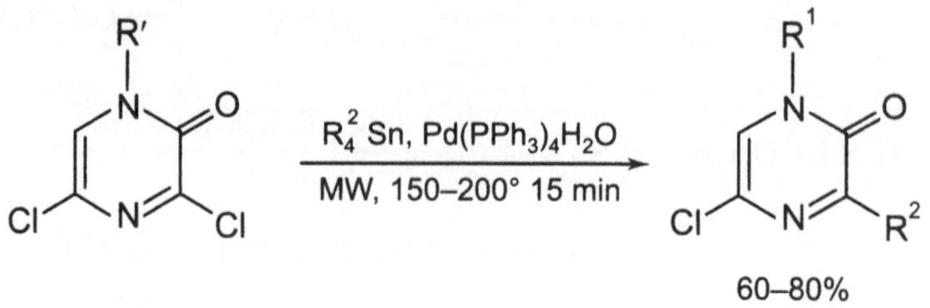

R^1 = MeO–Bn, Bn, Ph
R^2 = n–Bu, Me, Ph

(Scheme–5)

2.2.13.2 Stille Coupling in SC–CO$_2$

Stille coupling mediated by fluorous tagged phosphine (PTP-1 and PTP-2) has been reported[10] in SC–CO$_2$ using $(n–Bu)_4NCl$ as catalyst (Scheme–6).

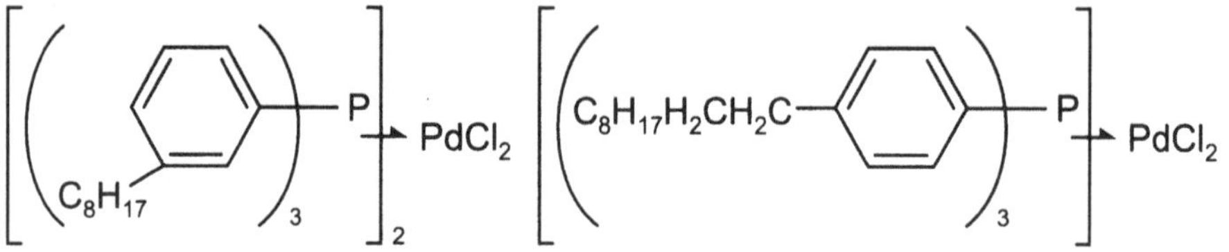

PTP–I   PTD–II

(Scheme–6)

2.2.13.3 Stille Coupling in Ionic Liquids

It has been found[11] that use of palladium complexes immobilized in ionic liquid offer great advantage over the classical organic solvents used for stille coupling reaction. A large number of stille coupling reactions with Pd(O) or Pd(II) catalyst associated with Ph$_3$As in the presence of CuI has been developed in [Bmin][BF$_4$] (Scheme–7)[11]

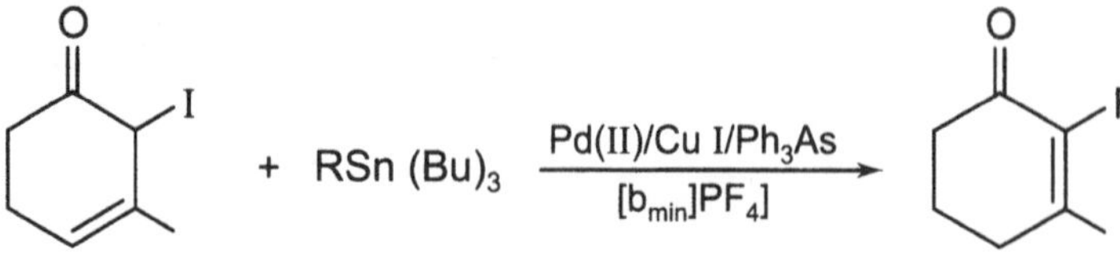

(Scheme–7)

2.2.13.4 Stille Coupling using Fluorous Phase Technique

Stille coupling has been achieved (Scheme–8) with perfluoro-tagged tin compounds[12,13].

(Scheme–8)

References

1. J.K. Stille, Angew. Chem. Int. Ed. Engl. 1986, **25**, 508; T.N. Mitchell, Synthesis, 1992, 803.

2. F.K. Sheffy, J.P. Godschalx and J.K. Stille, J. Am. Chem. Soc., 1984, **106**, 4833; I.P. Beltskaya, J. Organomet. Chem., 1983, **250**, 551; J.K. Stille and B.L. Grot, J. Am. Chem. Soc., 1987, **109**, 813.

3. M. Kosugi, K. Sasazawa, Y. Shimizu and T. Migata, Chem. Lett., 1977, 301.

4. D.R. Mckean, G. Parrinello, A.F. Renaldo and J.K. Stille, J. Org. Chem., 1987, **52**, 422.

5. J.K. Stille and B.L. Groh, J. Am. Chem. Soc., 1987, **109**, 813.

6. T. R. Bailey, Tetrahedron Lett., 1986, **27**, 4407.

7. J. Malm, P. Bjork, S. Gronowitz and A.–B. Hörnfeldt, Tetrahedron Lett., 1992, **33**, 2199.

8. C.R. Jhonson, J.P. Adams, M.P. Braun and C.B.W. Senarayake, Tetrahedron Lett., 1992, **33**, 919.

9. N. Kaval, K. Bisztray, W. Deharen, C.O. Kappe and E. Vander Eycken, Molecular Diversity, 2003, **7**, 125.

10. S. Schneider and W. Bannwarth, Angew. Chem. Int. Edn., 2000, **39**, 4142.

11. S.T. Handy and X. Zhang, Org. Lett., 2001, **3**, 233.

12. M. Hosino, P. Degenbolb and D.P. Curran, J. Org. Chem., 1997, **62**, 8341.

13. J.K. Stille, Angew. Chem. Int. Ed. Engl., 1986, **25**, 508.

2.2.14 Sakurai Reaction

It is the titanium mediated inter or intramolecular addition of allylic silanes to α, β-unsaturated ketones[1] or to aldehydes[2] (Scheme–1).

(Scheme-1)

It is possible to perform Sakurai reaction between α, β-unsaturated ketones with allyltrimethylsilane in presence of $InCl_3$ using $[C_4min][PF_6]$ or $[C_4min][BF_4]$ as solvent[3] (Scheme-2).

(Scheme-2)

References

1. H. Sakurai, J. Am. Chem. Soc., 1977, **99**, 1673; Pure Appl. Chem. 1982, **54**, 1; Tetrahedron Lett., 1976, 1295

2. D. Seabach, Argew. Chem. Int. Ed., 1985, **24**, 765.

3. J. Howarth, P. James and D. Dais J. Mol. Catal. A. 2004, **214**, 143.

2.2.15 Wittig Reaction

The reaction[1,2] of carbonyl compounds (aldehydes or ketones) with phosphorus ylides (or phosphorane) (commonly known as Wittig reagent) yield alkenes (Scheme–1). It is found that cis olefins predominate in aliphatic systems and trans in conjugated (Scheme–1).

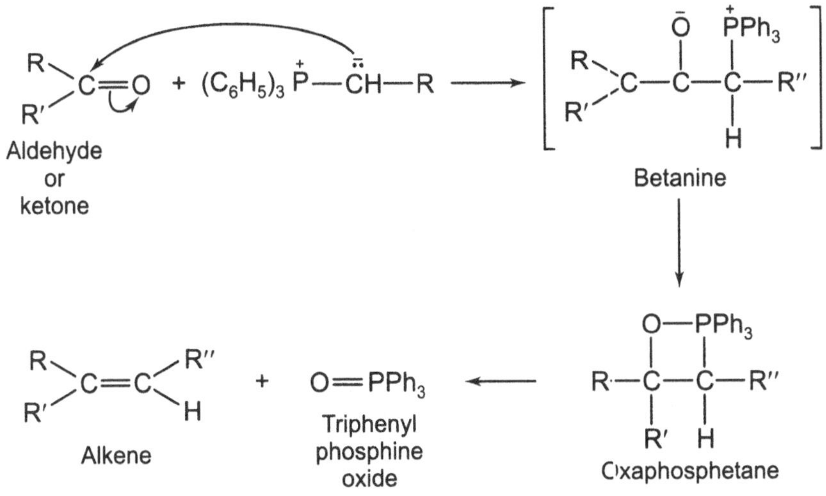

(Scheme–1)

The phosphorous ylides are obtained by the reaction of an alkyl halide with triphenylphosphine. The formed phosphonium salt is treated with a strong base like sodium hydride or phenyl lithium to give phosphorous ylide (Scheme–2).

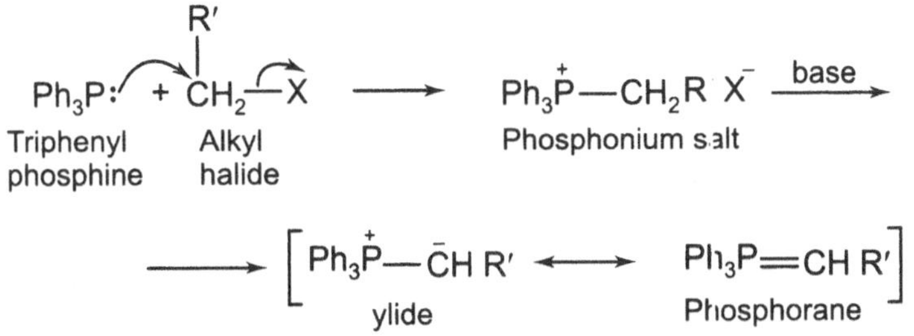

(Scheme–2)

2.2.15.1 Wittig Reaction using Phase Transfer Catalysts

It has been shown[3-5] that phase transfer catalysts, *viz.*, alkyl triphenyl phosphonium salts react with aqueous sodium hydroxide to generate ylides, which combine with organic phase aldehydes to produces olefins (Scheme–3).

$$\overset{+}{Ph_3P}-CH_2C_6H_5\overset{-}{Cl} \; + \; NaOH(aq.) \; \xrightarrow[\text{Solution}]{CH_2Cl_2} \; \left[Ph_3P{=}CH\,C_6H_5 \right]$$

Benzyltriphenyl
phosphonium chloride

$$\text{ylide}$$

$$\Big\downarrow R\,CHO$$

$$R\,CH{=}CHC_6H_5 \; + \; Ph_3PO$$

(Scheme–3)

The yield of olefin increases with the increase in concentration of alkali up to maximum and then decreases. The yield also depends on the alkyl group attached to the triphenylphosphonium salt. The quaternary phosphonium salts are better than the quaternary ammonium salts.

The PTC catalysed Wittig reaction is limited to only aldehydes. Thus, the method is very convenient for the synthesis of olefins the type $R\,CH{=}CHR'$.

2.2.15.2 Wittig Reaction using Microwaves

Phosphorus ylides undergo Wittig reaction with ketones in the absence of solvent on MW irradiation to give improved yields compared to conventional procedures[6]. The preparation of several phosphonium salts is possible using domestic MW oven by the reaction of neat triphenylphosphine with organic halides showing remarkable rate enhancement in a pressurized vessel[7] (Scheme–4).

$$Ph_3P \; + \; R_1CH_2X \; \xrightarrow{MW} \; \overset{+}{Ph_3P}-CH_2R_1\overset{-}{X} \; \xrightarrow[MW]{base}$$

Triphenyl   Alkyl   Phosphonium
phosphine   halide   salt

$$\longrightarrow \left[\overset{+}{Ph_3P}-\overset{-}{C}H\,R_1 \; \longleftrightarrow \; Ph_3P{=}CH\,R_1 \right] \; \xrightarrow[MW]{O=C\overset{R}{\underset{R}{\big\langle}}}$$

                    ylide   Phospharane

$$\longrightarrow \; R_1{-}CH{=}CR_1$$
                    Ofelin

(Scheme–4)

2.2.15.3 Modification of Wittig Reagent

Several modification of the Wittig reagent have been made in order to improve the reactivity of ylides.

(a) Horner-Wadsworth Modification

In this procedure, the ylides are obtained from phosphonate ester (instead of triphenyl phosphonium salt), which in turn is readily available from alkyl halide and triethylphosphite via an **Arbunov rearrangement[8]**. These ylides (being more reactive than the corresponding phosphoranes) react with ketones that are inert to phosphoranes. Thus, the reaction of ethyl bromoacetate with triethylphosphite gave the phosphonate ester, which on treatment with base (NaH) and subsequent reaction with cyclohexanone gives α, β-unsaturated ester, cyclohexylidene acetate in 70 per cent yield (only 20% yield in obtained with triphenyl phosphorane). This method is commonly known as **Horner-Wadsworth Emmons Modification of the Wittig reaction** (Scheme–5).

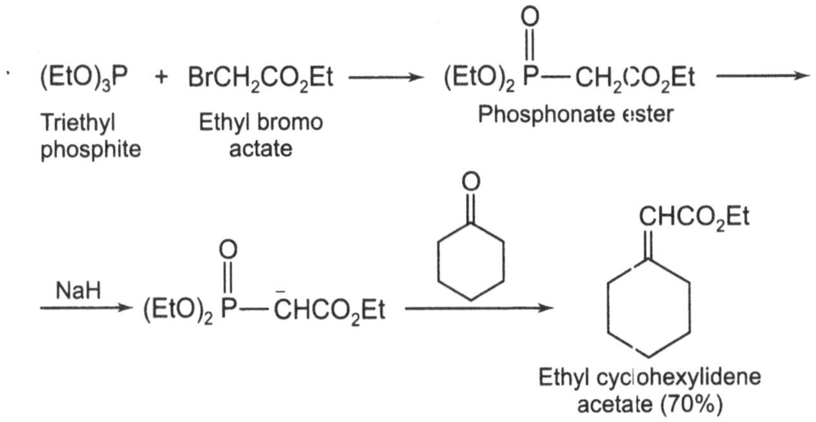

$$(EtO)_3P \ + \ BrCH_2CO_2Et \longrightarrow (EtO)_2 \overset{\overset{O}{\parallel}}{P} - CH_2CO_2Et \longrightarrow$$

Triethyl phosphite   Ethyl bromo actate   Phosphonate ester

$$\xrightarrow{NaH} (EtO)_2 \overset{\overset{O}{\parallel}}{P} - \bar{C}HCO_2Et \longrightarrow$$

Ethyl cyclohexylidene acetate (70%)

(Scheme–5)

The major product obtained in Horner-Wadsworth-Emmons modification is usually the (E) alkene compared to a mixture of (E) and (Z) alkenes obtained in the usual Wittig reaction.

Similar reaction of the appropriate phosphonate ester with benzaldehyde in presence of NaH gives trans stilbene in 84% yield[9] (Scheme–6).

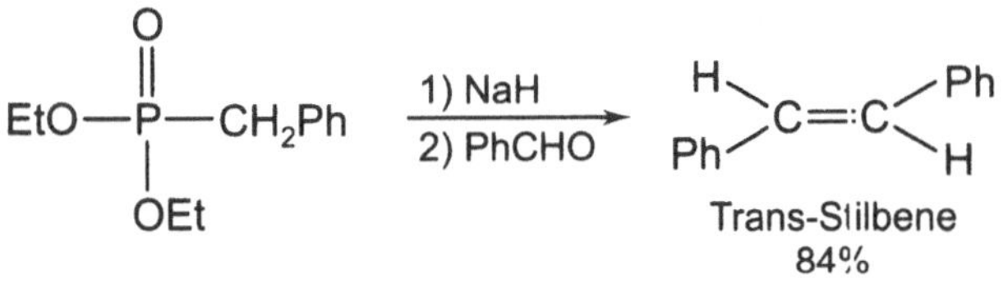

$$EtO - \overset{\overset{O}{\parallel}}{\underset{OEt}{P}} - CH_2Ph \xrightarrow[\text{2) PhCHO}]{\text{1) NaH}} \overset{H}{\underset{Ph}{\diagdown}} C=C \overset{Ph}{\underset{H}{\diagup}}$$

Trans-Stilbene
84%

(Scheme–6)

The above reaction (Scheme–5) is sometimes performed in an organic/ water biphase system[10,11] and in place of strong base like NaH, a PTC can be used in aqueous NaOH with good results. Even much weaker base like K_2CO_3 or $KHCO_3$ can be used. Substrates with base and acid sensitive functional groups can be directly used.

Water has been shown to be an efficient medium for the Wittig reaction employing stabilized ylides and aldehydes. Yields ranging from 80–90 per cent and high E-selectivities (up to 99%) are obtained. Typical example are given below (Scheme–7).

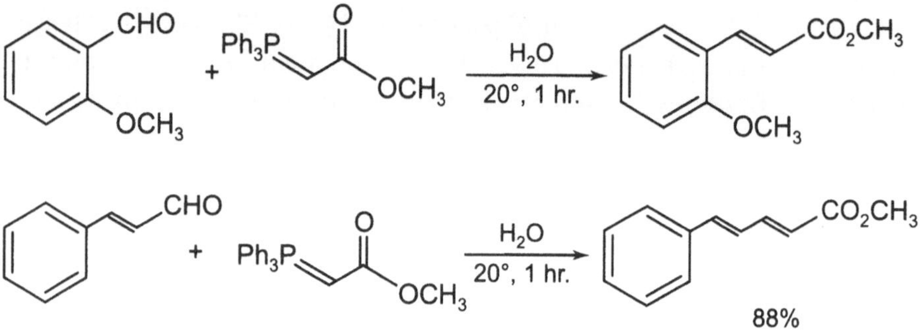

(Scheme–7)

(*b*) The Wittig-Horner Reaction

In the Wittig-Horner reaction, which is a modification of Wittig reaction, the readily available phosphine oxide, $Ph_2 \overset{\overset{O}{\|}}{P}-CH_2R$ is used. Its Lithio derivative on reaction with aldehydes or ketones gave α-hydroxy phosphine oxide, which on treatment with NaH eliminate water to give the corresponding alkene (Scheme–8).

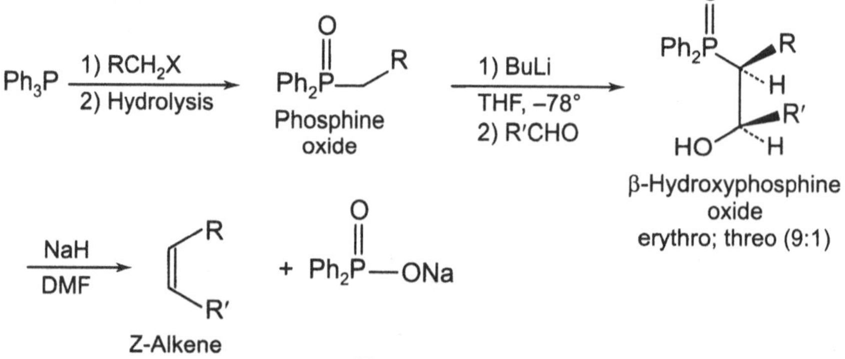

(Scheme–8)

The PTC catalyzed Wittig Hormer reaction using aqueous sodium hydroxide and either tetraalkylammonium salts or crown ether as catalyst gives olefins (Scheme–9).

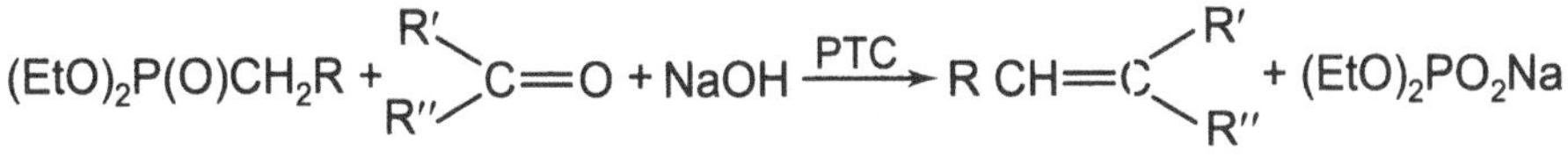
(EtO)2P(O)CH2R +
R′
R″
C=O + NaOH
PTC
R CH=C
R′
R″
+ (EtO)2PO2Na

R=CN, CO_2Et,S Ph etc.

(Scheme–9)

It has been shown[11] that with aldehydes, the reaction may be carried out without a typical PTC, suggesting that the starting phosphonates are themselves able to catalyse two phase reaction.

2.2.15.4 Wittig Reaction in Solid Phase

The Wittig reaction has been reported[12] to occur in solid phase. In this procedure, a 1 : 1 mixture of finely powdered inclusion compound of 4-methylcyclohexanone and (–)–B (Derived from tartaric acid[12] and a catalytic amount of benzyltrimethylammonium hydroxide) is heated at 76° with Wittig reagent (Carbethoxymethylene triphenyl phosphorane) to give optically active 1-(carbethoxymethylene) cyclohexane or the corresponding 4-methyl or 3, 5-dimethyl compound (Scheme–10).

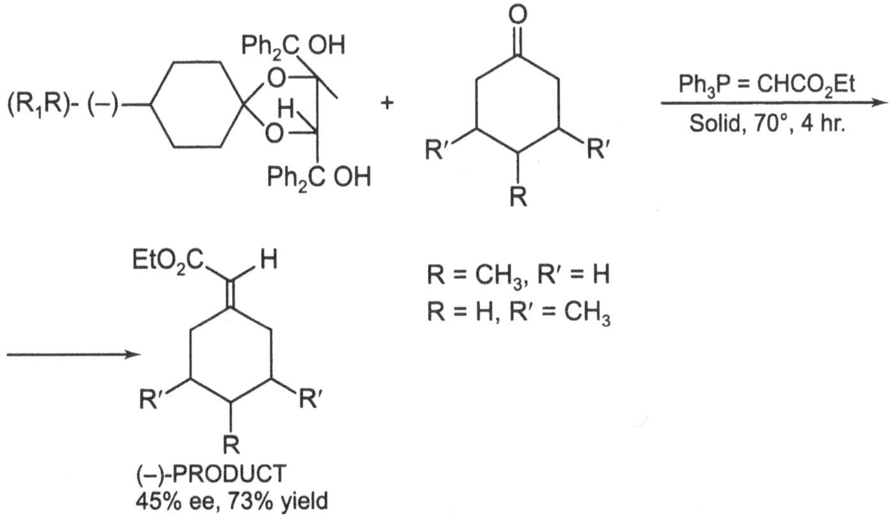
Ph2C OH
(R,R)- (–)-
O
H
O
Ph2C OH
+
O
R′
R′
R
Ph3P = CHCO2Et
Solid, 70°, 4 hr.
EtO2C
H
R′
R′
R
(–)-PRODUCT
45% ee, 73% yield
R = CH3, R′ = H
R = H, R′ = CH3

(Scheme–10)

2.2.15.5 Wittig Reaction in Ionic Liquids

In Wittig reaction, the separation of the alkene from the byproduct (Ph$_3$PO) is a classical problem, which is normally done by crystallization or chromatography. The ionic liquid, [bmin][PF$_4$] can be used as a medium to perform the reaction[13]. The advantage is easy separation of the alkene from Ph$_3$PO and also the recycling of the solvent. In ionic liquids, E-stereoselectivity was observed, similar to that observed in organic solvents.

References

1. G. Wittig, Liebigs Ann., 1949, **562**, 187, Chem. Ber., 1961 **94**, 1373, R. Ketcham, J. Org. Chem., 1962, **27**, 4666.

2. For a general treatise, see Cadogen. Organophosphorus reagents in Organic Synthesis, Academic Press, N.Y. 1970. For a monograph, See Johnson, ylid chemistry, Academic Press, N.Y. 1966. Reviews, Bestmann and Vostrowsky, Top. Curr. Chem., 1983, **109**, 85, 165.

3. G. Markl and A. Merz, Synthesis,1975, 295.

4. S. Hung and I Stemmler, Tetrahedron Lett., 1974, 3151.

5. W. Tagaki, I. Inouse, Y. Yano and T. Okonogi, Tetrahedron Lett., 1974, 2587.

6. A. Spinella, T. Fortunati and R. Sorienti, Synlett, 1997, 93.

7. J.K. Kiddle, Tetrahedron Lett., 2000, **41**, 1339.

8. Prbuzov, Pure Applied Chem., 1964, **9**, 307.

9. W.S. Wadsworth and W.D. Emmons, J. Am. Chem. Soc., 1961, **83**, 1733.

10. C. Pilchuri, Synthesis, 1976, 187.

11. M. Milolajezk, S. Grzejszezk, W. Midura and A. Zatoria, Synthesis, 1976, 397.

12. F. Toda and H. Akai, J. Org. Chem., 1900, **55**, 3446.

13. V. Le Boulaive and R. Gr'ee, Chem. Commun. 2000, 2195.

2.2.16 Miscellaneous C = C Bond Forming Reactions

2.2.16.1 Organometallic Reagents

(*a*) Grignard Reagents, particularly vinyl magnesium chloride react with α, β-unsaturated ketones to give the corresponding vinyl compounds (Scheme–1).

(Scheme–1)

Benzylmagnesium bromide on reaction with benzaldehyde gives stilbene (Scheme–2).

$$C_6H_5CHO \ + \ \xrightarrow[\substack{2) \ H_3^+O}]{1) \ C_6H_5CH_2MgBr} \ C_6H_5CH\!\!=\!\!CHC_6H_5$$

Benzyldehyde   Stilbene

(Scheme–2)

(*b*) Organo lithium reagents particularly aryl lithium on reaction with allylic halides give 3-phenyl propene (Scheme–3).

$$\longrightarrow \ PhCH_2 \ CH\!\!=\!\!\overset{*}{C}H_2$$

3-Phenyl propene

(Scheme–3)

(*c*) Organosilanes, particularly allylic silanes on reaction with carbonyl compounds give the substituted alkene (Scheme–4).

$$R_2CO + CH_2\!\!=\!\!CHCH_2Si(CH_3)_2 \ \xrightarrow[\substack{or \\ BF_3}]{TiCl_4} \ R_2\overset{\overset{\displaystyle OH}{|}}{C}\!\!-\!\!CH_2CH\!\!=\!\!CH_2$$

(Scheme–4)

(*d*) Organostannes particularly alkenyl stannane on reaction with aldehydes in presence of acetyl chloride give substituted alkenes (Scheme–5).

$$R \ CHO + CH_2CHCH_2Sn \ (C_4H_9)_3 \ \xrightarrow[\substack{CH_3 \ COCl}]{n \ Bu_2SnCl_2} \ R \ \overset{\overset{\displaystyle OCOCH_3}{|}}{C}HCH_2CH\!\!=\!\!CH_2$$

(Scheme–5)

(*e*) Organo copper reagents, particularly lithium dimethyl cuprate on reaction with vinyl halides give the corresponding alkenes (Scheme–6).

$$R_2 \ Cu \ Li \ + \ \diagup\!\!\!\!=\!\!\!^{X} \ \longrightarrow \ \diagup\!\!\!\!=\!\!\!^{R}$$

(Scheme–6)

In place of lithium dialkyl cuprate, lithium divinyl cuprate can be used to synthesise alkenes (Scheme–7).

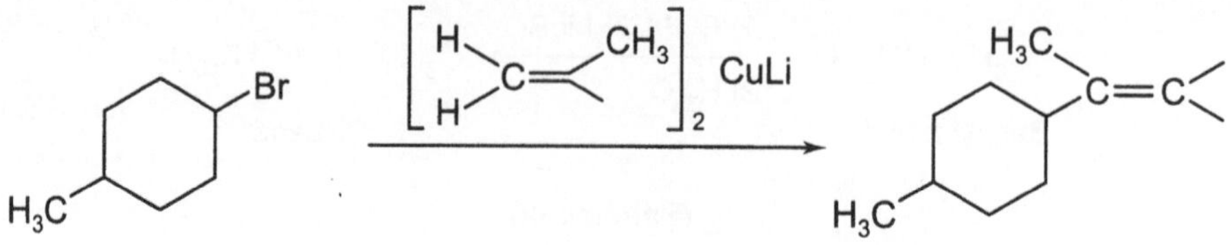

(Scheme–7)

(*f*) Organo zinc compounds, particularly iodomethylenezinc iodide (generated *in situ*) on reaction with carbonyl compounds give alkene (Scheme–8).

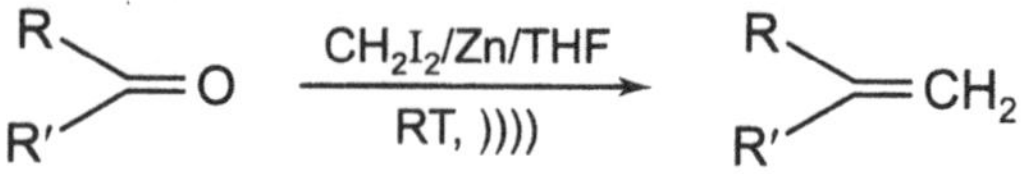

(Scheme–8)

See also section 2.1.29.

2.2.16.2 Michael Addition

The Michael addition of active nitriles to acetylenes, catalysed by PTC (quaternary ammonium chloride) given the substituted alkenes (Scheme–9).

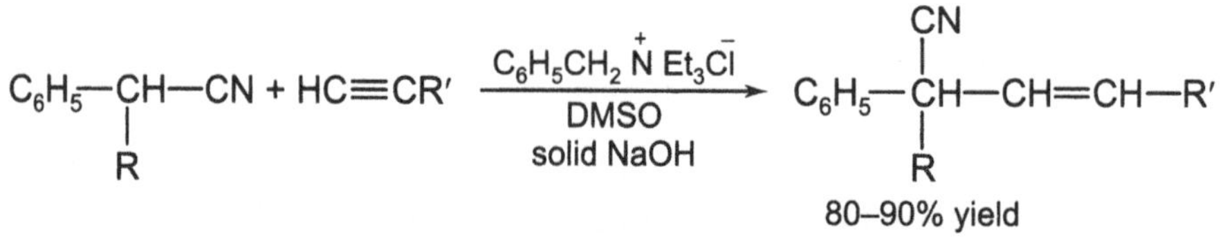

80–90% yield

(Scheme–9)

See also section 2.1.26.

2.2.16.3 Dehydrohalogenations

Phase transfer catalysts have been used for dehydrohalogenations of alkyl halides to give alkenes (Scheme–10).

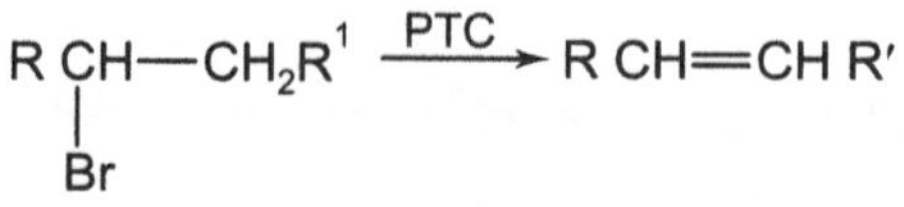

(Scheme–10)

2.2.16.4 Dehydration of Aldols

Aldols on dehydration give unsaturated compounds (See Aldol condensation, section 2.1.4).

2.2.16.5 Henry Reaction

(See section 3.1.12).

2.2.16.6 Stobbe Condensation

Condensation of dialkyl succinates with carbonyl compounds in presence KOH or Pot. tert. Butoxide or NaOH give the salts of α, β-unsaturated half esters (H. Stobbe, Ber, 1983, 26, 2312; Ann. 1894, 282, 280; W.S. Johnson, C.H. Doub, Org. Reactions, 1951, 6, 1.

$$R_2CO + \underset{\underset{CH_2CO_2Et}{|}}{CH_2CO_2Et} \xrightarrow[HOCMe_2]{KOMe_3} R_2C{=}\underset{\underset{CH_2COOH}{|}}{C}{-}CO_2Et$$

2.2.16.7 Dehydrogenation

Generally dehydrogenation results in the formation of carbon-carbon double bonds. Thus, androst-4-en-3, 17-dione on heating with DDQ in dioxane in presence of gaseous HCl gives 72% yield of androsta-4, 6-diene-3, 17-dione (A.B. Turner and H.J. Ringold. J. Chem.. Soc., 1967, 1720.

Androst-4-en-3-17-dione   Androsta-4-6-diene-3, 17-dione

2.3 CARBON-CARBON TRIPLE BOND (C ≡ C) FORMATION

2.3.1 Introduction

Compounds containing $C \equiv C$ are only obtained by indirect procedures, some of them are given as follows:

(*i*) **From alkenes** by reaction with bromine followed by dehydrohalogenation of the formed vic-dibromides.

$$R—CH=CH—R \xrightarrow{Br_2} R—\overset{\overset{\displaystyle H}{|}}{\underset{\underset{\displaystyle Br}{|}}{C}}—\overset{\overset{\displaystyle H}{|}}{\underset{\underset{\displaystyle Br}{|}}{C}}—R$$

vic. Dibromide

$$2\,R—\overset{\overset{\displaystyle H}{|}}{\underset{\underset{\displaystyle Br}{|}}{C}}—\overset{\overset{\displaystyle H}{|}}{\underset{\underset{\displaystyle Br}{|}}{C}}—R \;+\; 2NH_2^- \longrightarrow RC\equiv CR + 2\,NH_3 + 2\bar{Br}$$

(*ii*) **From alkynes** by treatment with a strong base followed by reaction with alkyl halides

$$R—C\equiv CH + NaNH_2 \longrightarrow RC\equiv \overset{\ominus}{C}:$$

$$\downarrow R' I$$

$$R—C\equiv C—R'$$

This is example of S_N2 reaction

$$RC\equiv C: + \overset{R'}{\underset{H}{C}}—\overset{..}{\underset{..}{Br}}: \xrightarrow{S_N2} R—C\equiv C—CH_2R'$$

(*iii*) **From Grignard reagent** by treatment with other alkynes followed by S_N2 raction with alkyl halides.

$$R—MgX + CH_3C\equiv CH \longrightarrow CH_3C\equiv CMgI + MgI_2$$

Propyne

$$R'I \Big| S_N2$$

$$CH_3C\equiv C—R'$$

Alkynes

The Grignard reagent obtained from alkynes (R — C ≡ C MgI) can be transformed into other compounds as in the case of Grignard reagents.

(*iv*) **From organo lithium reagents:** Appropriately substituted organo lithium reagents react with epoxides to give substituted alkynes.

$$CH_2\text{---}CH \underset{O}{\diagup} \quad \xrightarrow[\text{2) } H_3O^+]{\text{1) } CH_3C\equiv C\ Li} \quad CH_3\equiv C\ CH_2\ CH_2\ OH$$
$$40\%$$

(*v*) **Sonogastira reaction** provides a general method for the preparation of unsymmetrical alkynes.

2.3.2 Sonogashira Reaction

It is a powerful method for the creation of carbon-carbon bonds[1] and is a palladium and copper cocatalysed coupling of terminal alkynes with aryl and vinyl halides (Scheme–1). It is a general method for the preparation of unsymmetrical alkynes.

Cl—⟨ ⟩—Br + Ph≡H $\xrightarrow[\Delta]{\text{Pd/Cu}}$ Cl—⟨ ⟩≡—Ph

| Aryl halide | Phenylacetylene (terminal alkyne) | | unsymmetrical alkyne |

(Scheme–2)

2.3.2.1 Sonogashira Reaction in Water

A aqueous sonogashira type coupling reaction proceeds in water[2] as the sole solvent, without the need for copper (I) or any transition-metal phosphine complex, which overcome the problem of intrinsic toxicity and air-sensitivity of transition metal complexes and the use of phospane ligands (Scheme–2).

+ ≡—⟨ ⟩ $\xrightarrow[\text{H}_2\text{O, MW 175°}]{\text{TBAB, Na}_2\text{CO}_3}$

| 2-Bromo naphthalene | Phenyl acetylene | Coupled product |

(Scheme–2)

Another example of sonogashira coupling of aryl bromides and iodides with phenyl acetylene using polymeric complex (A) or the monomeric (B) catalyst, pyrolidine as base and TBAB as additive[3] has been performed under microwave heating. Use of catalyst (B) gave better yields (Scheme–3).

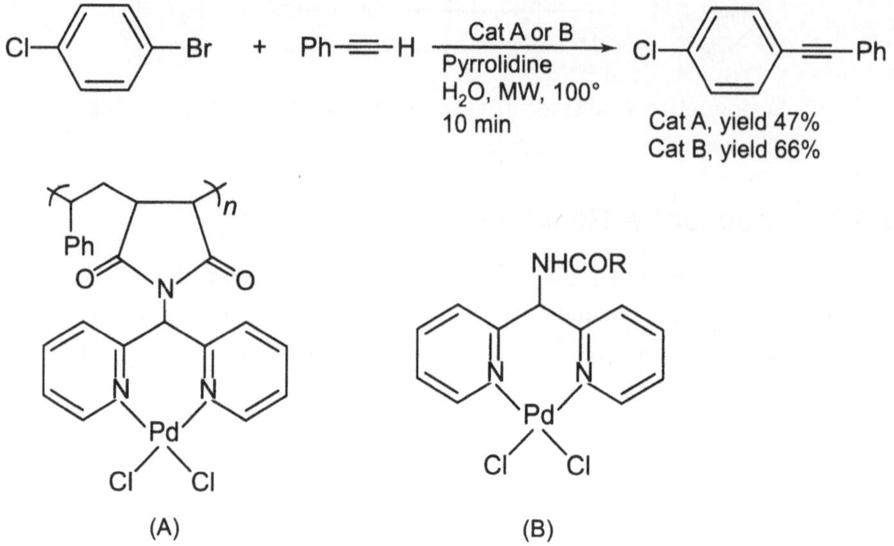

(Scheme–3)

Sonogashira coupling of aryl and heteroaryl bromides and iodides with phenyl acetylene occurred in the presence of heterogeneous Pd catalyst[4] (C) under microwave heating to give the coupled product (Scheme–4)

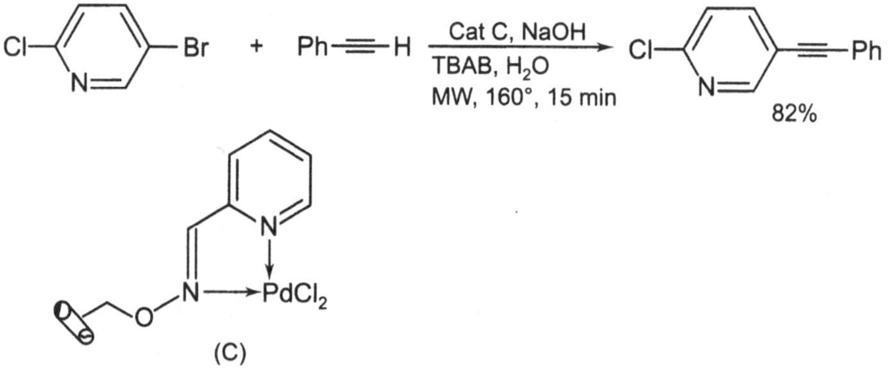

(Scheme–4)

Sonogashira coupling under transition – metal – free conditions[5] involving ultralow Pd concentrations as contaminations[6] and using NaOH as base, polyethylene glycol (PEG) as phase – transfer catalyst and water as solvent under microwave heating gave the coupled products in good yield (Scheme–5)

$$R'{-}\langle\bigcirc\rangle{-}X \quad + \quad R^2{\equiv}H \xrightarrow[\substack{MW,\ 170^\circ \\ 5\ min}]{\substack{Pd,\ NaOH \\ PEG,\ H_2O}} \quad R^1{-}\langle\bigcirc\rangle{\equiv}R^2$$

80–90%

R_1 = H, Me, COMe, OMe
R_2 = Alkyl, aryl
X = Br, I

(Scheme–5)

2.3.2.2 Sonogashira Reaction in Ionic Liquids

Copper and ligand free Sonogashira reaction catalysed by Pd (O) nanoparticles proceed under ultrasound irradiation in ionic liquid [bbmin] [BF$_4$] (Scheme–6)[7]

$$R_1{-}\langle\bigcirc\rangle{-}X \quad + \quad R_2{\equiv} \xrightarrow[\substack{\text{)))),}\ Pd\ Cl_2,\ 30^\circ}]{[bbmin]\ [BF_4]} \quad R_1{-}\langle\bigcirc\rangle{\equiv}R_2$$

R_1 = H, CH$_3$, NO$_2$, CHO
R_2 = Alkyl, Cyclohexyl
X = I, Br

(Scheme–6)

References

1. R. Chinchilla and C. Nájera, Chem. Rev., 2007, **107**, 874.
2. P. Appukkutan, W. Dehaen and E.V. der Eycken, Eur. J. Org. Chem., 2003, 4713.
3. J. Gil – Molto', S. Karlström and C. Nájera, Tetrahedran, 2005, **61**, 12168.
4. K.M. Dawood, W. Solodenko, A. Kirschning, ARKIVOC, 2007, 104; K.M. Dawood and A. Kirshing, Tetrahedron, 2005, **61**, 12121.
5. N.–E. Leadbeater, M. Marco and B.S. Tominack, Org. Lett., 2003, **5**, 3919.
6. R.K. Arvela, N.E. Leadbeter, M.S. Sangi, V.A. Williams, P. Grandos and R.D. Singer, J. Org. Chem., 2005, **70**, 161.

Carbon-Oxygen and C=O Bond Formation

3.1 CARBON-OXYGEN BOND FORMATION

3.1.1 Introduction

Next to carbon-carbon bond formation, carbon-oxygen bond formation is important for organic synthesis. A large numebr of compounds containing carbon-oxygen bonds are present in a large number of natural products like esters, ethers, glycosides. Some of the well-known reactions which are useful for carbon-oxygen bond formation have already been discussed in chapter 2 in reactions for carbon-carbon bond formation. These include aldol condensation, acyloin condensation, Arndt-Eistert synthesis, Baylis-Hillman Reaction, Benzoin condensation, Darzen reactior, Pinacol coupling, Reformatsky reaction.

Following are given some of the methods/procedures which are useful for carbon-oxygen bond formation.

3.1.2 Acetylation

Phenols, amines and quinones can be conveniently acetylated. Depending on the nature of compound to be acetylated, appropriate acetylating reagent can be used[1]. These include:

- Acetic anhydride-sulphuric acid (catalytic amount).
- Acetic anhydride-sodium hydroxide (aqueous solution).

- Acetic anhydride-sodium acetate.
- Acetic anhydride-zinc chloride.
- Acetic anhydride-acetic acid-zinc dust.
- Acetic anhydride-zinc dust.
- Acetyl chloride-pyridine.

3.1.2.1 Acetylation using Ionic Liquids

Simple alcohols like 2-naphthol and *t*-butyl alcohol can be acetylated at room temperature using dicyanamide ionic liquid and acetic anhydride even in the absence of any other catalyst[2]. Using ionic liquids, glycosylation and acetylation of nucleotides[3] can also be effected.

3.1.2.2 Acetylation using Microwaves

Using microwave technique asperin is prepared by the reaction of salicylic acid and acetic anhydride by heating in a microwave oven.

Salicylic acid + $CH_3COOCOCH_3$ (Acetic anhydride) $\xrightarrow[\text{90 sec}]{\text{MW}}$ Aspirin

References

1. V.K. Ahluwalia and Renu Aggarwal, Comprehensive Practical Organic Chemistry, Universities Press, 2004, Page 3–8.
2. S.A. Forsyth, D.R. MacFarlane, R.J.Thomson and M. Itzstein, Chem. Commun., 2002, 714.
3. M.C. Uzagare, Y.S. Sanghvi and M.M. Salunkhe, Green Chem. 2003, 370.
4. A.K. Bose, B.K. Banik, N. Lavlinskaia, M. Jayaraman and M.S. Manhas, Chemtech. Sept., 1977, 18.

3.1.3 Algar-Flynn-Oyamada Reaction

Oxidation of 2′-hydroxy chalcones with hot alkaline hydrogen peroxide gives[1] flavanols.

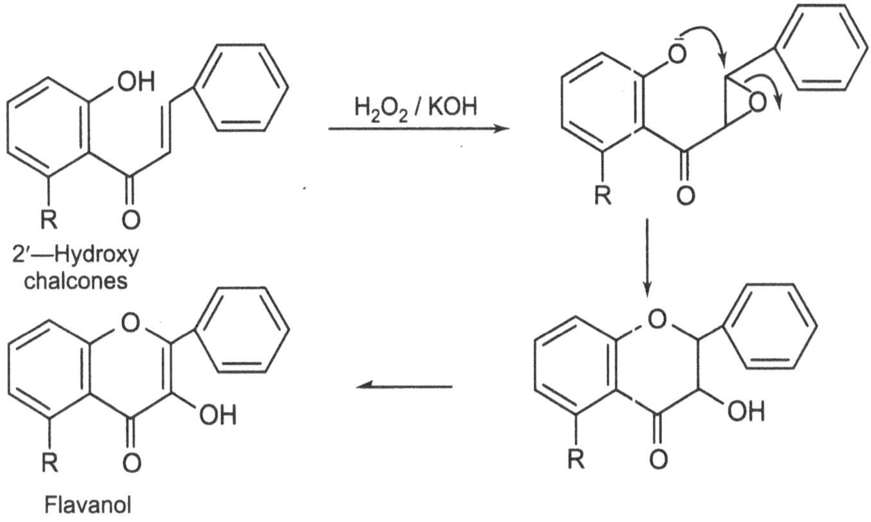
OH
H_2O_2 / KOH
R
O
2′—Hydroxy chalcones
O
OH
Flavanol

See also Baker-Venkataraman Rearrangement (section 2.1.7).

Reference

1. J. Algar, J.P. Flynn, Proc. Raj. Iris. Acad., 1934, **42B**, 1; B. Oyamada, J. Chem. Soc. Japan, 1934, **55**, 1256.

3.1.4 Baeyer-Villiger Oxidation

Ketones on oxidation with hydrogen peroxide or with peracids (RCO_3H) give esters[1] (Scheme–1).

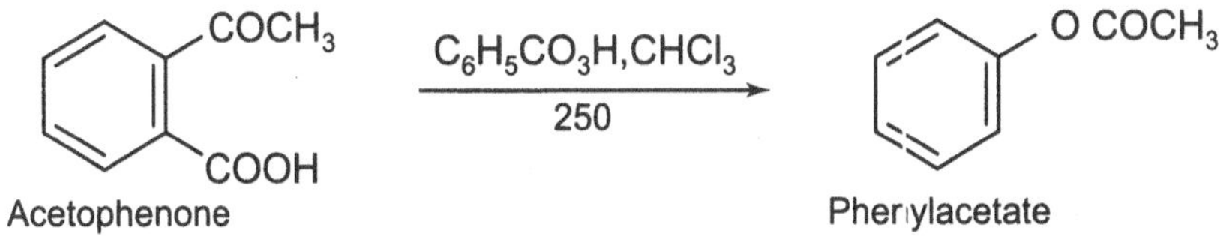
$COCH_3$
COOH
$C_6H_5CO_3H, CHCl_3$
250
O $COCH_3$
Acetophenone
Phenylacetate

(Scheme–1)

The Baeyer-Villiger Oxidation can not be used for substrates which contain C = C, – S – or >N—R groups.

3.1.4.1 Baeyer-Villiger Oxidation in Aqueous Phase

Baeyer-villiger oxidation of ketones has been satisfactorily carried out in aqueous heterogeneous medium with m-chloroperbenzoic acid[2] (MCPBA) (Scheme–2).

R—⟨O⟩—COMe $\xrightarrow[\text{H}_2\text{O, 0.5–1 hr}]{\text{MCPBA, 80°}}$ R—⟨O⟩—O COMe

R = H, Cl, OMe

(Scheme–2)

Baeyer villiger oxidation can be conveniently performed by using polymer supporter peracids (*e.g.*, polystyrene peracid)[3].

3.1.4.2 Baeyer-Villiger Oxidation in Solid State

Some Baeyer-villiger oxidation of ketones with MCPBA proceed much faster in the solid state than in solution. In this procedure, a mixture of powdered ketone and 2 mol equivalent of MCPBA is kept at room temperature to give the product[4] (Scheme–3).

$$\overset{\overset{\displaystyle O}{\|}}{\text{MeC}}\text{—}⟨O⟩\text{—Br} + \text{MCPBA} \xrightarrow[\text{Solid State}]{\text{RT, 5 days}} \text{MeOCO—}⟨O⟩\text{—Br}$$

p-Bromoacetophenone   *p*-Bromophenylacetate, 64%

$$\text{PhCOCH}_2\text{Ph} + \text{MCPBA} \xrightarrow[\text{Solid State}]{\text{RT, 24 hr}} \text{PhCOOCH}_2\text{Ph}$$

97%

$$\text{PhCOPh} + \text{MCPBA} \xrightarrow[\text{Solid State}]{\text{RT, 24 hr}} \text{PhOCOPh}$$

85%

(Scheme–3)

3.1.4.3 Enzymatic Baeyer-Villiger Oxidation

Phenylacetone on oxidation with the enzyme, cyclohexanone oxygenase in presence of NAD/reduced nicotinamide adenine dinucleotide gives[5] benzyl acetate (Scheme–4).

$$\text{Ph CH}_2\text{ COCH}_3 \xrightarrow[\text{O}_2\text{, ENZ–FAD, NADPH, H}^+]{\text{Cyclohexanone oxygenase}} \text{Ph CH}_2\text{OCOCH}_3$$

Phenyl acetone   Benzylacetate

References

1. A.V. Baeyer and V. Villager, Ber., 1899, **32**, 3625.
2. F. Fringulli, R. Germani, E. Pizzo and G. Savalli, Gazz. Chem. Ital., 1989, **119**, 249.
3. J.M.J. Frechet, and K.E. Macromolecules, J. Am. Chem., Soc., 1975, **8**, 130.
4. K. Tanka and F. Toda, Chem., Rev., 2000, **100**, 1028–29.
5. B.P. Branchaud and C.T. Walsh, J.Am. Chem. Soc., 1985, **107**, 2153.

3.1.5 Barbier Reaction

The reaction of an organo metallic reagent (obtained *in situ*) with the ketone give the corresponding alcohol. This reaction is known as Barbier reaction[1] (Scheme–1).

(Scheme–1)

Same products are also obtained by using Grignard reagent (see section 2.1.29.1), which is prepared separately and then reacted with the ketone. In the normal Barbier reaction only reactive alkyl halides can be used.

3.1.5.1 Barbier Reaction under Sonication

Using sonication[2], any alkyl halide can be used. The reaction is carried out in THF and even imperfectly dried alkyl halides give excellent yields of the alcohol. Even allylic or benzylic aldehydes, which generally give Wurtz coupling product can also be used (Scheme–2).

70–100%

(Scheme–2)

Some other examples of sonochemical Barbier reaction are given as follows (Scheme–3).

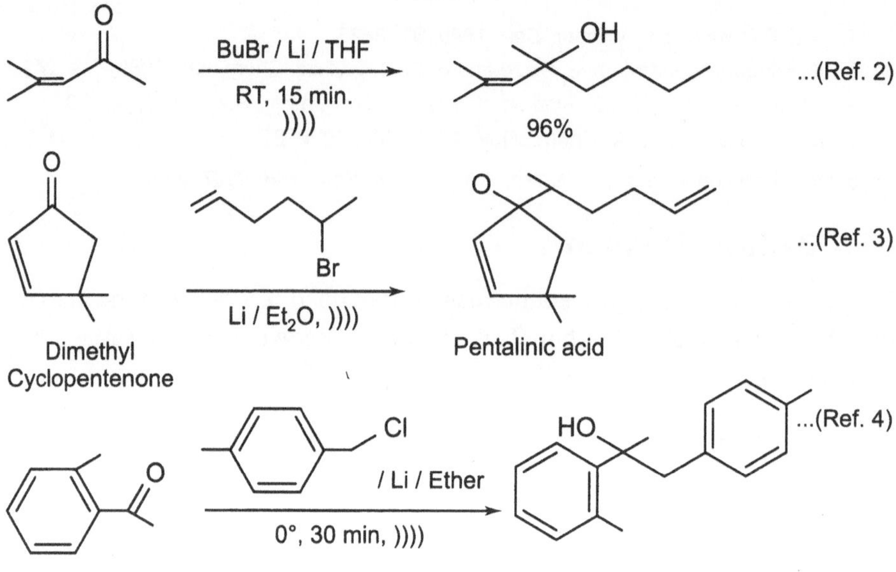

...(Ref. 2)

...(Ref. 3)

...(Ref. 4)

Dimethyl
Cyclopentenone

Pentalinic acid

(Scheme–3)

References

1. P. Barbier, C.R. Acad. Sci., 1898, **128**, 110.
2. J. L. Luche and J.C. Damiano, J.Am. Chem. Soc., 1980, **102**, 7926.
3. M. Ihara, M. Katogi, K. Fukumoto and T. Kametani, J.Chem. Soc., Chem. Commun., 1987, 721.
4. I.C. Burkow, L.K. Sydnes and D. C.N. Ubeda, Acta. Chem. Scand-Ser. B, 1987, **B41**, 235; S.B. Singh and G.R. Pettit, Syn. Commun., 1987, **17**, 877.

3.1.6 Bouveault-Blanc Reduction

Reduction of esters, aldehydes or ketones with sodium and alcohol give[1] alcohols (Scheme–1). However, acids do not undergo reduction by this procedure.

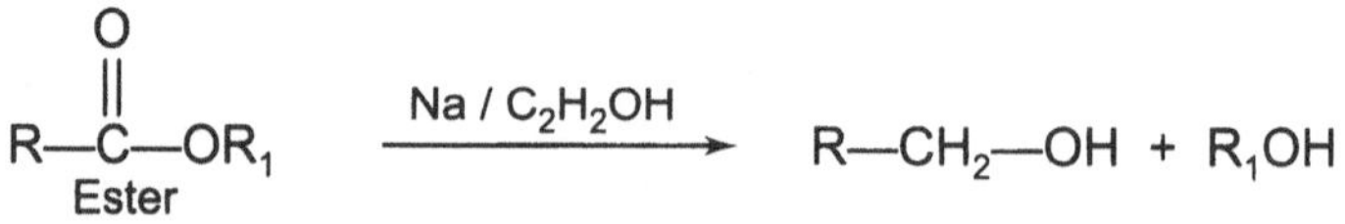

(Scheme–1)

Using this method even unsaturated ester can be reduced[2](Scheme–2).

$$CH_3 (CH_2)_7 CH = CH (CH_2)_7CO_2C_4H_9 \xrightarrow{Na / C_2H_2OH}$$

Butyl oleate

$$\longrightarrow CH_3 (CH_2)_7 CH = CH (CH_2)_7CH_2OH$$

Oleyl alcohol

(Scheme–2)

Better yields are obtained by sonication.

References

1. G. Bouveault and G. Blanc, Compt. Red., 1903, **136**, 1676; Bull. Soc. Chem. France, 1904, **(3)31**, 666; H.O. House, Modern Synthetic Reactions, W.A. Benjamin, California, 1972, p. 150.

2. L.A. Paquette, J. Org. Chem., 1962, **27**, 2274; K. Ruhmann, Synthesis, 1972, 236.

3.1.7 Benzil-Benzilic Acid Rearrangement

α-Diketones (benzils) undergo a base catalysed reaction called Benzil-Benzilic acid rearrangement[1] (Scheme–1).

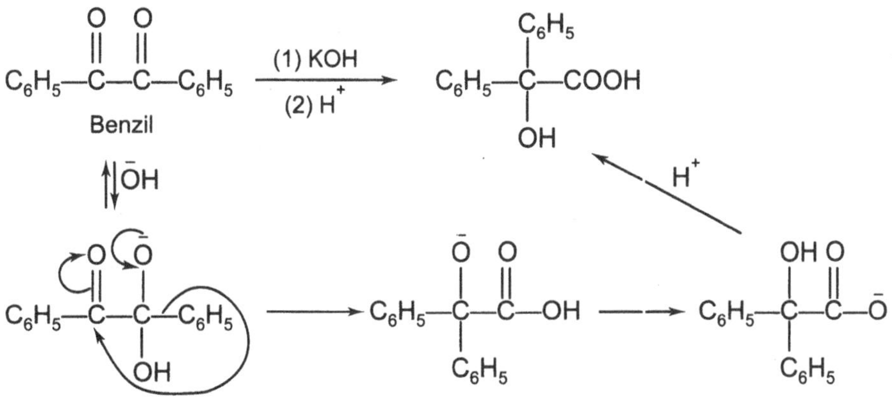

(Scheme–1)

3.1.7.1 Benzil-Benzilic Acid Rearrangement under Microwave Irradiation

Benzil-Benzilic rearrangement proceeds more efficiently and faster in solid state[2] and it takes 0.1 to 6 hr. for completion yielding 70–90 per cent of the rearranged product. It is best to conduct the rearrangement in solid state by MW irradiation[3] in much shorter time and better yields (Scheme–2).

$$
\underset{\text{Ar}-\overset{\overset{\displaystyle O}{\|}}{C}-\overset{\overset{\displaystyle O}{\|}}{C}-\text{Ar}'}{} \quad \xrightarrow[\text{3-5 min,}]{\text{KOH / MW}} \quad \underset{\underset{>95\%}{\overset{\text{Ar}'}{|}}}{\text{Ar}\!-\!\!\overset{\text{Ar}'}{\underset{\text{OH}}{|}}\!\!-\!\text{COOH}}
$$

(Scheme–2)

References

1. S. Selman and J.F. Eastham, Quart, Rev., 1960, **14**, 221.
2. F. Toda, Y. Tamaka, Y. Kagaura and Y. Sakaino, Chem. Lett., 1990, 373.
3. H.M. Yu, S. T. Chem., M. J. Tseng and K.M. Wang J. Chem. Res(S), 1994, 62.

3.1.8 Cannizzaro Reaction

Aldehydes without α-hydrogen(s) on treatment with concentrated aqueous alkali undergo self oxidation and reduction to give an alcohol and the salt of the corresponding carboxylic acid[1] (Scheme–1).

$$
\underset{\text{Benzaldehyde}}{C_6H_5CHO + C_6H_5CHO} \quad \xrightarrow{\text{KOH}} \quad \underset{\text{Benzylalcohol} \quad \text{Pot. benzoate}}{C_6H_5CH_2OH + C_6H_5C\overset{-}{O}\overset{+}{O}K}
$$

(Scheme–1)

Cannizzaro reaction best proceeds with aromatic aldehydes devoid of α-hydrogen. Some aliphatic aldehydes like formaldehyde and dimethyl acetaldehyde (which do not have α-hydrogen) also undergo cannizzaro reaction (Schem–2).

$$
\underset{\text{Formaldehyde}}{\text{HCHO} + \text{NaOH}} \quad \xrightarrow{\Delta} \quad \underset{\text{Methylalcohol} \quad \text{Sod. formate}}{\text{CH}_3\text{OH} + \text{HCOONa}}
$$

$$
\underset{\text{Dimethylacetaldehyde}}{2(CH_3)_2\,CHCHO + NaOH} \xrightarrow{\Delta} \underset{\substack{\text{2–methyl–1–} \\ \text{–propanol}}}{(CH_3)_2\,CH\,CH_2OH} + \underset{\substack{\text{Sod. 2–methyl–} \\ \text{–1–propionate}}}{(CH_3)_2\,CHCOONa}
$$

(Scheme–2)

3.1.8.1 Crossed Cannizzaro Reaction

The cannizzaro reaction between two different aldehydes (one of which is formaldehyde) gives the alcohol (Scheme–3).

$$\mathrm{RCHO + CH_2OH \xrightarrow{-OH} RCH_2OH + HCO\bar{O}}$$

(Scheme–3)

3.1.8.2 Intramolecular Cannizzaro Reaction

Certain compounds, which contain two carbonyl groups undergo internal cannizzaro reaction (Scheme–4).

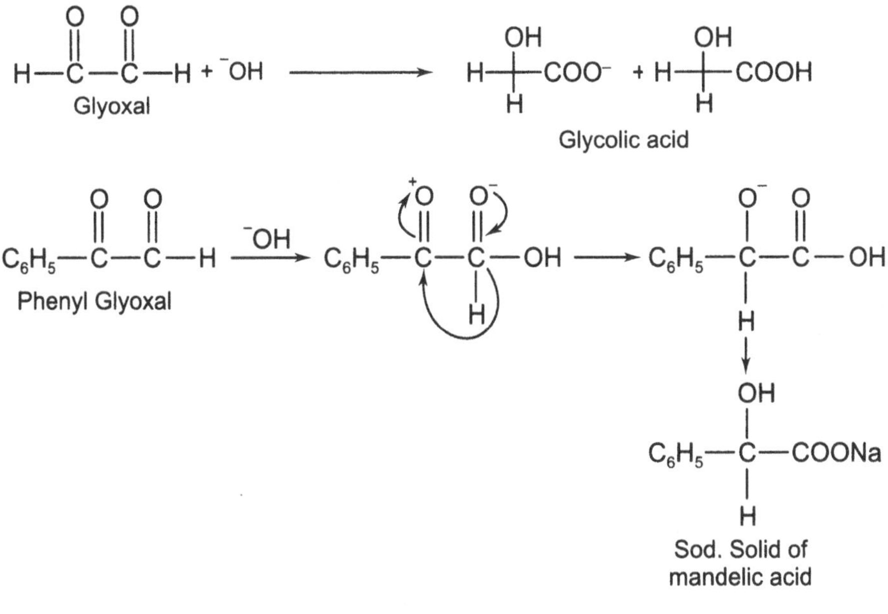

(Scheme–4)

3.1.8.3 Cannizzaro Reaction in Solid State

Cannizzaro reaction proceeds rapidly on a barium hydroxide, $Ba(OH)_2.8H_2O$ surface. Thus, a mixture of benzaldehyde and paraformaldehyde on mixing with barium hydroxide octaacetate and then irradiation in a microwave oven (100–110°) or heating in an oil bath gives major amount of the alcohol (Scheme–5)[2].

$$\underset{\text{Benzaldehyde}}{\mathrm{PhCHO}} + \underset{\text{Paraformaldehyde}}{\mathrm{(CH_2O)_2}} \xrightarrow[\mathrm{Ba\ (OH)_2.8H_2O}]{\text{MW or oil bath}} \underset{80\text{-}90\%}{\mathrm{PhCH_2OH}} + \underset{1\text{-}20\%}{\mathrm{PhCOOH}}$$

(Scheme–5)

3.1.8.4 Cannizzaro Reaction under Sonication

The cannizzaro reaction under heterogeneous conditions catalysed by barium hydroxide is considerably accelerated by low intensity ultrasound[3](cleaning bath). No reaction is observed without the use of ultrasound (Scheme–6).

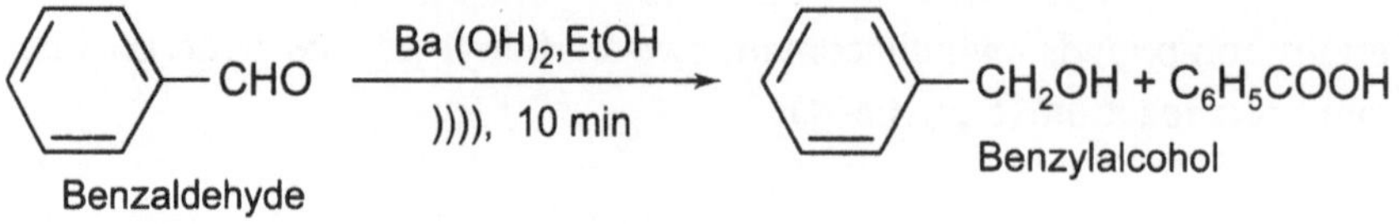

(Scheme–6)

See also Tischeno reaction (section 3.1.15).

References

1. S. Cannizzaro, Ann., 1853, **88**, 129; K. List and H. Limprecht, Ann., 1854, **90**, 180.
2. R.S. Varma, G.W. Kabalka, L.T. Evans and R.M. Pagni, Synth. Commun., 1985, **15**, 279.
3. A. Fuentes and J.V. Sinisterra, Tetrahedron Lett., 1986, 27, 2967.

3.1.9 Dakin Reaction

It involves[1] the oxidation of aldehyde or acyl group in phenolic/aldehydes or ketones to give dihydroxy compound (the –CHO or –COCH$_3$ group is replaced by OH) (Scheme–1).

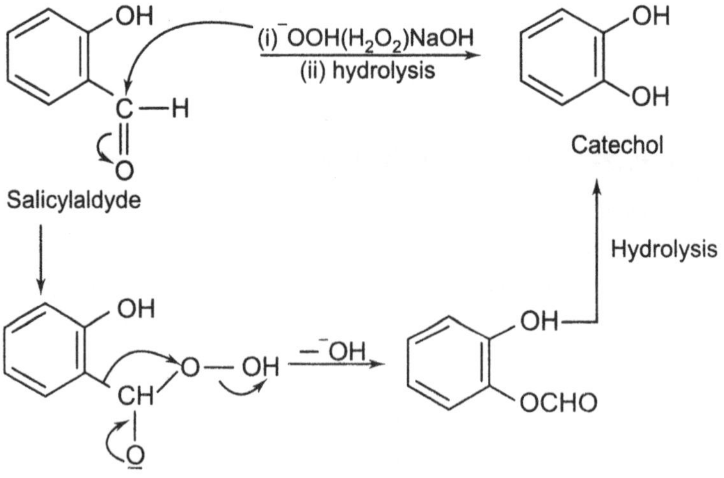

(Scheme–1)

Generally in Dakins oxidation, the yields are low.

3.1.9.1 Dakin Reaction in Solid State

Solid state oxidation of hydroxy benzaldehydes and acetophenones with urea-formaldehyde adduct is a super alternative[2] in terms of shorter reaction time, cleaner product formation, ease of manipulation and excellent yield.

The reagent, hydrogen peroxide-urea complex $\left[\begin{array}{c} H_2NCONH_2 \\ | \\ HOOH \end{array} \right]$ (UPH) is

commercially available and can also be easily prepared. The oxidation in carried out by heating a mixture of hydroxy aldehyde or ketone and urea-formaldehyde adduct (UPH) in the molar ratio (1:2) at 85° for 20 min to 1.5 hr. The product is isolated by extraction the reaction mixture with ethyl acetate (Scheme–2).

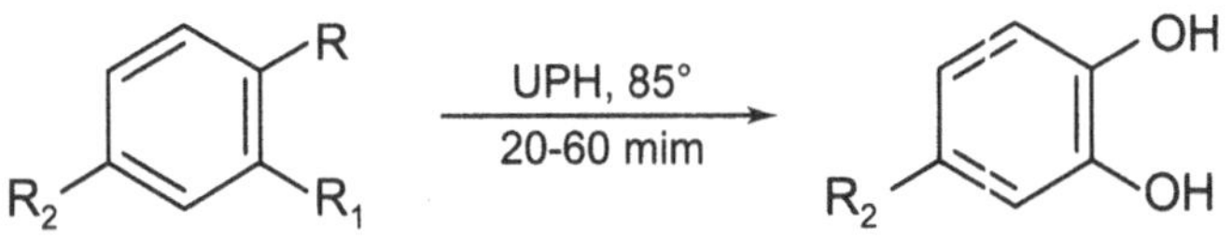

R = CHO, R_1 = OH, R_2 = H

R = CHO, R_1 = H, R_2 = OH

R = COCH₃, R_1 = OH, R_2 = H

R = COCH₃, R_1 = H, R_2 = OH

R = CHO, R_1 = OH, R_2 = NO₂

R – CHO, R_1 = H, R_2 = OMe

(Scheme–2)

3.1.9.2 Dakin Reaction under Ultrasonic Irradiation

Dakin reaction has been carried out in high yields using sodiumpercarbonate (SPC, Na_2CO_3. $1.5 H_2O$) in H_2O_2 –THF under ultrasonic irradiation[3] giving 85 to 80 per cent yields (Scheme–3).

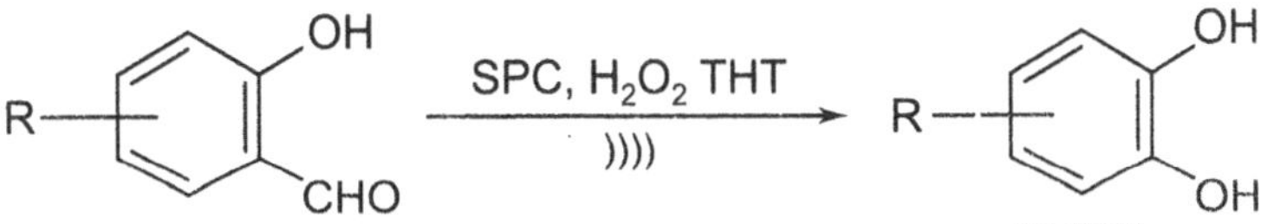

(Scheme–3)

References

1. H.D. Dakin, OS, 1941, **1**, 149; J.E. Lettler, Chem., Rev., 1949, **45**, 385; H.D. Dakin, J. Am. Chem. Soc., 1909, **42**, 477; H.D. Dakin, Org. Synth. Coll. Vol. I, 1941.

2. R.S. Varma and K.P. Naicker, Org. Lett., 1991, **1**, 189.

3. G.W. Kabalka, N.K. Reddy and C. Narayana, Tetrahedron Lett., 1992, **33**, 865.

3.1.10 Elb Persulfate Oxidation

Phenols on oxidation with potassium persulfate[1] in alkaline medium give hydroxy phenols. The –OH group enters the *para* position with respect to the original hydroxy group to give a hydroquinone derivative. However, if the *para* position is blocked, the –OH group goes to the *ortho* position (Scheme–1).

(Scheme–1)

Reference

1. E. Elbs, J. Praki. Chem., 1983, **48**, 179; S.M. Sethna, Chem. Rev., 1951, **91**, 49.

3.1.11 Knorr Quinoline Synthesis

Condensation of β-ketoesters with arylamines at 110° give the corresponding amides, which on cyclisation with conc. H_2SO_4 give α-hydroxy quinolines[1] (Scheme–1).

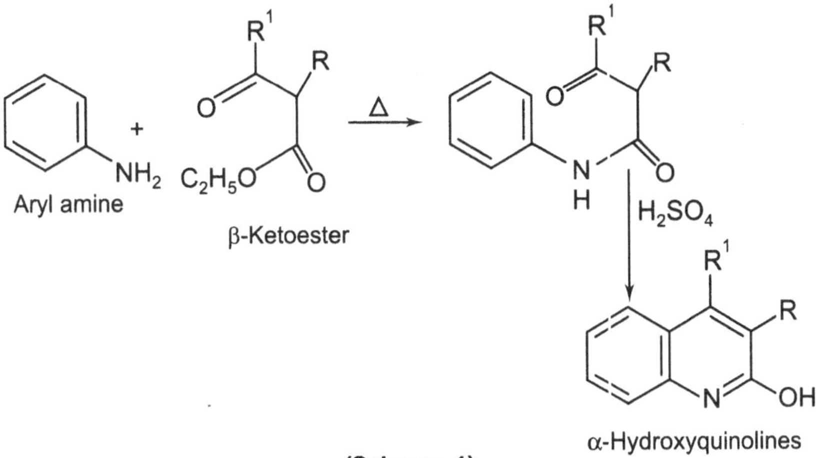

(Scheme–1)

Reference

1. L. Knorr, Ann., 1886, **236**, 69; 1888, **245**, 378.

3.1.12 Henry Reaction

It is the aldol condensation of nitroalkanes with aldehydes[1](Scheme–1).

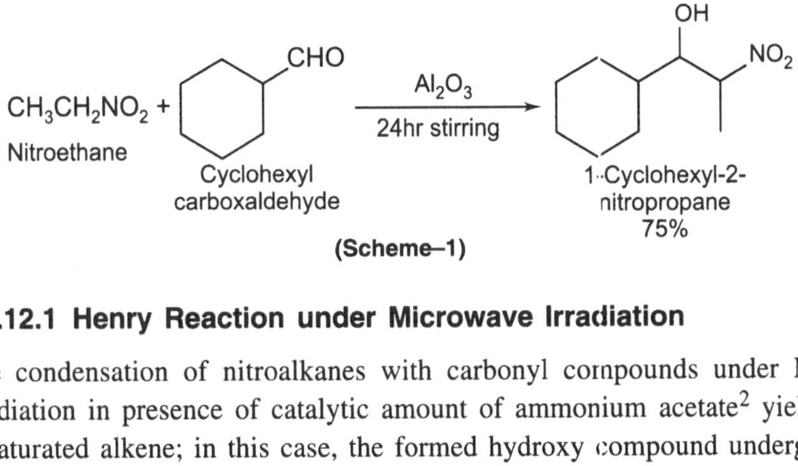

(Scheme–1)

3.1.12.1 Henry Reaction under Microwave Irradiation

The condensation of nitroalkanes with carbonyl compounds under MW irradiation in presence of catalytic amount of ammonium acetate[2] yielded unsaturated alkene; in this case, the formed hydroxy compound undergoes dehydration. This procedure avoids the use of large excess of polluting nitrohydrocarbons (Scheme–2).

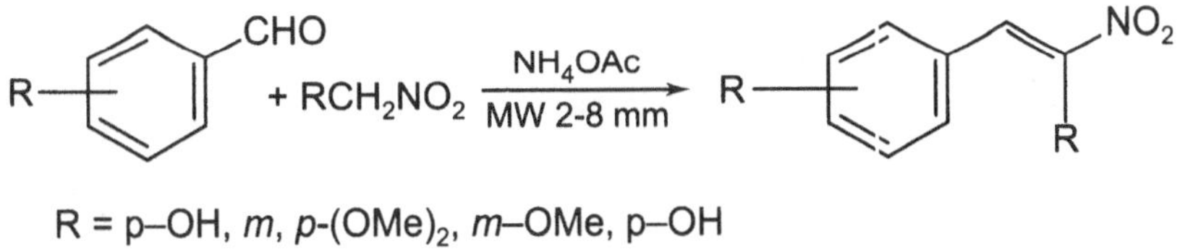

R = p–OH, *m, p*-(OMe)$_2$, *m*–OMe, p–OH

(Scheme–2)

3.1.12.2 Henry Reaction using Ionic Liquids

Henry reaction can be conducted in chloroaluminate ionic liquids[3]. The tetramethyl guanidine (trifluoroacetate) based ionic liquids has been reported as recyclable catalyst for Henry reaction to produce 2-nitroalkanols[4] (Scheme–3).

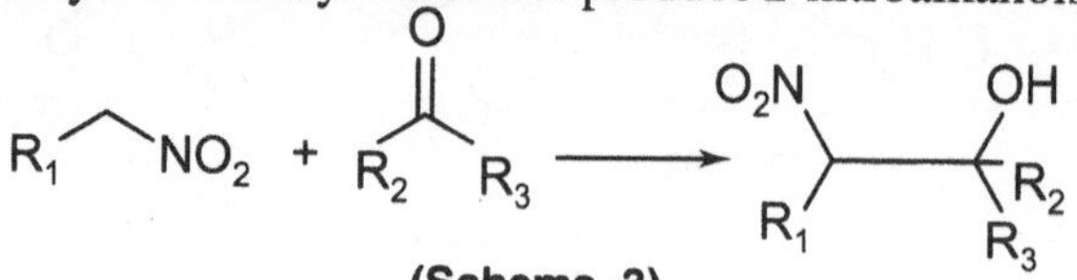

(Scheme–3)

References

1. L. Henry, Compt. Rend., 1894, **122**, 1265.
2. R.S. Varma, R. Dahiya and S. Kumar, Tetrahedron Lett., 1997, **38**, 5131.
3. A. Kumar and S.S. Pawar, J. Mol. Catal. A, 2005, **235**, 244.
4. T. Jiang, H.Gao, B. Han, G. Zhau, Y. Chang, W. Wu, L. Gao and Y. Yang, Tetrahedron Lett., 2004, **45**, 2699.

3.1.13 Stetter Reaction

The reaction of aldehydes with olefins in ionic liquids, [C$_4$ min] [BF$_4$], [C$_4$ min] [PF$_6$] and [C$_4$ min] [NTf$_2$] give[1] 1, 4-dicarbonyl compounds. For a range of aldehydes and olefins, both thiazolium salts and Et$_3$N could be employed giving good yield of 1,4-adducts (Scheme–1).

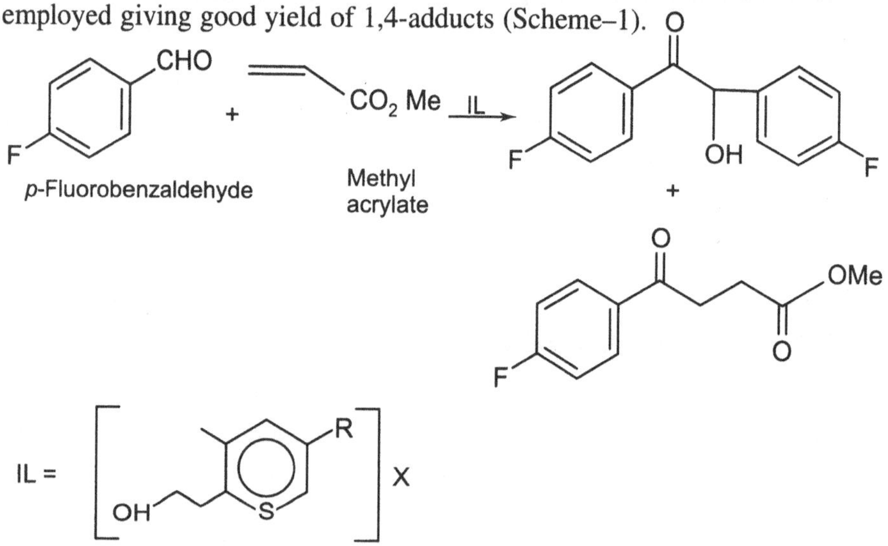

p-Fluorobenzaldehyde

Methyl acrylate

IL =

R = CH$_2$Ph, X = Cl
R = Et, X = Br

(Scheme–2)

Reference

1. S. Anjaiah, S. Chandrasekhar and R. Gree, Adv. Synth. Catal., 2004, **346**, 1329.

3.1.14 Thiele Acetylation

The reaction of quinones with acetic anhydride in presence of H_2SO_4 or BF_3 gives[1] triacetoxy aromatic compounds (Scheme–1).

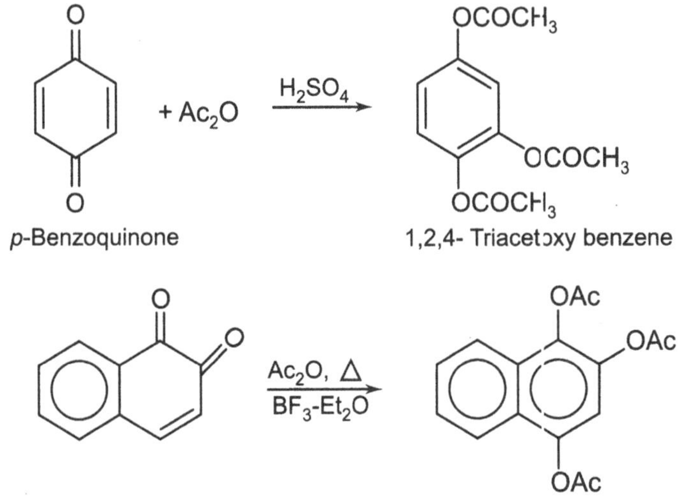

p-Benzoquinone   1,2,4- Triacetoxy benzene

(Scheme–1)

Reference

1. J. Thiele, Ber, 1893, **31**, 1247; J. Thiele and E. Winter, Ann; 1900, **311**, 341; J.M.W. Mc Omie and J.N. Blatchy, Org. React., 1972, **19**, 200.

3.1.15 Tischeno Reaction

Aldehydes on reaction with aluminium alkoxides give[1] esters (Scheme–1). All aldehydes, *i.e.*, with or without α-hydrogen undergo this reaction.

$$2RCHO \xrightarrow{Al(OC_2H_5)_3} R\ COOCH_2R$$

Aldehydes   Esters

(Scheme–2)

In **Cannizzaro reaction** (section 3.1.8), one molecule of the aldehydes is oxidised and the other molecule is reduced. However, in Tischeno reaction, the alcohol and acids combine to form esters. Thus, acetaldehyde and propionaldehyde give ethylacetate and ethyl propionate, respectively.

Reference

1. V. Tischeno, J.Russ. Phys. Chem. Soc., 1966, **58**, 355, 482, 540, 547; Chem. Zetr., 1906, **II**, 1309; 1522, 1555.

3.1.16 Williamsons Ether Synthesis

Alkyl halides on reaction with alkoxides or phenoxide give[1] ethers. This is a simple method for the formation of mixed or simple ethers (Scheme–1).

$$R^1{-}X + \overline{O}R \longrightarrow R^1{-}O{-}R + X^-$$

(Scheme–1)

3.1.16.1 Phase Transfer Catalysed Williamson Ether Synthesis

PTC technique provides a simple and convenient method for the Williamson ether synthesis by use of excess of alcohol or alkyl halide, lower temperature and larger alcohols ($C_7H_{18}OH$) give higher yield of ethers[2,3] (Scheme–2).

$$C_8H_{17}OH + C_8H_9Cl \xrightarrow[\substack{\text{NaOH}\\ \text{Solution}}]{\text{PTC}} \underset{\text{Major}}{C_8H_{17}OC_4H_9} + \underset{\text{Byproduct}}{C_8H_{17}OC_8C_{17}}$$

(Scheme–2)

Best results are obtained[3] by using five fold excess of aqueous sodium hydroxide (over alcohol), excess alkyl chloride (also used as solvent) and tetrabutylammonium bisulphate[3] (1-5 mol) as PTC at 25°–75°. Primary alcohols require more time or greater amount of catalyst.

It is well knoun that most alcohols do not react with dimethyl sulphate in presence of alkali or even by using alkali metal alkoxides. However, the reaction proceed easily[4] with tetrabutylammonium salts as catalysts. Activated alcohols and primary alcohols give high yields of ethers, but secondary alcohols react very slowly and tertiary alcohols do not react at al.

In case of phenol, K_2CO_3 is mostly used in presence of dimethyl sulphate. However, such ethers can be obtained[3] in quantitative yield by using catalytic amount of 18–crown 6 (Scheme–3).

(Scheme–3)

Some aromatic ethers are obtained by using phenol and aryl halide in presence of copper (Scheme–4).

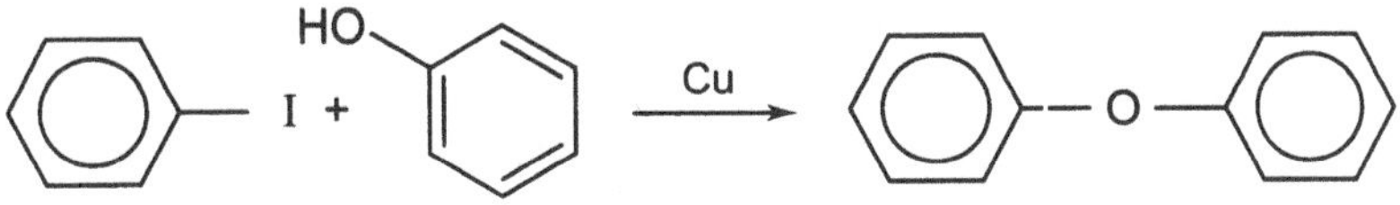

| Iodo benzene | Phenol | | Diphenyl ether (60%) |

(Scheme–4)

A variation of Williamsons ether synthesis is by using thallium (I) ethoxide. In this procedure, even substrates containing an additional oxygen function, such as –OH, –COOR, –CONH$_2$ can also be used (Scheme–5).

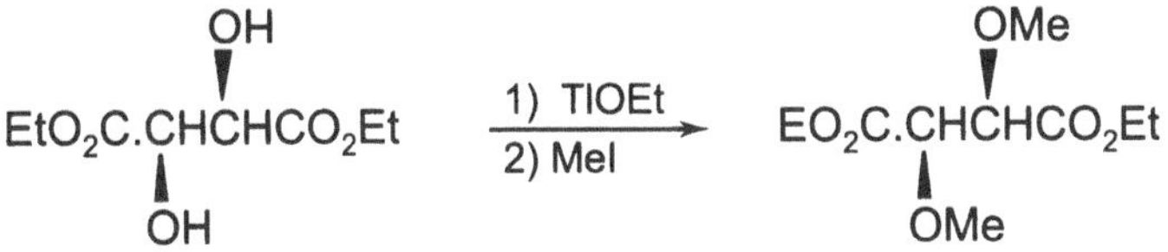

(Scheme–5)

References

1. R. Williamsons, J. Chem. Soc., 1952, **4**, 229; O.C. Dermer, Chem., Rev., 1934, **14**, 409.
2. J. Jarrouse, C.R. Hebd, Scances Acad. Sci. Ser.C, 1951, **232**, 1429.
3. H. H. Freeman and R.A. Dobois, Tetrahedron Lett., 1975, 3251; A. Merz, Angew. Chem., Int. Ed., Engl., 1973, **12**, 846.

3.1.17 Miscellaneous Methods

3.1.17.1 From Alkenes

(*i*) **Acid Catalysed Hydration of Alkenes:** Treatment of alkene with cold H$_2$SO$_4$ followed by hydrolysis of alkyl hydrogen sulphate gives alcohol (Markovnikov addition).

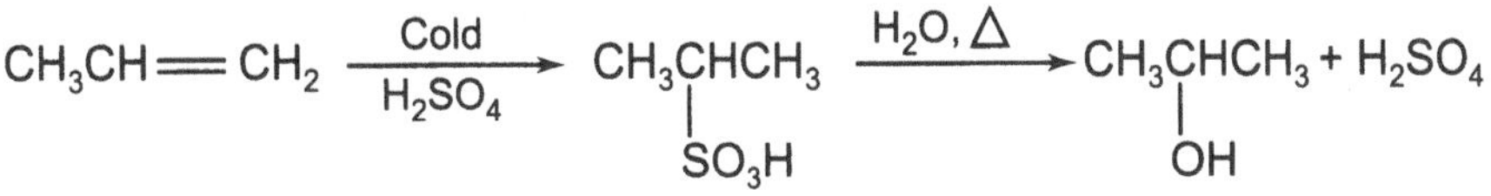

Alternatively, alkenes can be treated with water in presence of acid.

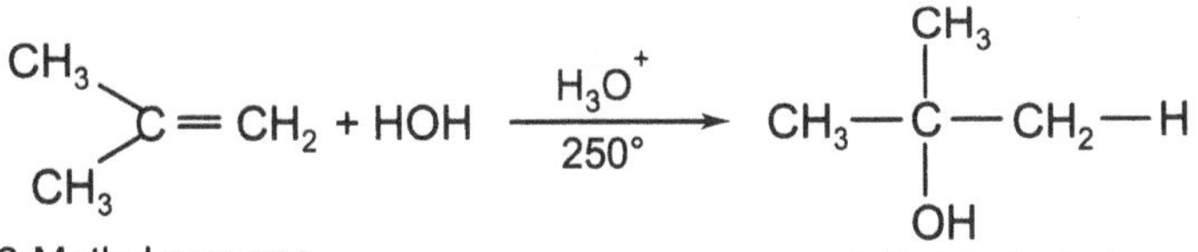

2-Methyl propene   tert. Butyl alcohol

(*ii*) Addition of acids to alkenes give the corresponding ester in SO_3H-functionalised ionic liquids (I), which serve as catalyst and solvent (J. Fraga-Dubreeuil, K. Bourahla, M. Rahmouni, J.P. Bazureau and J. Hamelin, J. Catal. Commun., 2002, 3, 185).

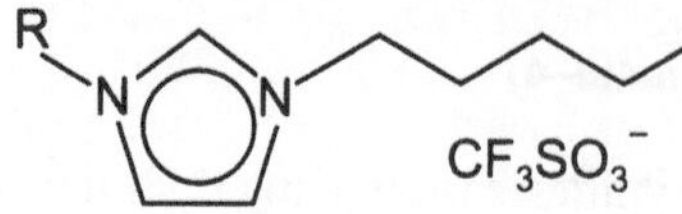

R = Me, Et, Bu, HX

I

In place of simple alkenes, cyclic olefines can also be used.

(*iii*) Alkenes can also be converted into alcohols by treatment with water in polyethylene glycol (PEG) (N.F. Leininger, J.L. Gainer and D.J. Kirwan, AICHEJ, 2004, 50, 511).

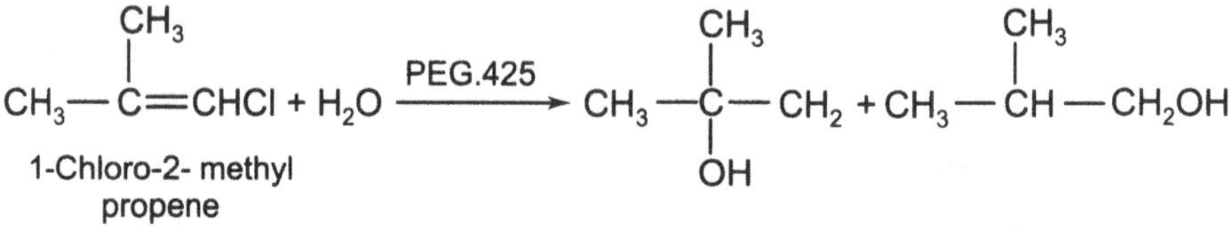

1-Chloro-2- methyl propene

(*iv*) **Oxymercuration of Alkenes:** Oxymercuration of olefines can be performed in one pot under sonication in $THF-H_2O$ with HgO and t-$BuCO_2H$ to give alcohols.

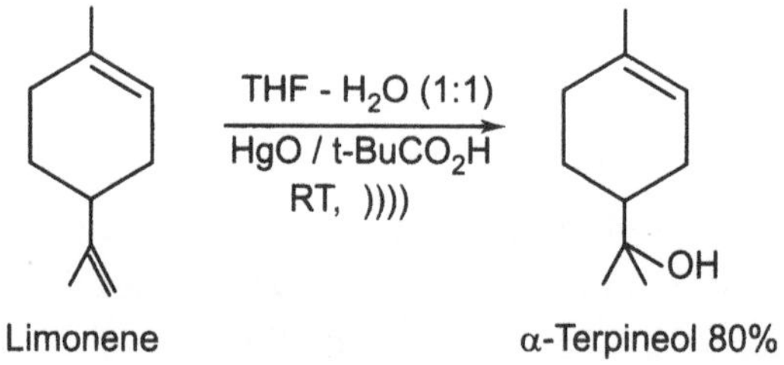

THF - H_2O (1:1)
HgO / t-$BuCO_2H$
RT,))))

Limonene   α-Terpineol 80%

(*v*) **Oxymercuration-Demercuration:** Alkenes react with mercuric acetate in $THF-H_2O$ to produce (hydroxyalkyl) mercury compounds, which on reduction ($NaBH_4$) give alcohols.

$$\underset{}{\overset{}{C}} = \underset{}{\overset{}{C}} + H_2O + Hg\ (OCOCH_3)_2 \xrightarrow[\text{Oxymercuration}]{\text{THF}} \overset{|}{\underset{|}{C}} - \overset{|}{\underset{|}{C}} - + CH_3COOH$$

$$\underset{\text{OH HgOCOCH}_3}{}$$

$$\overset{-}{OH} \Bigg| \begin{array}{l} \text{Demercuration} \\ \text{NaBH}_4 \end{array}$$

$$-\overset{|}{\underset{|}{C}} - \overset{|}{\underset{|}{C}} - Hg + CH_3COO^-$$

$$\underset{\text{OH H}}{}$$

(Markownikoff product)

(*vi*) **Dihydroxylation of Alkenes:** It is usually carried out with O_sO_4 in presence of an oxidising agent like BuOOH, $H_2O/\ ^-OH$ or $NaHSO_3$ to give cis diols.

Cyclooctene   cis-1, 2-Cyclooctanediol, 50%

Use of PEG as solvent and catalysed by N-methyl morpholine (NMO) give 95% yield of the diol.

$$C_6H_5CH{=}CH{-}C_6H_5 \xrightarrow[\substack{OsO_4 \\ PEG\ 400 \\ 2\text{-}4\ hr.}]{NMO} C_6H_5{-}\underset{\underset{OH}{|}}{CH}{-}\underset{\underset{OH}{|}}{CH}{-}C_6H_5$$

Stilbente   Diol (95%)

For details see section 5.2.2 (Inter conversion of Functional Groups: alkenes).

Trans diols are obtained from alkenes by treatment with peracid followed by hydrolysis of the formed epoxide.

$$\underset{}{X} \xrightarrow{RCO_3H} \underset{}{\overset{O}{X}} \xrightarrow{H_2O} \underset{HO}{\overset{OH}{X}}$$

trans Diol

3.1.17.2 From Alkyl Halides

Alkyl halides can be transformed into alcohols by treatment with alkali

$$R-X \quad \xrightarrow{\;^{-}OH\;} \quad R-OH$$

Alkyl halides   Alcohols

The displacement of halogen by hydroxyl group is best accomplished by the use of betanine quaternary salts, $R_3N^+CH_2CO_2^-$; these salts exhibit 10–50 fold greater activity than ordinary tetraalkylammonium salt (Charles M.Stakes. and Charles L. Liotta, Phase Transfer catalysts, Principles and Techniques, Academic Press Inc. Ny, 1918, p. 127).

$$IC_8H_{17}Br + (C_{12}H_{25})_2\overset{+}{N}CH_2COO^- \longrightarrow IC_8H_{17}OH$$

Alkyl halides

See also Grignard reagents (section 2.1.29.1) and

Williamson ether synthesis (section 3.1.15).

3.1.17.3 From Aldehydes, Ketones and Esters

For reduction of aldehydes with $NaBH_4$ to alcohols, (see Functional Group Transformation (section 5.2.10); with sodium alcohol, See Bouveault Blanc-reduction, (section 5.2.10); ketones on reduction give secondary alcohol (For detail, see Transformation of Functional Groups section 5.2.11). Even enzymatic reduction of ketones has been used. Esters can also be reduced to alcohols (Functional Group Transformation section 5.2.13).

3.2 C=O BOND FORMATION

Following procedures are used for the formation of $C = O$ bond.

1. In a modified Reformatsky reaction (section 2.1.36), using nitriles in place of aldehydes and ketones, the product obtained is an imine, which undergoes readily hydrolysis to give the ketones. This reaction is called **Blaise Reaction.**

2. Cyclic ketones are obtained in the Dieckmann condensation (section 2.1.13) and also by oxidation of cyclic alcohols with Jones reagent.

Cyclohexanol   Cyclo octanone

3. The reaction of secondary arylamines with sodium bisulphite addition compound of glyoxal gives oxindole. The reaction is known as **Hinsberg oxindole synthesis** (O. Hinsberg, Ber., 1888, 21, 110; 1892, **25**, 2545; 1908, **41**, 1367).

Oxindole

In place of sodium bisulphite adduct of glyoxal, α-haloacyl halides can also be used (R. Stolle, Chem. Ber., 1913, **46**, 3915; W. Sumter, Chem., Rev., 1944, **34**, 396).

4. Conjugated aldehydes and ketones on reaction with Grignard reagent give two products 1, 4-addition and 1, 2-addition.

CH$_3$CH=CH—C—CH$_3$
         ‖
         O
3-Penten-2-one

$\xrightarrow[\text{2) H}_3\text{O}^+]{\text{1) C}_2\text{H}_5\text{MgI}}$

1,4-addition

CH$_3$—C—CH$_2$—C—CH$_3$
       | ‖
       C$_2$H$_5$ O
4-Methyl-2-hexanone

1,2-addition

                              OH
                              |
CH$_3$—CH=CH—C—CH$_3$
                              |
                              C$_2$H$_5$
3-Methylhex-4-en-3-ol

5. **Pinacol-Pinacolone Rearrangement:** CR. Fittig, Ann., 1859, **110**, 17; 1860, 114, 54) is a useful reaction for C = O bond formation.

        CH$_3$ CH$_3$
         | |
H$_3$C—C—C—CH$_3$ $\xrightarrow{\text{H}^+}$ H$_3$C—C—C—CH$_3$ + H$_2$O
         | | | ‖
        OH OH   CH$_3$ O
         Pinacol   Pinacolone

The above rearrangement can also be performed in ionic liquid.

        OH OH   Ph
         | | |
Ph—C—C—Ph $\xrightarrow[180°, 2M]{\text{IL}}$ Ph—C—C—Ph
         | | | ‖
        Ph Ph   Ph O

IL = $\overset{+}{\text{P}}$(CH$_2$)$_3$SO$_3$H   p— CH$_3$C$_6$H$_4$SO$_3^-$

6. Aryl bromides on heating with amines in MW oven in the presence of catalyst, give the corresponding benzamides. M. Larhed, Org. Lett., 2005, 7, 3327).

Aryl bromide   +   Amine $\xrightarrow[\substack{\text{H}_2\text{O, MW,}\\170°, 10\text{min}}]{[\text{Pd}], \text{Mo(CO)}_6}$ Benzamides

7. Alcohols on refluxing with Al(OCHMe$_2$)$_3$ in acetone or benzene or toluene gave the corresponding ketones **(oppenauer oxidation)** (H. Meerwein, and R. Schmidt, Ann., 1925, **444**, 221; W. Ponndorf, Angew. Chem., 1929, **39**, 138).

$$R-\underset{\underset{OH}{|}}{CH}-R^1 + CH_3-\underset{\underset{O}{||}}{C}-CH_3 \underset{\Delta}{\overset{Al(OCHMe_2)_3}{\rightleftharpoons}} R-\underset{\underset{O}{||}}{C}-R^1 + CH_3-\underset{\underset{OH}{|}}{CH}-CH_3$$

(For other methods for the oxidation of secondary alcohols to ketones see Functional Group Transformation, Alcohol, section 5.2.6.6).

Secondary alcohols can also be oxidised to ketone with CrO$_3$/C$_5$H$_5$N **(Sarett Oxidation)** (G.I. Poos, G.E. Arth, R.E. Beyler and L.H. Sarett, J. Am. Chem. Soc., 1953, **75**, 422; J.R. Holum, J. Org. Chem., 1961, **26**, 484).

8. The reaction of olefins or secondary alcohols with nitriles in strongly acidic medium gives amides **(Ritter Reaction)** (Boehm., Schumann and Hansen, Arch. Pharm., 1933, **271**, 490).

$$(CH_3)_2C=CH_2 + CH_3CN \underset{H_2O}{\overset{H^+}{\longrightarrow}} CH_3\underset{\underset{O}{||}}{C}\ NHC\overset{CH_3}{\underset{CH_3}{\diagdown}}$$

$$R_3COH + HCN \underset{H_2O}{\overset{H^+}{\longrightarrow}} HC\underset{\underset{O}{||}}{}\ NHCR_3$$

9. The reaction of adiponitrile with sodium ethoxide gives 2-cyanocyclopentanone. This is an intramolecular version of Thorpe reaction and is called **Thorpe-Ziegler Method** (J.J. Bloomfield and V. Fennessey, Tet. Lett., 1964, 2273).

Adiponitrile   2-Cyano cyclo pentanone

10. Arenes on oxidation with $KMnO_4$, impregnated alumina in MW oven gives ketones (A. Oussaid and L. Loupy, J. Chem., Res (S), 1997, 342).

11. Oxidation of alkenes with $PdCl_2$ immobilised in [b$_{min}$] [PF$_4$] and [b$_{min}$] [PF$_6$] using H_2O_2 as oxidant give ketones (R.S. Varma, E. Shale-Demessce and H.R. Pillai, Green. Chem., 2002, 107). This is **Wacker Type Oxidation Reaction**.

$$\text{Vinyl benzene} \xrightarrow[\text{[b}_{min}\text{] [PF}_6\text{]}]{PdCl_2.H_2O_2,\ 60°} \text{Acetophenone}$$

Vinyl benzene   Acetophenone

12. Nitriles can be hydrolysed to amides by using urea hydrogen peroxide complex (UHP) under solvent free conditions (H. Tanka, S. Kishigami and F. Toda, J. Chem., Soc., 1991, 56, 4333).

$$\xrightarrow[\text{85°, 1-1.5 hr}]{UPH}$$

80%

□□□

Carbon-Nitrogen (Single, Double and Triple) Bond, Carbon-Halogen, Carbon-Sulphur, Nitrogen-Nitrogen Double Bond and Nitrogen-Oxygen Bond Formation

4.1 CARBON-NITROGEN SINGLE BOND FORMATION

Following reactions have been used for C—N bond formation. As will be seen, in most of the reactions, along with C—N, C—C bonds are also formed. The nitrogen containing heterocyclic compounds contain both C—C and C — N bonds.

4.1.1 Aza-Michael Reaction

It is an important carbon-nitrogen bond forming reaction. It consists[1] in the reaction of amines with olefinic compounds in presence of polystyrene supported sulfonic acid (PSSA) as catalyst (Scheme–1).

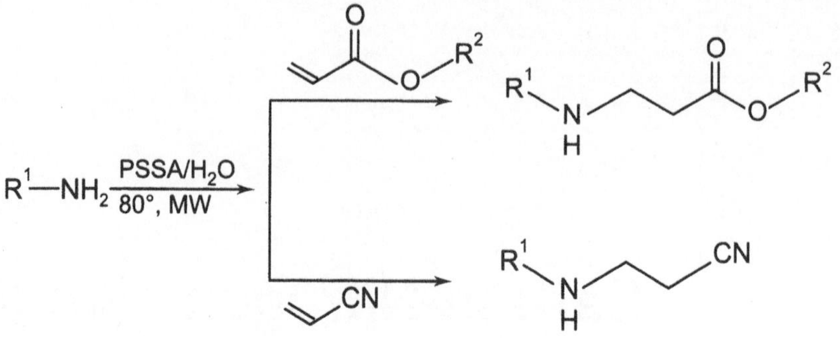

(Scheme–1)

Reference

1. V. Polshettiwar and R.S. Varma, Tetrahedron Lett., 2007, **48**, 8735.

4.1.2 Backmann Rearrangement

Normally, in Beckmann rearrangement, the oximes of ketones on heating with acidic reagents like PCl_5, HCOOH, $SOCl_2$ *etc.*, are converted into anilides[1]. However, a solid-state microwave assisted Beckmann rearrangement is also possible[2,3]. In this procedure, the oxime of a ketone is mixed with montmorillonite K–10 clay in dry media and the mixture heated for 7–10 min in a microwave oven to give the corresponding anilide in 91 per cent yield (Scheme–1).

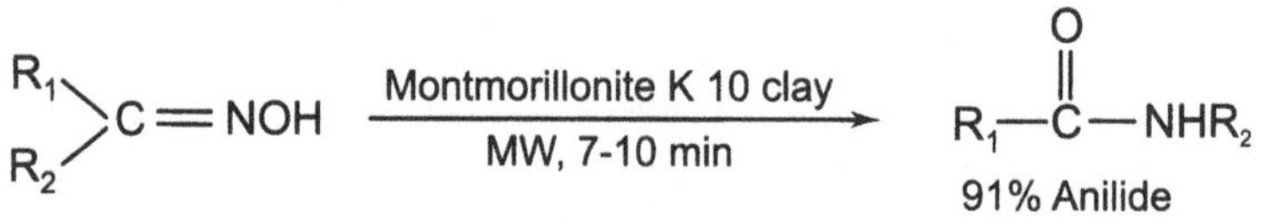

(Scheme–1)

References

1. E. Beckmann, Chem., Ber., 1886, **19**, 988. R.T. Conley, J. Org. Chem., 1963, **28**, 210.

2. I. Almena, A. Diaz-Ortiz, E. Diaz-Barra and A. Loupy, Chem. Lett., 1996, 333; S. Caddick, Tetrahedron, 1995, 10400.

3. A.I. Bosch, P. de La Cruez, E. Diez-Barra and F. Langa, Syntlett., 1995, 1259.

4.1.3 Biginelli Reaction

Acid catalysed condensation of an aldehyde, a β-ketoester and urea gives tetrahydropyrimidones[1] (Scheme–1).

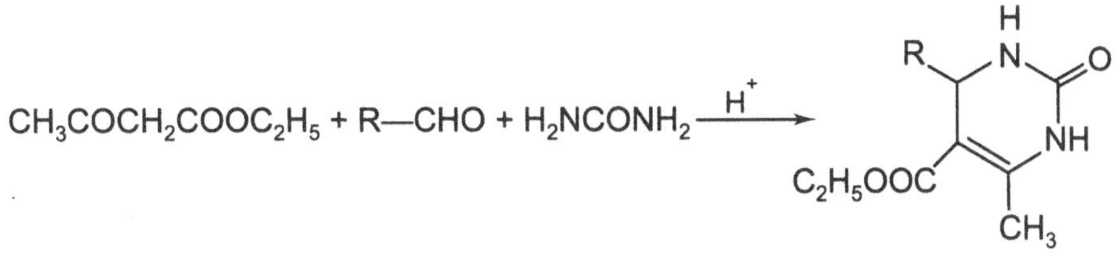

$CH_3COCH_2COOC_2H_5$ + R—CHO + H_2NCONH_2
H^+
R
H
N
O
NH
C_2H_5OOC
CH_3

(Scheme–1)

It is more convenient to perform the Biginelli reaction by mixing the reagents in presence of catalytic amount of clay and heating in a microwave oven[2]. The yield is 65–95 per cent.

Ionic liquids, such as $[b_{min}]$ $[BF_4]$ and $[b_{min}]$ $[PF_6]$ have also been used as catalysts for the Biginelli condensation reaction under solvent free conditions[3] (Scheme–2).

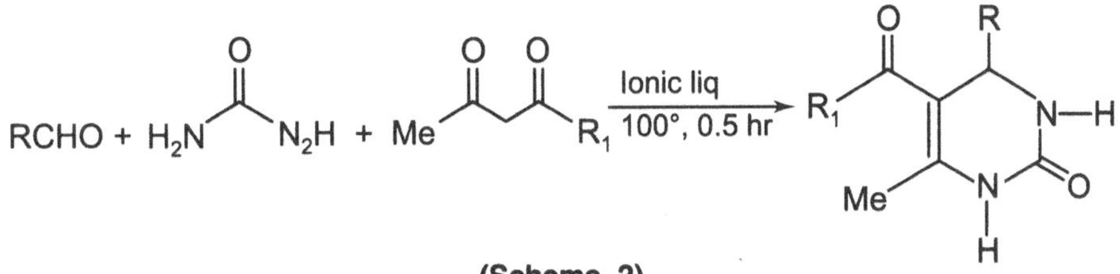

RCHO + H_2N
O
N_2H + Me
O O
R_1
Ionic liq
100°, 0.5 hr
R_1
O
R
N—H
O
Me
N
H

(Scheme–2)

References

1. P. Biginelli, Ber., 1891, **24**, 1317, 2962; 1893, **26**, 447; Zaugg, Martin, Organic Reactions, 1965, **14**, 88.
2. C.O. Kappe, D. Kumar and R.S. Varma, Synthesis, 1999, 1799; M. Kidwai, S. Saxena, R. Mohan and R. Venkataraman, J. Chem., Soc., Perkin Trans I, 2002, 1845.
3. J. Peng and Y. Deng, Tetrahedron Lett., 2001, **42**, 5917.

4.1.4 Fischer-Indole Synthesis

It consists in heating aryl hydrazones of aldehydes and ketones in presence of $ZnCl_2$ or polyphosphoric acid (PPA) or BF_3 to give indoles[1] (Scheme–1).

R_1
H_2C
C—R
N
N
H
$ZnCl_2$
Δ
R^1
N
H
R

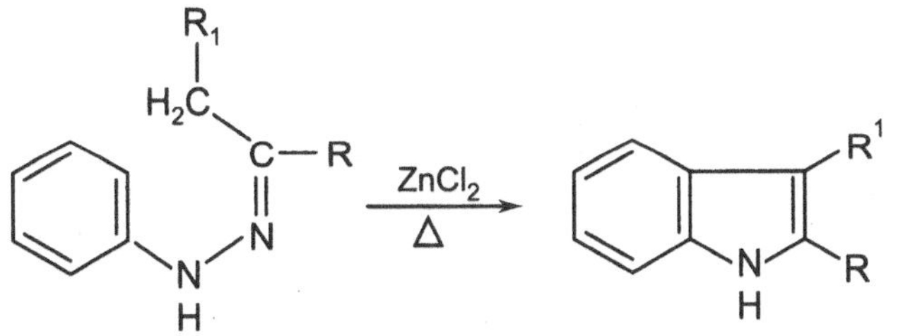

(Scheme–1)

Indole itself cannot be prepared by the cyclisation of phenyl hydrazone of acetaldehyde. However, it is conveniently prepared by the decarboxylation of indole 2-acetic acid, (by heating with copper chromite in quinoline solution) which is prepared by the cyclisation of pyruvic acid phenylhydrazone (Scheme–2).

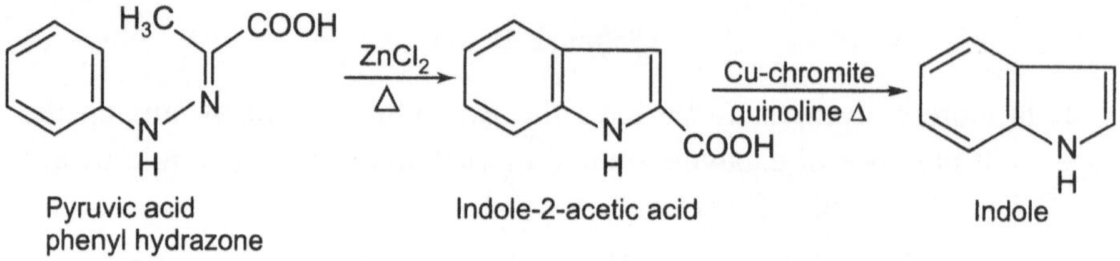

Pyruvic acid
phenyl hydrazone

Indole-2-acetic acid

Indole

(Scheme–2)

The last step of decarboxylation can be conveniently affected by MW irradiation by using quinoline as solvent[2].

4.1.4.1 Fischer-Indole Synthesis in Solid State

The reaction of cyclohexanone and phenyl hydrazine in presence of montomorillonite KSF clay using microwave irradiation give carbozole[3] (Scheme–3).

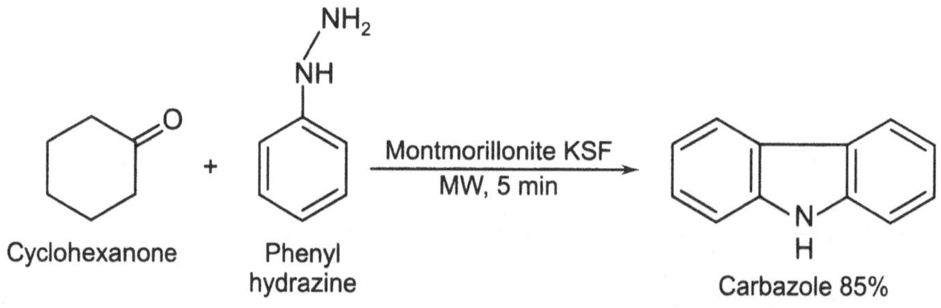

Cyclohexanone

Phenyl
hydrazine

Carbazole 85%

(Scheme–3)

4.1.4.2 Fischer-Indole Synthesis in Water

The reaction of phenyl hydrazine and ethyl methyl ketone in water at 220° gave 2,3-dimethyl indole[4] (Scheme–4).

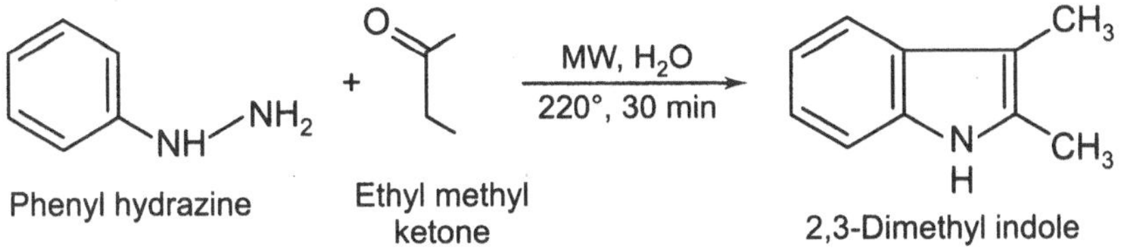

Phenyl hydrazine

Ethyl methyl ketone

2,3-Dimethyl indole

(Scheme–4)

References

1. E. Fischer and F. Jourdan, Ber., 1883, **16**, 2241; E. Fischer and O. Hess, Ber., 1884, **17**, 559.
2. G.B. Jones and J. Chapman, J. Org. Chem., 1993, **58**, 5558.
3. D. Villemin, B. Labiad and Y. Ouhilal, Chem. and Ind., 1989, 607.
4. C.R. Strauss and R.W. Trainer, Aust. J. Chem., 1998, **51**, 703.

4.1.5 Friedlander Synthesis

It involves[1] base catalysed condensation of 2-aminobenzaldehydes with ketones to form quinoline derivatives (Scheme–1).

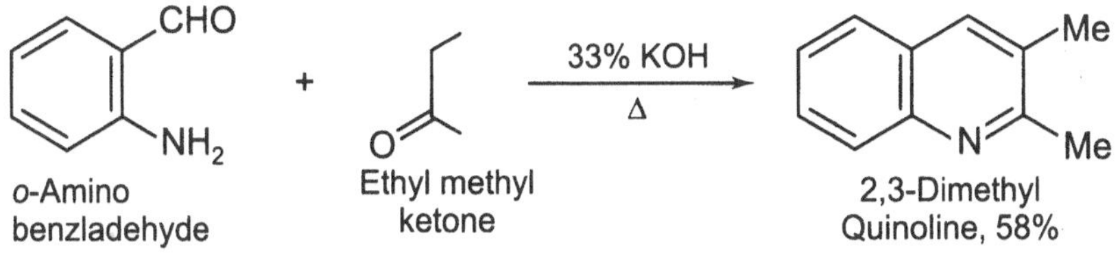

o-Amino benzladehyde

Ethyl methyl ketone

2,3-Dimethyl Quinoline, 58%

(Scheme–1)

4.1.5.1 Friedlander Synthesis under Microwave Irradiation

KSF day catalysed Friendlander synthesis involving 2-aminoaldehydes or ketones with carbonyl compounds containing α-methylene group has been achieved in solvent free conditions under MW irradiation to give quinoline derivatives[2] (Scheme–2).

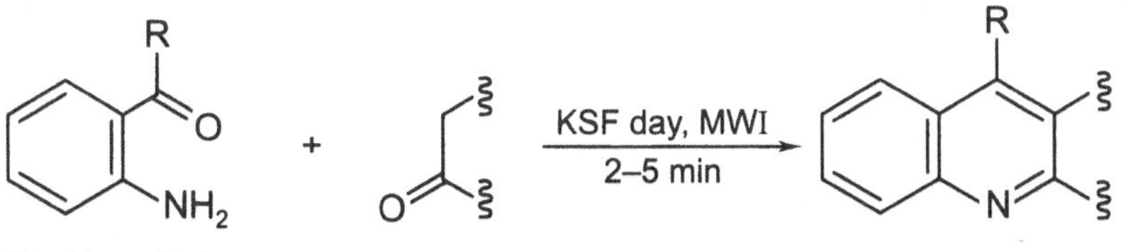

R = H or alkyl

(Scheme–2)

References

1. P. Friedlander, Chem. Ber., 1882, **15**, 2572; P. Friendlander and C.P. Gohring, Ber., 1883, **16**, 1833; Manske, Chem. Rev., 1942, **30**, 124.

2. G. Sabitha, R.S. Babu, B.V.S. Reddy and J.S. Yadav, Synth. Commun., 1999, **29(24)** 4403.

4.1.6 Graebe-Ullman Synthesis

2-Aminodiphenylamine on treatment with nitrous acid gives benzotriazole, which on decomposition gives[1] carbazole (Scheme–1).

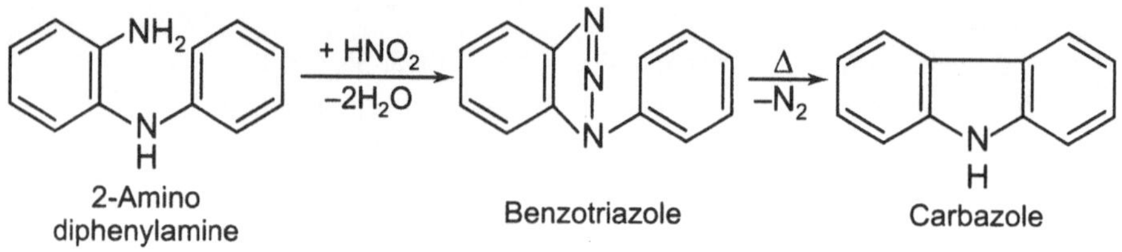

2-Amino diphenylamine   Benzotriazole   Carbazole

(Scheme–1)

4.1.6.1 Graebe-Ullman Synthesis under Microwave Irradiation

A mixture of Benzotriazole and chloropyridines on irridiation with microwave (in the presence pyrophosphoric acid gave the product (A), which on further heating in microwave oven gave the corresponding β-carboline[2] (Scheme–2).

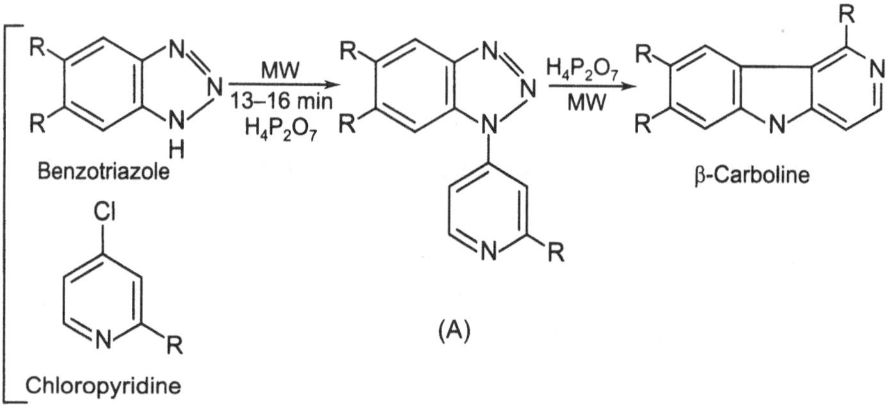

(Scheme–2)

References

1. C. Graebe and F. Ullmann, Ann., 1896, **291**, 16; F. Ullmann. Ann., 1904, **332**, 82.

2. A. Mohna, J.I-Vaguero, J.J. Garefa and J. Alvarez-Builla, Tetrahedron Lett., 1933, **34**, 2673.

4.1.7 Hantzsch Pyridine Synthesis

Condensation of two molecules of ethyl acetoacetate with an aldehyde in presence of ammonia followed by oxidation of the resultant dihydropyridine derivative gives[1] pyridine derivatives (Scheme–1).

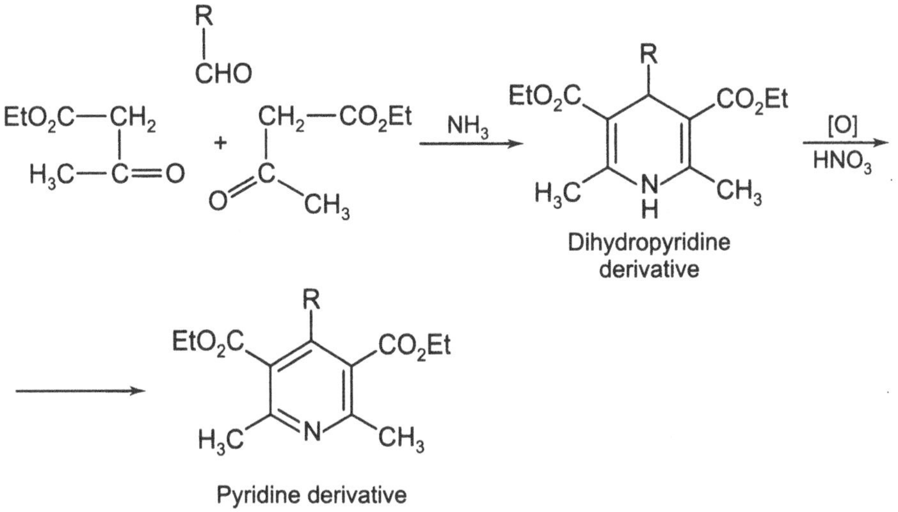

Dihydropyridine derivative

Pyridine derivative

(Scheme–1)

4.1.7.1 Hantzsch Pyridine Synthesis under MW Irradiation

The reaction of 1,3-dicarbonyl compounds and arylaldehyde and ammonium nitrate (an ammonia source) under MW irradiation gives pyridine derivatives[2] (Scheme–2).

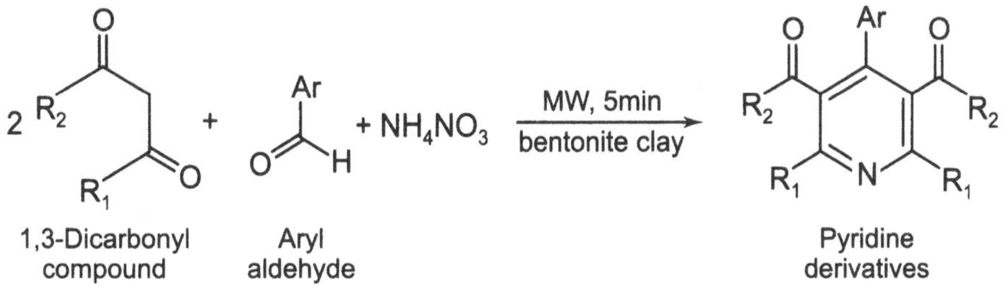

| 1,3-Dicarbonyl compound | Aryl aldehyde | | Pyridine derivatives |

(Scheme–2)

References

1. A. Hantzsch, Ann., 1882, **215**, 1, 72; Ber., 1885, **18**, 1744; 1886, **19**, 289.
2. I.C. Cottrill, A.Y. Usyatinsky, J.M. Arnold, D.S. Clark, J.S. Dornick, P.C. Michels and Y.L. Khmelnitsky, Tetrahedron Lett., 1998, **39**, 1117.

4.1.8 Hantzsch Pyrrole Synthesis

The reaction of β-ketoesters with α-chloroketones in presence of ammonia or primary amine gives pyrrole derivatives[1] (Scheme–1).

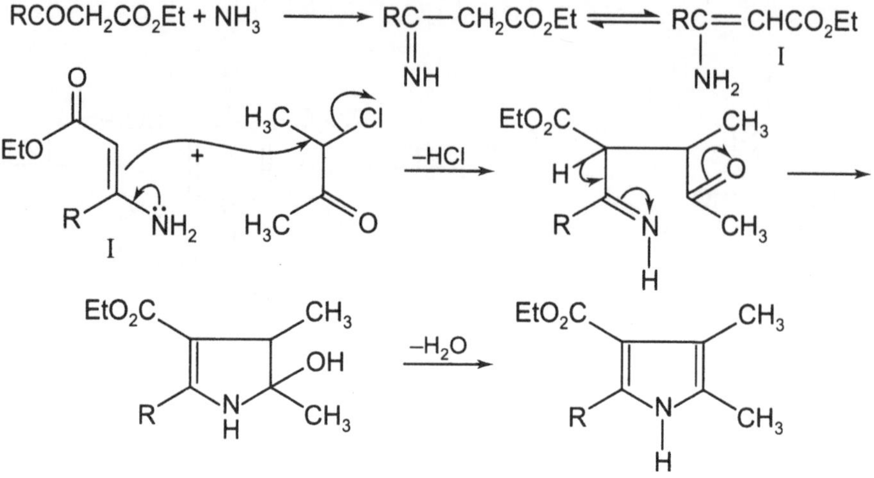

(Scheme–1)

Reference

1. A. Hentzsch, Ber., 1890, **23**, 1474.

4.1.9 Hetro-Diels-Alder Reaction

It is also known as **Aza-Diels-Alder Reaction**[1]. The reaction consists in the reaction of 2-aza butadienes. Thus, 1-dimethyl aza butadienes react with dienophiles giving pyridines or dihydropyridines[2] (Scheme–1).

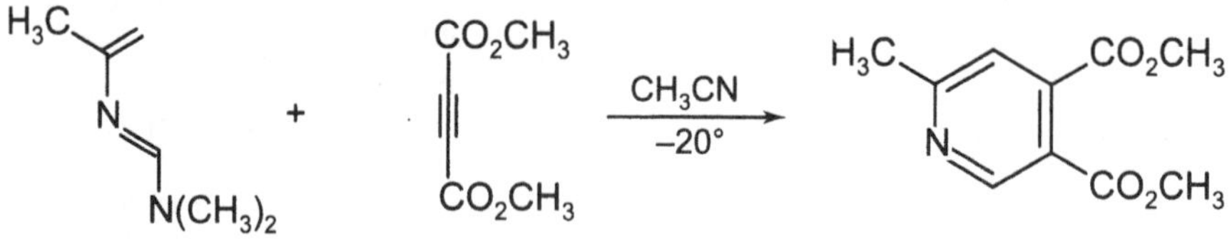

| 1-Dimethylamino-
-2-aza-isoprene | Dimethyl
acetylene
dicarboxylate | 58% |

(Scheme–1)

1-Azabutadienes also react with methyl vinyl ketone to give the products (Scheme–2)[3].

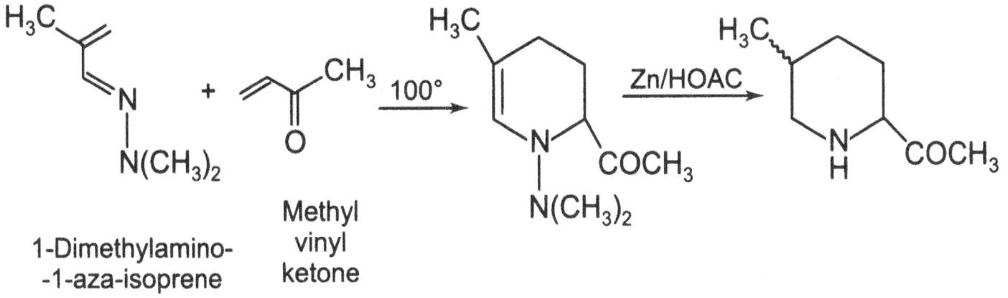

(Scheme–2)

The hetero Dields-Alder reaction can also be carried out in water. Thus, the reaction of iminium salts (generated *in situ* under Mannich-like conditions) reacted with dienes to give Aza-Diels-Alder reaction products[4] (Scheme–3).

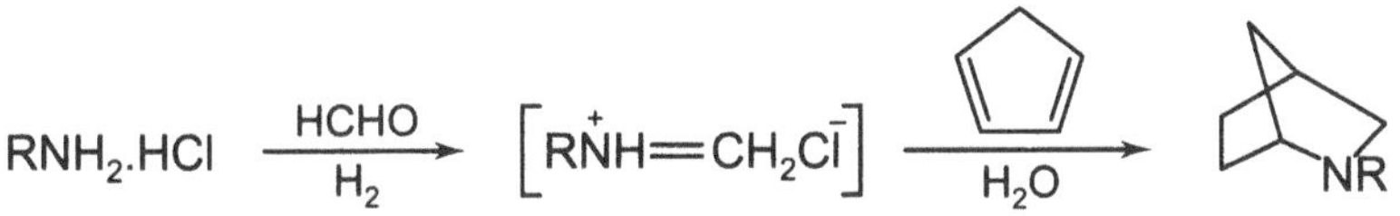

(Scheme–3)

In the above method, different dienes like 1,3-hexadiene, 2,3-dimethylbutadiene, 2-methyl butadiene, 2,4-hexadiene, 4-methyl-2,4-pentadiene *etc*. can be used. The iminium salts can be generated by reacting formaldehyde with amines like $BnNH_2$ and $MeNH_2$.

4.1.9.1 Intramolecular Aza-Diels-Alder Reaction

Intramolecular aza-Diels-Alder reaction also occurs[5] in aqueous medium to give fused ring systems with bridgehead nitrogen (Scheme–4).

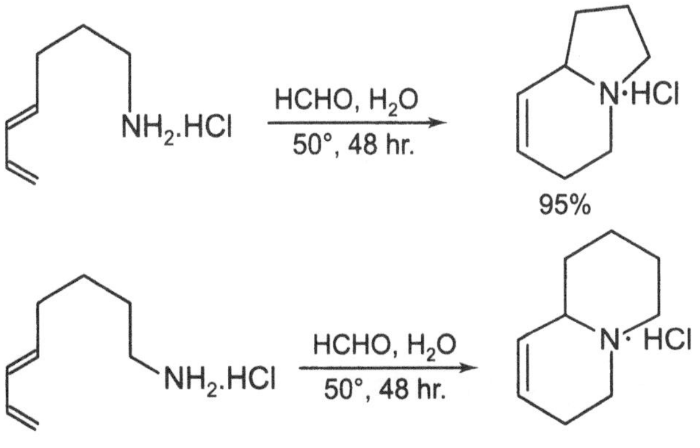

(Scheme–4)

An aza-Diels-Alder reaction has also been conducted[6] in SC—CO_2 using scandiumperfluaroalkanesulfonate [Sc(OSO$_2$C$_8$H$_{17}$)$_3$] as catalyst (Scheme–5).

(Scheme–5)

References

1. D.L. Boger, Tetrahedron, 1983, **39**, 2869.
2. A. Demoulin, H. Gorissen, A.–M. Hesbain-Prizque and L. Ghosez, J.Am. Chem. Soc., 1975, **97**, 4409.
3. B. Serckx-Poncin, A.–M. Hesbain-Frisque and L. Ghosez, Tetrahedron Lett., 1982, **23**, 3261.
4. P.A. Grieco and S.D. Larsen, J.Am. Chem. Soc., 1985, **107**, 1768.
5. W. Oppolzer, Angew. Chem. Int. Ed. Engl., 1972, **11**, 1031.
6. J. Matsu, T. Tsuchiya, K. Odashima and S. Kobayashi, Chem. Lett., 2000, 178.

4.1.10 Knorr Pyrrole Synthesis

Condensation of α-amino ketones with carbonyl compounds containing active methylene group (such as β-diketones or β-ketoesters) gave pyrrole derivatives[1] (Scheme–1).

Ethyl α-aminoacetate   Acetyl   Pyrrole
                       acetone   derivative

(Scheme–1)

Reference

1. L. Knorr, Ber., 1884, **17**, 1635; Ann., 1886, **236**, 290; L. Knorr and H. Lanse, Ber., 1902, **35**, 2998; A.H. Corwin in R.C. Elderfield Heterocyclic Compounds, 1950, 287.

4.1.11 Paal-Knorr Synthesis

Pyrroles are synthesised[1] by the condensation of 1,4-dicarbonyl compounds with ammonia, primary amines or hydrazines (Scheme–1).

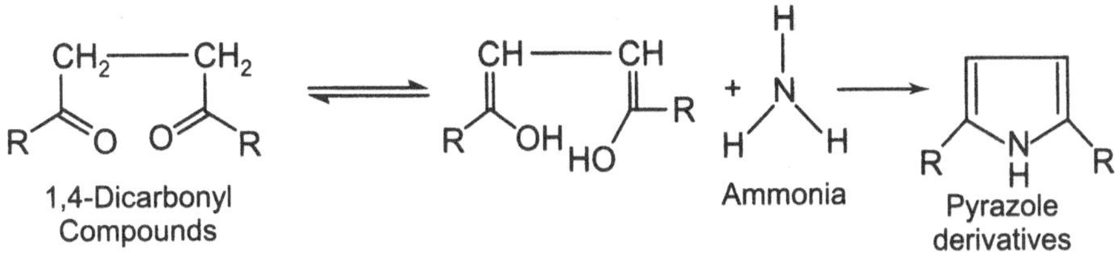

(Scheme–1)

Reference

1. C. Paal, Ber., 1885, **18**, 367; L. Knorr, Ber., 1985, **18**, 299.

4.1.12 Miscellaneous Procedures for C — N Bond Formation

1. Amines can be obtained by the following methods:

(*i*) Using **Gabrial Synthesis**

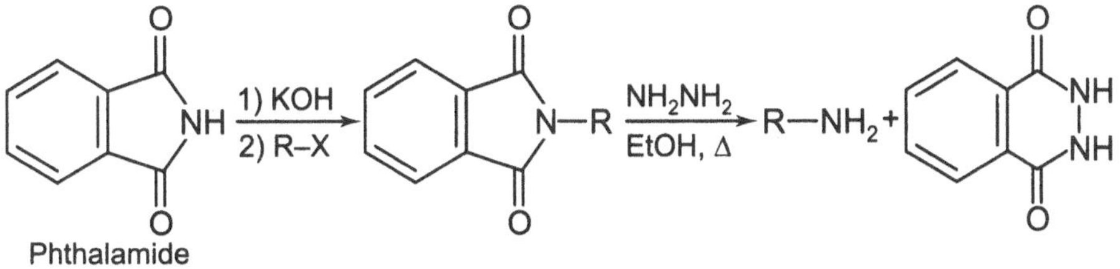

Phthalamide

(*ii*) Reduction of alkyl azides

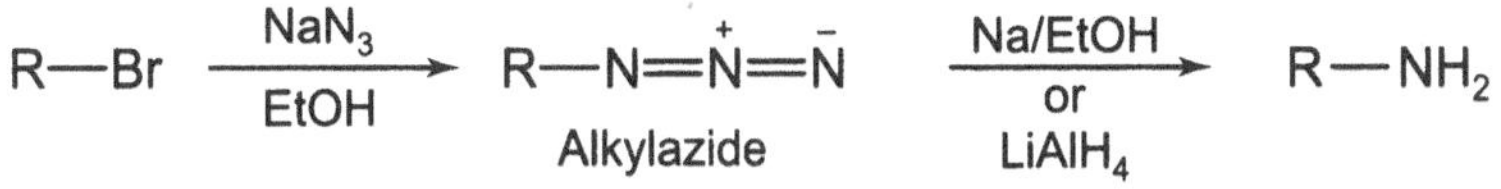

(*iii*) Reduction of nitroarenes

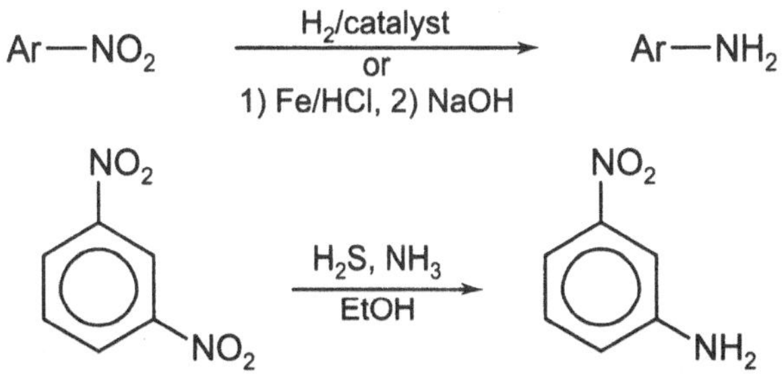

(*iv*) Reductive amination of aldehydes or ketones

$$\text{R—CH—NH}_2 \quad \overset{\circ}{1} \text{ Amine}$$

With the reactions:

$$\underset{\text{R}}{\overset{\text{R}'}{\diagdown}}\text{C}{=}\text{O} \quad \text{Aldehyde or Ketone}$$

$$\xrightarrow{\text{NH}_3/\text{H}_2} \quad \underset{\text{R—CH—NH}_2}{\overset{\text{R}'}{|}} \quad \overset{\circ}{1} \text{ Amine}$$

$$\xrightarrow{\text{R''NH}_2/\text{H}_2} \quad \underset{\text{R—CH—NHR''}}{\overset{\text{R}'}{|}} \quad \overset{\circ}{2} \text{ Amine}$$

$$\xrightarrow{\text{R''R'''NH/H}_2} \quad \underset{\text{R—CH—NR''R'''}}{\overset{\text{R}'}{|}} \quad \overset{\circ}{3} \text{ Amine}$$

(*v*) Reduction of nitriles, oximes and amides

$$\text{R—C}{\equiv}\text{N} \xrightarrow[\text{2) H}_2\text{O}]{\text{1) LAH/Et}_2\text{O}} \text{R CH}_2\text{NH}_2 \quad \overset{\circ}{1} \text{ Amine}$$

Nitrile

$$\underset{\text{R}\diagup\text{C}\diagdown\text{R}'}{\overset{\text{N—OH}}{\|}} \quad \text{Oxime} \xrightarrow{\text{Na/EtOH}} \underset{\text{R—CH—R}'}{\overset{\text{NH}_2}{|}} \quad \overset{\circ}{1} \text{ Amine}$$

$$\underset{\text{R}\diagup\text{C}\diagdown\text{NH}_2}{\overset{\text{O}}{\|}} \quad \overset{\circ}{1} \text{ Amide} \xrightarrow[\text{2) H}_2\text{O}]{\text{1) LiAlH}_4/\text{Et}_2\text{O}} \text{R—CH}_2{-}\text{NH}_2 \quad \overset{\circ}{1} \text{ Amine}$$

$$\underset{\text{R}\diagup\text{C}\diagdown\text{NH—R}'}{\overset{\text{O}}{\|}} \quad \overset{\circ}{2} \text{ Amide} \xrightarrow[\text{2) H}_2\text{O}]{\text{1) LiAlH}_4/\text{Et}_2\text{O}} \text{R CH}_2\text{NH—R}' \quad \overset{\circ}{2} \text{ Amine}$$

$$\underset{\text{R}\diagup\text{C}\diagdown\text{N}\diagdown\text{R}''}{\overset{\text{O}}{\|}} \quad \overset{\circ}{3} \text{ Amide} \xrightarrow[\text{2) H}_2\text{O}]{\text{1) LiAlH}_4/\text{Et}_2\text{O}} \underset{\overset{|}{\text{R}''}}{\text{R CH}_2\text{ N—R}'} \quad \overset{\circ}{3} \text{ Amine}$$

(*vi*) Hofmann Rearrangement

$$R-\overset{\overset{\displaystyle O}{\|}}{C}-Cl \xrightarrow{Br_2^-OH} R-NH_2 + CO_3^{2-}$$

(*vii*) Curtius Rearrangement

$$R-\overset{\overset{\displaystyle O}{\|}}{C}-Cl \xrightarrow[-NaCl]{NaN_3} R-\overset{\overset{\displaystyle O}{\|}}{C}-N_3 \xrightarrow[-N_2]{\Delta} R-N=C=O$$

$$R-N=C=O \xrightarrow{H_2O} R-NH_2 + CO_2$$

2. **Ullmann Reaction:** The reaction of acetanilide with bromobenzene in presence of K_2CO_3 gives diphenyl amine by refluxing in presence of copper bronze.

$$\underset{\text{Acetanilide}}{C_6H_5NHCOCH_3} + \underset{\substack{\text{Bromo} \\ \text{benzene}}}{C_6H_5Br} + K_2CO_2 \xrightarrow[\text{Reflux}]{\text{Cu}} \underset{\substack{\text{Diphenyl} \\ \text{amine}}}{C_6H_5NHC_6H_5} + CH_3COOK + KBr$$

3. **Mannich Reaction:** See section 2.2.10

4. N-Alkylation of secondary amines in presence of a PTC (polyethylene glycol monomethyl ether) with methyl iodide in presence of solid KOH on sonication gives N-methyl derivatives (R.S. Davidson, A.M. Patel, A. Safdar and D. Thornthwaite, Tetrahedron, 1983, 24, 5907).

Benzopyrrole

MeI/Solid/KOH
PEG methyl ether
20°, 30 min))))

65%

$$Ph_2NH \xrightarrow[\substack{\text{PEG methyl ether} \\ 20°, 1 \text{ hr. }))))}]{\text{PhCH}_2\text{Br/Solid KOH/toluene}} Ph_2N\,CH_2\,Ph$$

98%

5. The reaction of diethyl malonate with urea in presence of base gives barbituric acid.

Diethyl malonate + Urea $\xrightarrow[\text{EtOH}]{\text{C}_2\text{H}_5\text{ONa}}$ Barbituric acid

6. Amino acids are synthesised as follows:

$$CH_2(CO_2Et)_2 \; + \; HO\!-\!N\!=\!O \longrightarrow HO\!-\!N\!=\!C(CO_2Et)_2$$

Diethyl malonate   Nitrous acid

$$\xrightarrow[\text{Zn/HCl}]{\text{Reduction}} H_2N\,CH\,(CO_2Et)_2 \xrightarrow[-\text{HCl}]{\text{CH}_3\text{COCl}} CH_3CONHCH(CO_2Et)_2$$

1) KOH/H$_2$O
2) HCl
3) Δ, –CO$_2$

$$H_2NCH_2COOH$$
Glycine

7. Strecker Synthesis: Aldehydes on treatment with ammonia and HCN followed by hydrolysis of the formed α-amino nitriles gives α-amino acids.

$$R\!-\!CHO + NH_3 + HCN \longrightarrow R\,CH\,CN \xrightarrow[\text{H}_2]{\text{H}_3\text{O}^+,\,\Delta} R\,CH\,CO_2^-$$

$$\underset{\substack{NH_2 \\ \text{α-Amino} \\ \text{nitrile}}}{} \qquad \underset{\substack{+NH_3 \\ \text{α-Amino} \\ \text{acids}}}{}$$

Much better yields of α-amino nitriles are obtained by sonication. Also, the formation of α-aminonitriles can be accelerated by using ionic liquids as promotors (J.S. Yadav, B.V.S. Reddy, B. Eshwaiah, M. Srinivas and P. Vishnumurthy, New J. Chem., 2003, **27**, 462.

8. Synthesis of Symmetric Urea Derivatives: An effective procedure for the direct synthesis of symmetric urea derivative involves the reaction of amines using CO$_2$ in ionic liquids (F. Shi, Y. Deng, T. SiMa, J. Peng, Y. Gu and B. Qiao, Angew. Chem. Int. Ed., 2003, **42**, 3257).

9. Nucleophilic substitution of activated aryl halides with secondary amines has been affected in ionic liquids [b$_{min}$][PF$_4$] and [b$_{min}$] [PF$_6$] at room temperature (J.S. Yadav, B.V.S. Reddy, A.K. Basak and A.V. Naraish, Tetrahedron Lett., 2003, 44, 2217).

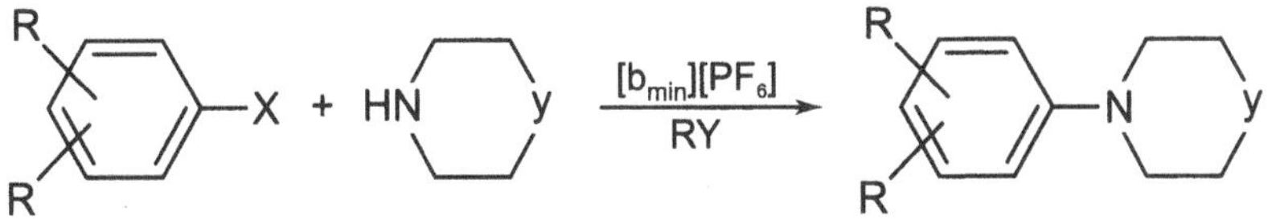

10. Coupling of amines with halides in presence of NaOH/H$_2$O gives the corresponding N-alkylderivatives (Y. Ju and R.S. Varma, Green Chem., 2004, 6, 219).

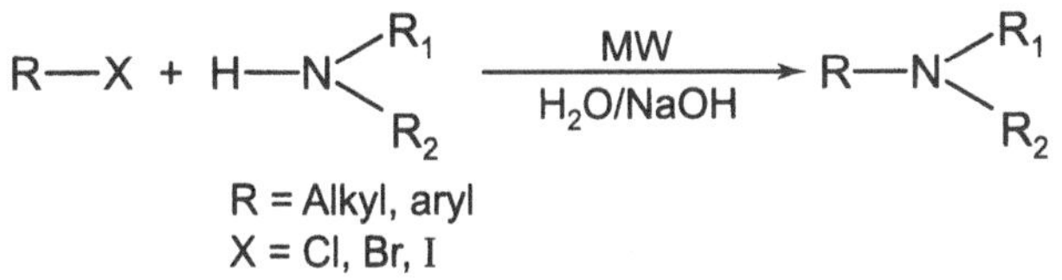

$$R—X + H—N\begin{matrix} R_1 \\ R_2 \end{matrix} \xrightarrow[\text{H}_2\text{O/NaOH}]{\text{MW}} R—N\begin{matrix} R_1 \\ R_2 \end{matrix}$$

R = Alkyl, aryl
X = Cl, Br, I

11. **Willgerodt Reaction:** The reaction of acetophenone by heating with sulphur, aqueous ammonia and pyridine at 180° (10 mm pr) gives amides (C.R. Strauss and K.W. Trainor, Aust. J. Chem., 1995, **48**, 1665).

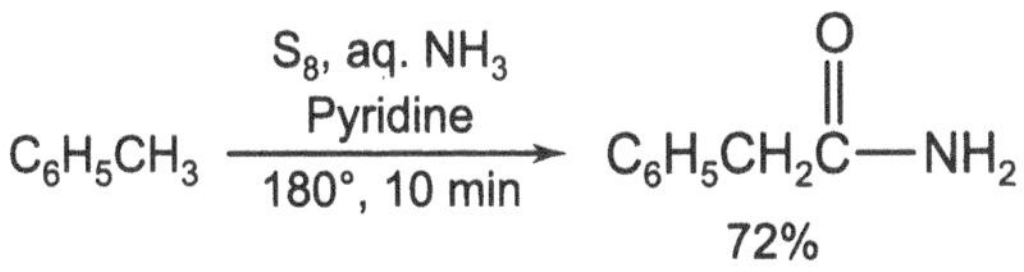

$$C_6H_5CH_3 \xrightarrow[\text{180°, 10 min}]{\substack{\text{S}_8\text{, aq. NH}_3 \\ \text{Pyridine}}} C_6H_5CH_2\overset{\overset{\displaystyle O}{\|}}{C}—NH_2$$

72%

4.2 CARBON-NITROGEN DOUBLE BOND (C = N) FORMATION

Following reaction/procedures are useful for C = N bond formation.

1. **Curtius Rearrangement** of tertiary alkyl azides, via the formation of nitrene intermediate (W. Lwowski, S. Linke and G.T. Tisue, J.Am. Chem. Soc., 1967, **89**, 6308) gives rise to the formation of imines.

$$\underset{\substack{\text{Tert. Alkyl}\\\text{chloride}}}{\overset{R}{\underset{R}{\overset{|}{>}}}C-Cl} + \underset{\substack{\text{Sod.}\\\text{azide}}}{Na\,N_3} \longrightarrow \underset{\substack{\text{tert. Alkyl azide}}}{\overset{R}{\underset{R}{\overset{|}{>}}}C-\ddot{\overset{..}{N}}-\overset{+}{N}\!\!\equiv\!\!N} \longrightarrow$$

$$\xrightarrow[-N_2]{\Delta}\quad \underset{\text{Nitrene}}{R-\overset{\overset{\displaystyle R}{|}}{\underset{\underset{\displaystyle R}{|}}{C}}-\ddot{\overset{..}{N}}} \longrightarrow \underset{\text{Imines}}{\overset{R}{\underset{R}{>}}C=N-R}$$

Imines can also be obtained from carbonyl compounds and aniline.

$$C_6H_5CHO + H_2NC_6H_5 \xrightarrow{-H_2O} C_6H_5CH=N\,C_6H_5$$

2. Oximes contain the C = N group and are obtained from carbonyl compounds and hydroxylamine by heating in a MW oven in presence of pyridine and absolute alcohol (R.N. Gedye, F.E. Smith and K.C. Westaway Can. J. Chem. 1988, 66, 17).

$$\overset{\cdot R}{\underset{R'}{>}}C=O + NH_2OH \xrightarrow[\text{MW, 2 min}]{\text{Pyridine, EtOH}} \underset{\text{Oxime}}{\overset{R}{\underset{R'}{>}}C=NOH}$$

3. **Doebner-Miller Synthesis:** The reaction of primary amines with aldehydes in presence of H_2SO_4 gives quinoline derivatives (A.A. Mortan, The Chemistry of Heterocyclic Compounds, McGraw. Hill, Inc. New York, N.Y., 1946, 263; Campbell and Schaffner, J.Am. Chem., Soc., 1945, **67**, 86.

Aniline   2-Methylquinoline

$$\xrightarrow[\Delta,\ H^+]{CH_3CHO}$$

4. **Skraup Synthesis:** A mixture of aromatic amines, glycerol is heated with H_2SO_4 in presence of nitrobenzene to give quinoline (Z.H. Skraup, Ber., 1880, **13**, 2086; R.H. Manske, M. Kulka, Org., Reactions, 1953, **7**, 80; Bergstrom. Chem. Rev., 1944, **35**, 152).

Aniline + Glycerol $\xrightarrow[C_6H_5NO_2]{H_2SO_4}$ Quinoline

5. **Thorpe (Zieglar) Reaction:** Base catalysed condensation of nitriles yield imines (H. Barton, F.G.P. Remfry and Y.F. Thorpe, J. Chem., Soc., 1904, **85**, 1726; K. Zieglar, E. Eberle and M. Ohlinger, Ber., 1933, **504**, 94; J.P. Schaefer, J.J. Bloomfield, Org. Reactions, 1967, **151**).

6. **Knorr Quinoline Synthesis:** Condensation of arylamines with β-ketoesters gives anilides, which an cyclisation with conc. H_2SO_4 gives α-hydroxy quinolines (L. Knorr, Ann., 1886, **236**, 69; 1888, **245**, 357, 378) (Scheme–1).

(Scheme–1)

7. **Dehydrogenation:** Hydrazocompounds on heating with diethylazodicarboxylate undergo dehydrogenation to give diazo compounds (F. Yoneda, K. Suzuki and N. Nitta, J. Org. Chem., 1967, **32**, 727).

$$C_6H_5\,NHNHC_6H_5 \xrightarrow[EtO_2,O°;\ few\ hrs.]{EtO_2CN = NCO_2Et} C_6H_5N{=}NC_6H_5$$

Hydrazabenzene   Azobenzene

4.3 CARBON-NITROGEN TRIPLE BOND (C≡N) FORMATION

The Cyanides or nitriles are the only compounds which have C≡N group. Following procedures are used.

1. **From Alkyl Halides:** The reaction (stirring) of alkyl halide with aqueous solution of NaCN or KCN in presence of a phase transfer catalyst yields alkyl cyanide in excellent yields [C.M. Starks, J.Am. Chem. Soc., 1971, **93**, 195; N. Sugimoto, T. Fujita, N. Shigematsu and A. Ayada, Chem. Pharm. Bull., 1962, **10**, 427 (Japanese patent 1961/1963)].

$$R-X + NaCN \xrightarrow[\text{Stirring, RT}]{\substack{\text{Benzyl triethyl} \\ \text{ammonium chloride}}} R-CN + NaX$$
aq.

R = $CH_3(CH_2)_7-$ or
   $C_6H_5CH_2CH_2$ or
   C_6H_5CO-

2. **From Aldehydes:** Aromatic aldehydes are rapidly converted into nitriles (89–95%) by heating with hydroxylamine hydrochloride supported on montmorillonite K10 clay in the absence of solvent. The reaction in carried out in MW owen (R.S. Varma and K.P. Naicker, Molecules on line, 1998, **2**, 94; R.S. varma, K.P. Naicker, D. Kumar, R. Dahiya and P.J. Liesen, J. Microwave Power Electromag. Energy, 1999, **34**, 113).

$$\xrightarrow[\text{MW, 1–1.5 min}]{\text{K 10 clay—NH}_2\text{OH. HCl}}$$

89–95%

R_1 = H; R_2 = H, OH, Br, Me, OMe, NO_2
R_1 = R_2 = OMe

3. **From Alkynes:** A mixture of ethyne and ammonia is passed over alumina catalyst at 573 K give acetonitrile.

$$HC{\equiv}CH + NH_3 \xrightarrow[\text{Al}_2\text{O}_3]{573 \text{ K}} H_3CCN + H_2$$
Ethyne   Acetonitrile

Alternatively addition of HCN to alkynes in presence of Ba(CN)₂ give nitriles

$$HC{\equiv}CH + HCN \xrightarrow{\text{Ba(CN)}_2} CH_2{=}CHCN$$
Ethyne   Acrylonitrile

4. **Knoevenagel Condensation** of ketones or aldehydes with cyanoacetic acid or the ester give substituted cyanides.

5. The reaction of Grignard reagents with cyanogen or cyanogen chloride yields alkyl cyanides.

6. **From Amines:** Amines on diazotisation give arenediazonium salts, which on treatment with CuCN give the corresponding cyanide.

$$Ar-NH_2 \xrightarrow[0-5°]{HONO} \underset{\substack{\text{Arene} \\ \text{diazonium} \\ \text{cation}}}{Ar-\overset{+}{N_2}} \xrightarrow[90-100°]{CuCN} Ar\ CN$$

7. Nitriles are conveniently obtained by the reaction of dichlorocarbene (generated *in situ* by the PTC technique [A.P. Kreshkov, E.N. Suguskima and B.A. Krozdov, J. Appl. Chem. USSR (Engl. Transl.), 1965, **38**, 2357] with amides, thioamides, aldoximes and amidines (V.K. Ahluwalia and Renu Aggarwal, Organic Synthesis) Narosa Publishing House, New Delhi, 2006, Page 23 and the references cited there in).

$$\left.\begin{array}{l} R\ CONH_2 \\ R\ CSNH_2 \\ R\ CH{=}NOH \\ R\ \underset{\underset{NH_2}{|}}{C{=}NH} \end{array}\right\} + CHCl_3 + NaOH\ \underset{aq}{\xrightarrow{C_6H_5CH_2\ \overset{+}{N}\ Et_3\overset{-}{Cl}}}\ R{-}C{\equiv}N$$

4.4 CARBON-HALOGEN BOND FORMATION

Following methods/procedures are useful for carbon-halogen bond formation

1. **From Alcohols:** Alcohols on reaction with dichlorocarbene generated *in situ* in phase catalysed system (I. Tabushi, Z. Yoshida and N. Takahashi, J. Am. Chem. Soc., 1971, **93**, 1820) gave good yields of chlorides.

$$ROH + CHCl_3 + NaOH \xrightarrow{C_6H_5CH_2 \overset{+}{N} Et_3\bar{Cl}} RCl + NaCl + H_2O$$

Other procedures include

$$R-CH_2OH + HX \longrightarrow RCH_2X + H_2O$$
$$X = Cl, Br, I$$

$$R-OH + SOCl_2 \longrightarrow RCl + SO_2 + HCl$$

$$RCH_2OH + PCl_5 \longrightarrow RCH_2Cl + POCl_3 + HCl$$

2. **From Carboxylic Acids:** Following sequence of reactions are followed.

$$R-COOH \xrightarrow{AgNO_3} R-\overset{O}{\overset{\|}{C}}-\bar{O} \overset{+}{Ag} \xrightarrow[CCl_4]{Br_2} R-Br + AgBr + CO_2$$

Carboxylic acid   Silver salt

The reaction is known as **Hunsdiecker reaction** (H.C. Hunsdiecker, Chem., Ber., 1942, **75**, 291).

Another procedure involves the photolytic reaction of carboxylic acid with bromine in presence of HgO.

$$Cl-C_6H_4COOH + Br_2 \xrightarrow{HgO, h\nu} Cl\ C_6H_4\ Br$$

p-Chloro benzoic acid   *p*-Bromo chlorobenzoe 80%

This reaction is known as **Hunsdiecker-Borodin-Cristol-Firth Reaction** (A.J. Meyers J. Org. Chem., 1979, **44**, 3405; A. Borodin, Ann., 1942, **75**, 291).

Carboxylic acids, having CH_2 group in α-position on treatment with P/Br_2 give α-bromo or chloro carboxylic acids **(Hell Volhard-Zelinsky Reaction)**.

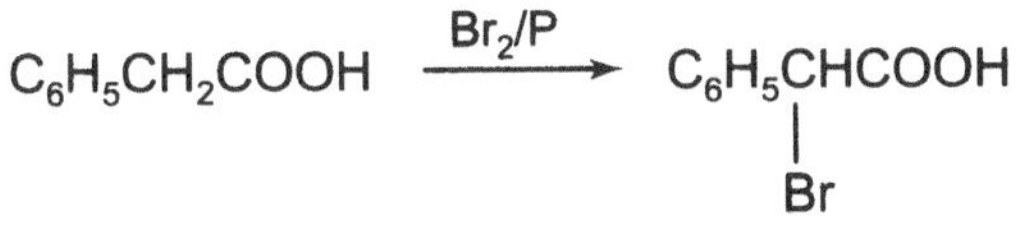

$$C_6H_5CH_2COOH \xrightarrow{Br_2/P} C_6H_5CHCOOH$$

$$\underset{\text{α-Bromophenylacetic acid}}{\overset{|}{Br}}$$

3. **From Arylamines:** Arylamines are diazotised to give arenediazonium salt (which is obtained *in situ*) and treated with CuCl, CuBr or KI to give the corresponding aryl halide.

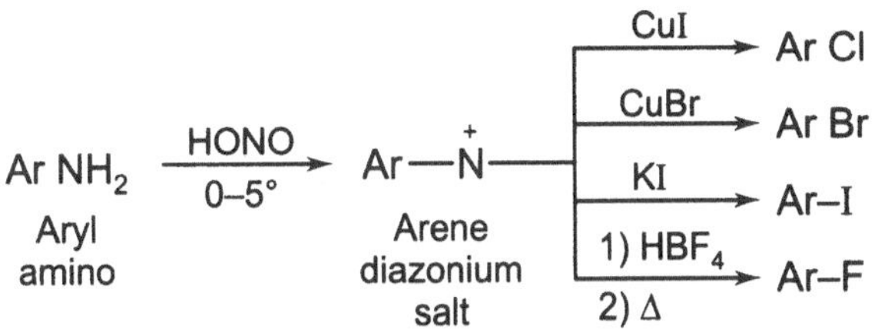

The above reaction is known as **Sandmeyer reaction**. However, the reaction of diazonium salt with HBF₄ to give the corresponding fluoro compound is known as **Schiemann reaction**.

4. **From Arenes:** Halogenation of arenes by halogen in presence of FeCl₃ gives the corresponding halides.

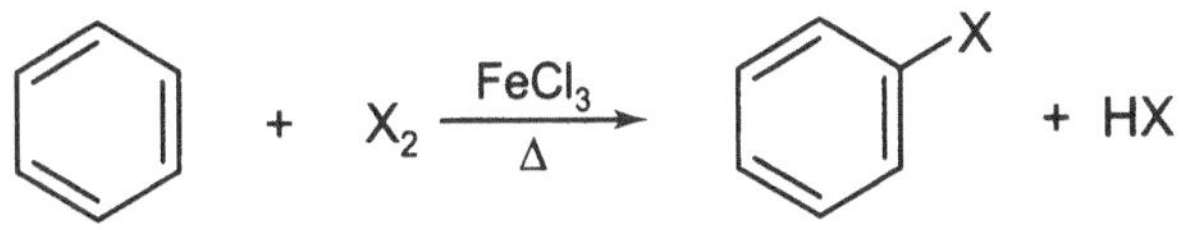

$$X = Cl, Br\ I$$

Chlorobenzene is obtained by passing a mixture of benzene vapours, air and HCl over copper chloride **(Rasching Process)**.

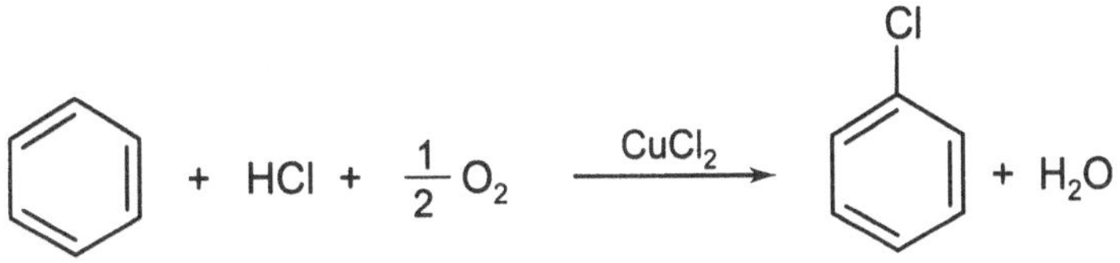

5. **From Alkenes:** Unsymmetrical alkenes on treatment with HBr give the bromo compound. The bromine adds on as per Markovnikov rule. However, in presence of peroxide, anti-Markonikov addition takes place.

$$\underset{\text{Propene}}{CH_3CH{=}CH_2} + HBr \longrightarrow \overset{\overset{\displaystyle Br}{|}}{\underset{\text{2-Bromopropane}}{CH_3CHCH_3}}$$

$$\xrightarrow{\text{Peroxide}} \underset{\text{Propyl bromide}}{CH_3CH_2CH_2Br}$$

4.5 CARBON-SULPHUR BOND FORMATION

Following are given some of the methods/procedures for carbon sulphur bond formation.

1. α-Diketones on heating with diethyl thiodiacetate in presence of base give thiophene derivative. The method is known as **Hinsberg Thiophene synthesis** (O. Hinsberg, Chem. Ber., 1910, **42**, 901; N. Wynberg, J. Org. Chem., 1964, **29**, 1919; N. Wynberg, J. Am. Chem. Soc., 1965, **87**, 1739).

Benzil + Diethyl thiodiacetate $\xrightarrow[\Delta]{t\text{-BuOK}}$ Thiophene derivative

2. **S-Alkylation:** Arylthiols and alkane thiols on treatment with alkyl halides in presence of K_2CO_3/DMF (Sonication) give S alkylated product (J.M. Khurana and P.K. Sahoo, Syn. Commun., 1992, 1691).

$$RSH + R'X \xrightarrow[))))]{K_2CO_3/DMF} RSR'$$

3. **From Epoxides:** The reaction of epoxides with thiophenol in presence of catalytic amount of NaOH (aqueous) on heating in MW oven for 5 min. yielded the corresponding β-hydroxy sulphides (V.Pirotic and S. Colonna, Green Chem., 2005, **7**, 43).

Cyclohexene
epoxide

β-Hydroxysulfide
(trans) 85–98%

The epoxides can be converted into thiranes in good yields by reacting with potassium thiocyante in [b$_{min}$] [PF$_6$] –H$_2$O (2 : 1) at room temperature (J.S. Yadav, B.V.S. Reddy, C.R. Reddy and K. Rajasekhar, J. Org. Chem., 2003, **68**, 2525).

Expoxides

Thiiranes

4. **Thiocyanation of Alkyl Halides:** Alkyl halides on treatment with the ionic liquid [b$_{min}$] [SCN] are converted into the corresponding thiocyanates at room temperature (A. Kamal and G. Chouhan, Tetrahedron Lett., 2005, **46**, 1489).

$$C_6H_5CH_2Cl \ + \ [b_{min}][SCN] \ \xrightarrow{RT} \ C_6H_5SCN + [bmin]]Cl]$$

(1.2 equiv)   80–95%

KCNS (1.5 eq.)

Alkyl halides on heating in MW oven with sodium phenylsulphinate, adsorbed onto alumina gave sulphones in 40–99 per cent yield (D.Villemin and A.B. Alloum, Synth. Commun., 1990, **20**, 925).

$$C_6H_5CH_2Cl \ \xrightarrow[\text{MW, 5 min}]{PhSO_2Na, Al_2O_3} \ C_6H_5CH_2SO_2Ph$$

40–99%

5. **Thionotion Reaction:** Thioketones, thioamides, thioesters and thioflavonoids can be obtained by heating of the corresponding oxygen containing compound with Lawesson's reagent (0.5 equiv.) in a MW oven under solvent free conditions. This is a ecofriendly procedure (R.S. Varma and D. Kumar, Org. Lett., 1999, **1**, 697).

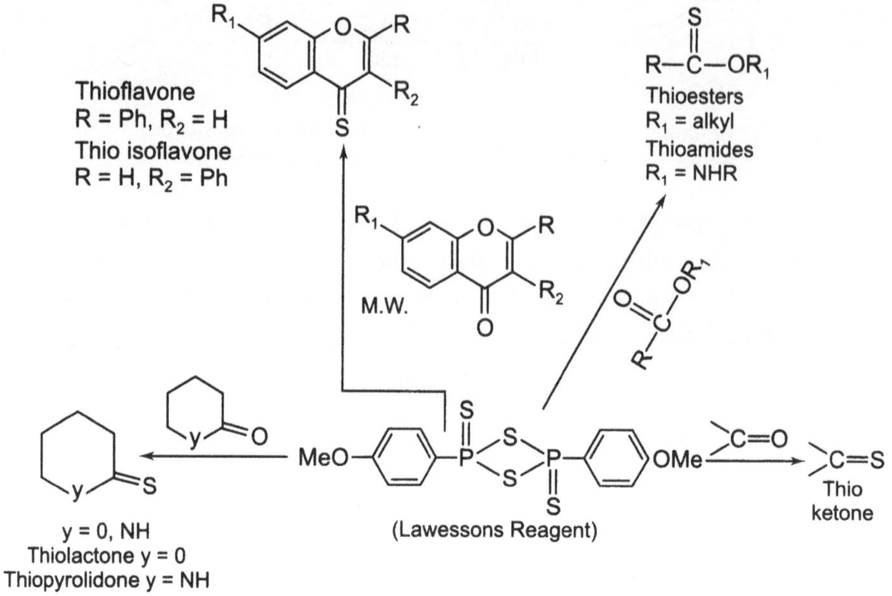

Thioflavone
R = Ph, R$_2$ = H
Thio isoflavone
R = H, R$_2$ = Ph

Thioesters
R$_1$ = alkyl
Thioamides
R$_1$ = NHR

M.W.

y = 0, NH
Thiolactone y = 0
Thiopyrolidone y = NH

MeO—⟨⟩—P(=S)—S—S—P(=S)—⟨⟩—OMe

(Lawessons Reagent)

Thio ketone

6. A one pot synthesis of β-ketosulfones involve condensation of ketones with relatively, benign [hydroxy (tosyloxy) iodo] benzene and sodium arene sulfinate in presence of a PTC catalyst (tetrabutyl ammonium bromide, TBAB) and sodium benzene sulfinate. Intimate grinding of the reaction mixture gave the β-ketosulfones in high yields (Dalip Kumar, Swapna Sundaree, V.S. Rao and R.S. Varma, Tetrahedron Lett., 2006, **47**, 4197).

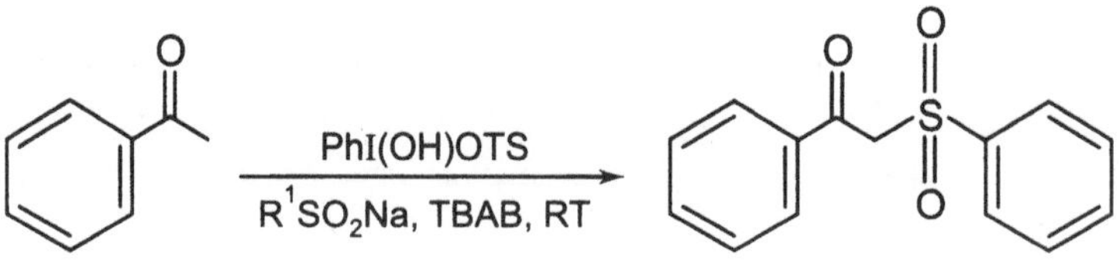

PhI(OH)OTS
R^{1}SO$_2$Na, TBAB, RT

7. The reaction of α, β-unsaturated ketones with thiols in the ionic liquid [b$_{min}$] [PF$_6$]/H$_2$O 2:1 in absence of any catalyst gave the corresponding Michael adducts in excellent yields (J.S. Yadav, B.S.V. Reddy and G. Baishya, J. Org. Chem., 2003, **68**, 7098).

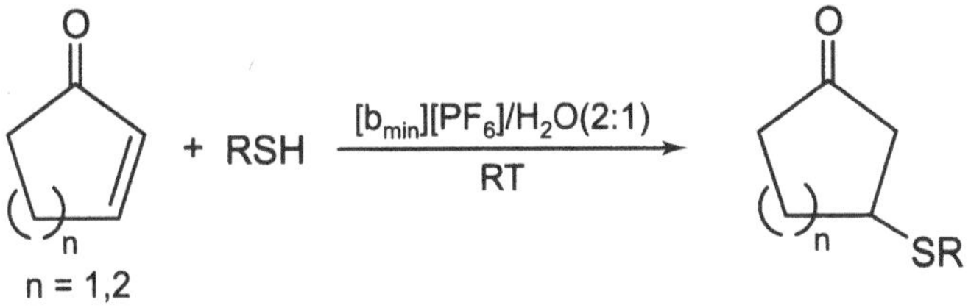

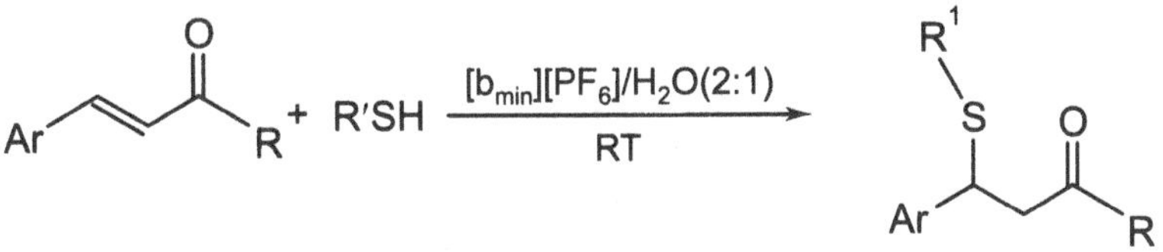

8. Thiols are obtained by the reaction of alkyl halides with NaSH.

$$C_2H_5Br + NaSH \longrightarrow C_2H_5SH + NaBr$$

Alternatively alkenes on reaction with H_2S in presence of acid catalyst give thiols.

$$CH_3-\underset{}{\overset{CH_3}{\overset{|}{C}}}=CH_2 + H_2S \xrightarrow{H_2SO_4} CH_3-\overset{CH_3}{\overset{|}{\underset{SH}{\underset{|}{C}}}}-CH_3$$

Alkyl halids are converted into grignard reagent, which on reaction with sulphur give the corresponding sulfphur containing grignard reagent; the latter on treatment with H^+ give thiols.

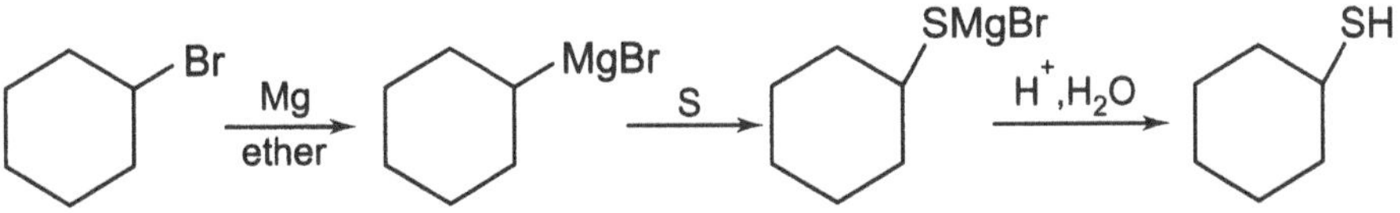

Cyclohexyl bromide | cyclohexane thiol

9. Thioethers or organic sulfides are readily obtained by displacement reactions between alkyl compounds and salts of thiols or by in addition of thiols to alkenes. Such reaction lead to anti markownikoff addition.

$$\underset{\text{Ethane-2-ol-1-thiol}}{HOCH_2CH_2SH} + (CH_3)_2SO_4 \xrightarrow[60\text{–}70°]{\text{aq. 25\% NaOH}} \underset{\text{2-(Methylthio) ethanol}}{HOCH_2CH_2SCH_3}$$

$$\underset{\text{Methanethiol}}{CH_3SH} + \underset{\text{Allylcyanide}}{CH_2{=}CHCH_2CN} \xrightarrow[\text{hr, 24 hr.}]{(C_6H_5CO_2)_2O} \underset{\text{4-(Methylthio)-butyronitrile}}{CH_3SCH_2CH_2CH_2CN}$$

10. Thiiranes can be used as synthons for a variety of products

4.6 NITROGEN-NITROGEN DOUBLE BOND (N = N) FORMATION

The N=N group is called the azo group. It is mostly formed by coupling reactions of arenediazonium salts with phenols or amines.

The azo compounds can also be synthesised from hydrazo derivatives by treatment with PEG. NO_2, used as an oxidant. This is a green route (R.Z. Qiao, Y. Zhang, X.P. Hui, P.F. Xu and Z.Y. Zhang, Green Chem. 2001, **3**, 186).

$$PEG + NO_2 \longrightarrow PEG.NO_2$$

$$R^1NH\,NH\,R^2 + PEGNO_2 \xrightarrow{\text{20–30 min}} R^1N{=}NR^2 + N_2$$

Hydrazo deriv.   Azo compound
                                                   75–80%

$$R^1 = R^2 = MeC_6H_4^-$$
$$R^1 = m{-}MeC_6H_4^-, R^2 = p{-}ClC_6H_4$$

4.7 NITROGEN-OXYGEN BOND FORMATION

Nitrogen-oxygen bonds are normally present in as N-oxides in nitrogen containing heterocyclic compounds or tertiary amines.

Trialkyl amines, such as tributylamine is oxidised with aq. methanolic 30–35 per cent H_2O_2 at 0° to room temperature.

$$(Bu)_3\,N \xrightarrow[\text{H}_2\text{O/MeOH}]{+\,\text{H}_2\text{O}_2} (Bu)_3\,\overset{+}{N}\!\!-\!\!\overset{-}{O}$$

95%

Pyridine on oxidation with the H_2O_2, peracids or dimethyl oxirane give 80–93% yield of the corresponding N-oxide.

50% H_2O_2, PhCN, MeOH
NaOH, pH8, 20–30°, 3 hr.

40%CH_3CO_3H,85°
50–60 min

Me_2C ⟨O O⟩ , H_2O,KOH

pH (7.5–8.0), RT, 2 hr.

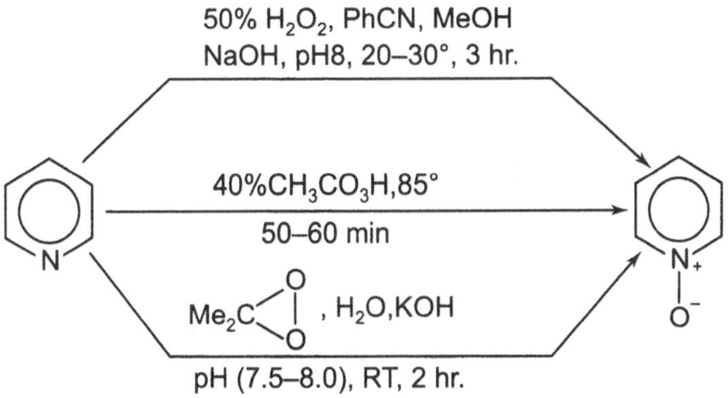

[G.B. Payne, P.H. Deming and P.H. Williams J. Org. Chem., 1961, 26, 659; H.S. Mosher, L. Turner, A. Carlsmith, Org. Syn., Collective Volume 1963, 4, 828; R.W. Murray and R. Jeyaraman, J. Org. Chem., 1985, 50, 2847].

Nitrogen containing compounds are best oxidised with urea formaldehyde adduct (UPH) to the corresponding N-oxides (R.S. Varma and K.P. Naicker, Org. Letters, 1999, **2**, 189).

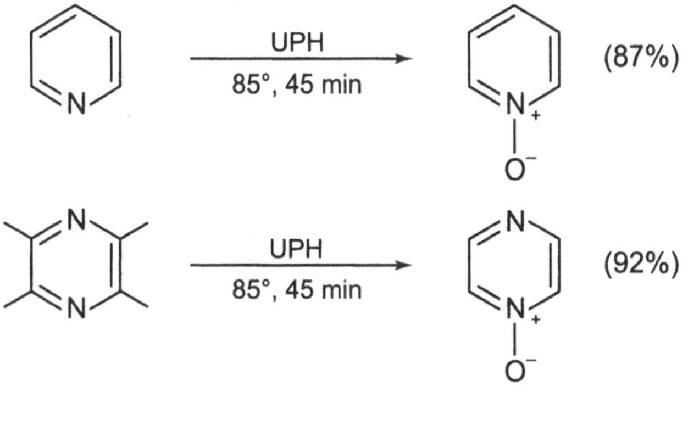

UPH
85°, 45 min
(87%)

UPH
85°, 45 min
(92%)

The reaction is carried out in solid state. Hydroxylamines on oxidation with diethylazodicarboxylate give nitroso compounds (E.C. Taylor and F. Yoneda, Chem., Commun., 1967, 199).

$$C_6H_5NHOH \xrightarrow[\text{Et}_2O, \ 0°, \ \text{two hrs.}]{EtO_2CN=NCO_2Et} C_6H_5NO$$

Phenyl   89%
hydroxylamine   Nitrosobenzene

❑❑❑

Interconversion of Functional Groups

5.1 INTRODUCTION

It is well known that organic compounds are divided into various families. Each family is characterised by the presence of groups called a functional group. In fact, the chemical properites of the families depend on the presence of the functional group it contains. Thus, as an example, the functional group for alkenes is carbon-carbon double bond, and that in alkynes, the functional group is carbon-carbon triple bond. However, there is no functional group in alkanes; these have carbon-carbon single bond and carbon-hydrogen bond, which are also present in molecules of almost all organic molecules.

A functional group is an arrangement of atoms and contains certain bonds other than carbon-carbon and carbon-hydrogen bonds. A exception to the above is the functional group of an aldehyde, $R—\overset{\overset{\textstyle H}{|}}{C}{=}O$, the functional group in this case contains a carbonyl group (C = O) and an C — H bond. A typical example is that of carboxylic acids, which contains a carboxyl group as the functional group.

$$R—\overset{\overset{\textstyle O}{\|}}{C}—OH$$

Carboxyl group
(Functional group)

The following table lists the name of the family, its general formula, the functional group present and typical example with their IUPAC names.

Table 5.1: Families of organic compounds

Family	General formula	Functional group	Examples (IUPAC name)
Alkanes	C_nH_{2n+2}	C—C and C—H bonds	CH_3 — CH_3 (Ethane)
Alkenes	C_nH_{2n}	$>C=C<$	$CH_2=CH_2$ (Ethene)
Alkynes	C_nH_{2n-2}	—C≡C—	HC≡CH (Ethyne)
Aromatic		Aromatic ring	(Benzene)
Haloalkanes[1] (Alkyl halides)	$C_nH_{2n+1}X$	—C—Ẍ:	CH_3CH_2Cl (Ethyl chloride)
Alcohols[2]	$C_nH_{2n+1}OH$	—C—ÖH	CH_3CH_2OH (Ethanol)
Ethers	$(C_nH_{2n+1})_2O$	—C—Ö—C—	$C_2H_5ÖC_2H_5$ (Ethoxy ethane)
Amines[3]	$C_nH_{2n+1}NH_2$	—C—N—H H	$C_2H_5-NH_2$ (Ethylamine)
Aldehydes	$C_nH_{2n+1}CHO$	:O: ‖ —C—H	$CH_3-\overset{O}{\overset{\|}{C}}-H$ (Ethanal)
Ketones	$(C_nH_{2n+1})_2CO$	:O: —C—C—C—	$CH_3\overset{O}{\overset{\|}{C}}CH_3$ (Propanone)
Carboxylic acids	$C_nH_{2n+1}COOH$	:O: ‖ —C—ÖH	$CH_3\overset{O}{\overset{\|}{C}}OH$ (Ethanoic acid)
Esters	$C_nH_{2n+1}COOR$	‖ —C—C—	$CH_3\overset{O}{\overset{\|}{C}}-OCH_3$ (Methyl ethanoate)
Amides[4]	$C_nH_{2n+1}CONH_2$	O ‖ —C—NR'R"	$CH_3\overset{O}{\overset{\|}{C}}NH_2$ (Ethane amide)
Nitriles	$C_nH_{2n+1}C≡N$	—C≡N	$CH_3C≡N$ Ethanenitrile

Notes

1. Haloalkanes or alkyl halides can be primary, secondary or tertiary depending on whether the halogen is attached to primary, secondary or tertiary carbon atom. These are represented as shown below:

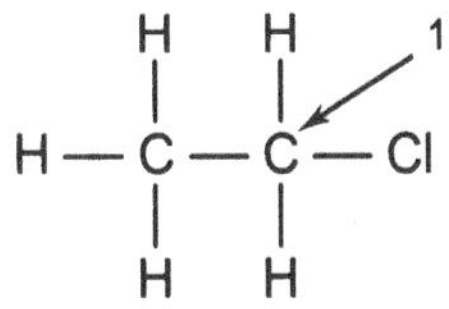

Primary alkyl chloride (Ethyl chloride)   Secondary alkyl chloride (2-Chloropropane)   Tertiary alkyl chloride (2-Chloro-2-methyl propane)

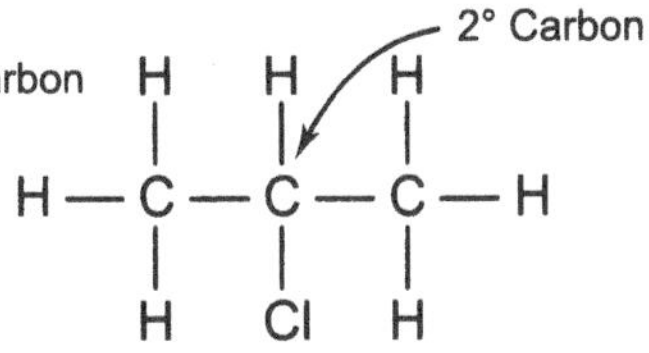

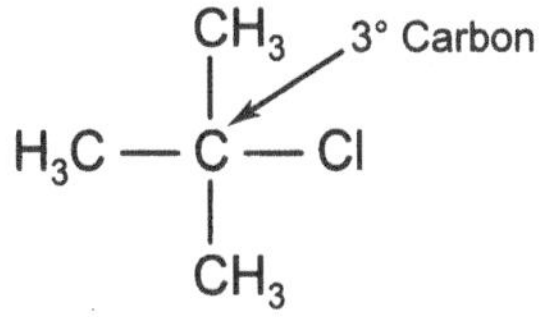

The functional groups in the 1°, 2° or 3° alkyl halides is $-\overset{|}{\underset{|}{C}}-\ddot{\underset{\cdot\cdot}{C}l}:$,

$-\overset{|}{C}H\ddot{\underset{\cdot\cdot}{C}l}:$ or $\overset{\diagdown}{\diagup}C\ddot{\underset{\cdot\cdot}{C}l}:$, respectively. In alkyl halides, the halogen can be chlorine, bromine or iodine.

2. The alcohols (as in the case of alkyl halides) can be primary, secondary or tertiary depending on whether the hydroxyl group is attached to primary, secondary or tertiary carbon atom. These are represented as shown below:

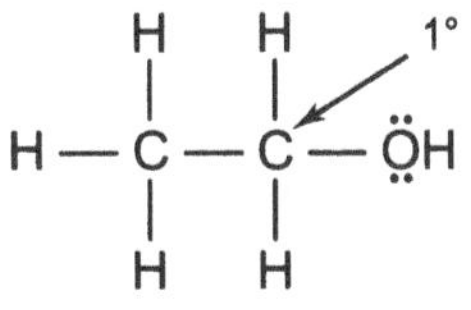

Primary alcohol (Ethyl alcohol)   Secondary alcohol (Isopropyl alcohol)   Tertiary alcohol (*tert.* Butyl alcohol)

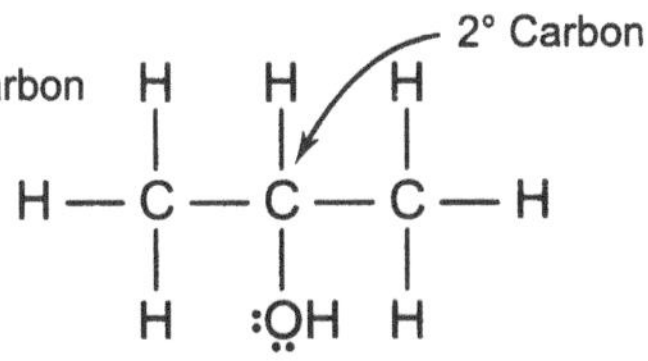

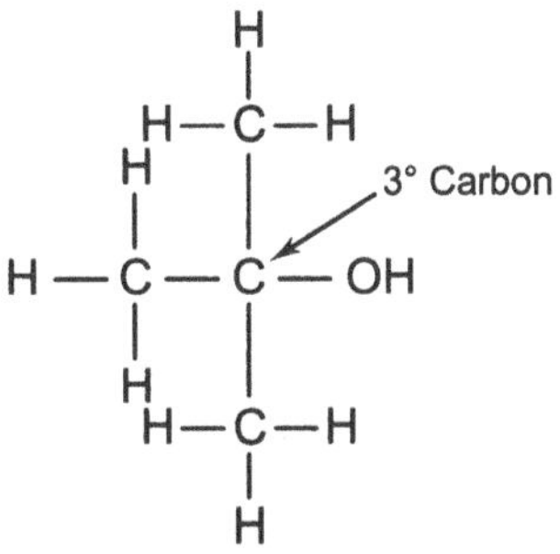

The functional group in 1°, 2° or 3° alcohols is $H-\overset{H}{\underset{|}{C}}-\ddot{\underset{\cdot\cdot}{O}}H$,

$-\overset{|}{C}H-\ddot{\underset{\cdot\cdot}{O}}H$ or $\overset{\diagdown}{\diagup}C-\ddot{\underset{\cdot\cdot}{O}}H$, respectively. In case of alcohol, if there is aryl group in place of alkyl group we get phenols.

3. The amines (as in the case of alkyl halides and alcohols) can also be primary, secondary or tertiary depending on the number of alkyl groups attached to nitrogen atom. These are represented as shown below:

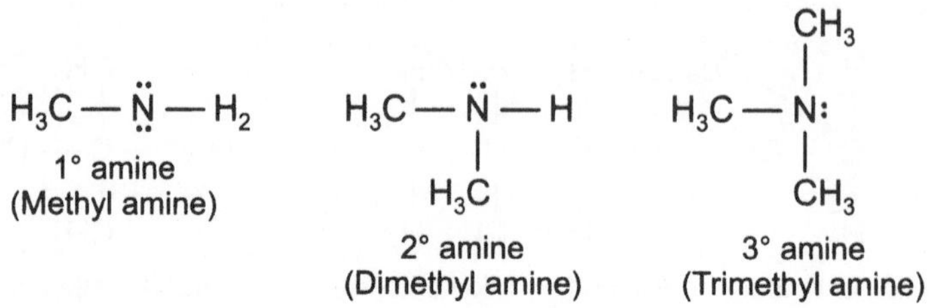

$$H_3C — \overset{..}{\underset{..}{N}} — H_2$$

1° amine
(Methyl amine)

$$H_3C — \overset{..}{N} — H$$
$$| $$
$$H_3C$$

2° amine
(Dimethyl amine)

$$\overset{\overset{CH_3}{|}}{H_3C — N:}$$
$$| $$
$$CH_3$$

3° amine
(Trimethyl amine)

Thus, in case of amines, the classification is quite different from alcohols and alkyl halides. For example, isopropylmine,

$$H_3C—\underset{\underset{NH_2}{|}}{CH}—CH_3$$, is a 1° amine, even though the amino group is

attached to a secondary carbon atom. In a similar way, piperidine

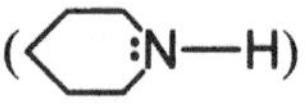

($\langle$ ⃝ :N—H) is a cyclic secondary amine.

4. Amides have the formulas $RCONH_2$, RCONHR or RCONR'R". Some examples are

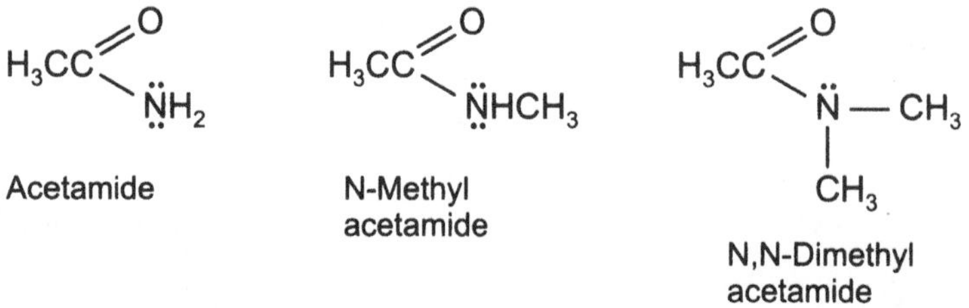

Acetamide

N-Methyl
acetamide

N,N-Dimethyl
acetamide

In these cases the functional groups are $—\overset{\overset{O}{||}}{C}—NH_2$, $—\overset{\overset{O}{||}}{C}—NHR'$ or

$—\overset{\overset{O}{||}}{C}—NR'R''$, respectively.

5.2 INTROCONVERSION OF FUNCTIONAL GROUPS

For organic synthesis, the interconversion of the functional groups is very important. Following are given interconvertions of various functional groups in various families of organic compounds. As far as possible, Green conditions have been given for the interconversion of functional groups.

5.2.1 Hydrocarbons

Hydrocarbons contain C — C and C — H bonds. The hydrogen of the C — H bond in hydrocarbons can be replaced with halogen. A typical example is the halogenation of alkanes.

$$CH_4 + Cl_2 \xrightarrow[\text{or heat}]{\text{light}} CH_3Cl + HCl$$

Excess   Methyl
chloride

$$CH_4 + Br_2 \xrightarrow{\text{light}} CH_3Br + HCl$$

Excess   Methyl
bromide

A typical reaction of C—H bond is the insersion of dichlorocarbene generated *in situ* by the reaction of chloroform and aqueous sodium hydroxide in presence of a phase transfer catalyst, benzyl trimethyl ammonium chloride. The reaction with adamantanes[1] and tetrahydrofuran[2] are given below.

$$C_6H_5CH_2N^+(CH_3)_3Cl^-$$

CHCl$_3$ + *aq.* NaOH

R=R'=H; yield 54%
R=CH$_3$=R'=H; yield 100%

Adamantanes

$$C_6H_5CH_2N^+(CH_3)_3Cl^-$$

CHCl$_3$ + *aq.* NaOH

THF   18%

The oxidation of adamantane occurs exclusively at the tertiary carbon C — I.

$$O_3, SiO_2$$

-78 to $60°C$
2 hr. to RT,

Adamantane   1-Adamantol
81–84%

The carbon-carbon bond in alkanes is quite strong and normally do not break.

5.2.2 Alkenes

The alkenes contain $C = C$ as the functional group. Following are given some typical reactions of $C = C$.

(*i*) **Addition of Dichlorocarbene:** Dichlorocarbene can be generated[3] by direct reaction between powdered sodium hydroxide and chloroform by sonication. The *in situ* generated dichlorocarbene adds to alkenes to give the adduct.

Styrene   Adduct, 96%

It is, however, found that dichlorocarbene can also be generated *in situ* by the reaction of chloroform with aqueous sodium hydroxide in presence of a phase transfer catalyst[4] (benzyl triethylammonium chloride). The *in situ* generated dichlorocarbene adds on to alkenes to give the adduct in 60–70 per cent yield.

Styrene   1-Phenyl-2,2-dichloro cyclopropane

(*ii*) **Cyclopropanation of Alkenes:** The reaction of alkenes with methylene iodide in presence of zinc (mossy zinc) give the corresponding cyclopropane derivative[5]. In this reaction, sonochemically activated zinc is used. The yield is 91 per cent compared to 51 per cent by the normal reaction. Ultrasonic source is a cleaning bath (50 KHz).

(*iii*) **Oxymercuration of Alkenes:** The oxymercuration of alkenes is a convenient procedure for carbon-oxygen bond formation. It is normally performed with mercuric acetate or trifluoroacetate due to non-availability of other mercury salts. It has been found that any mercury salt can be prepared from mercuric oxide and an organic acid under sonication[6]. The preparation of salt and oxymercuration can be performed in one pot under sonication. Thus, selective reaction of limonene, which is usually difficult can be affected in excellent yield (80%) compared to 40 per cent under usual condition.

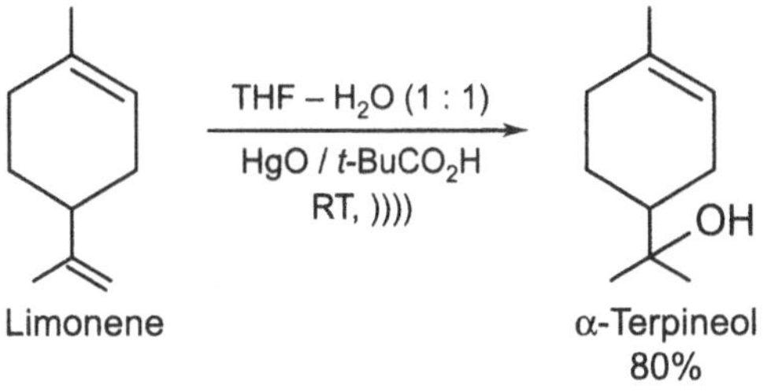

Limonene → α-Terpineol
80%

THF – H$_2$O (1 : 1)
HgO / t-BuCO$_2$H
RT,))))

(*iv*) **Epoxidation of Alkenes:** The epoxidation of alkenes is usually carried out with peracid (such as perbenzoic acid or m-chloroperbenzoic acid). It has been found that polymer supported peracid[7] give epoxides in those olefins which are unreactive towards organic peracids. The recovered polymer supported peracid can be reused for subsequents lots of epoxidation.

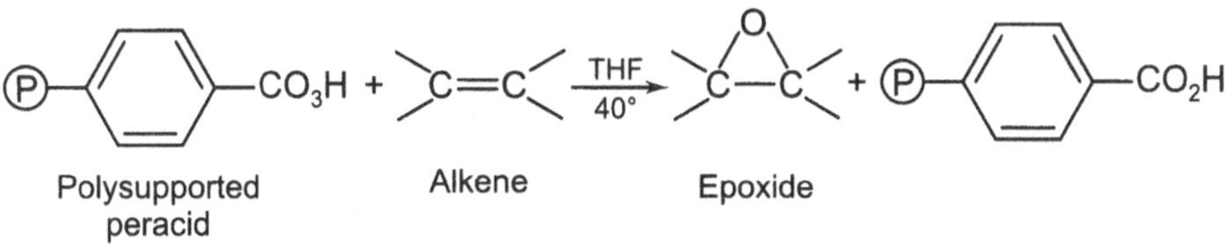

Polysupported peracid   Alkene   Epoxide

THF
40°

Epoxidation of alkenes with NaOCl using Jacobsens chiral Mn(III) Salen immobilised in a $[b_{min}][PF_6]$ catalyst is an efficient and recyclable procedure for asymmetric epoxidation[8]. In ionic liquid as solvent, the epoxidation proceeds *via* the formation of high valent manganese-oxo active intermediate, which was otherwise undetectable in organic solvents[9].

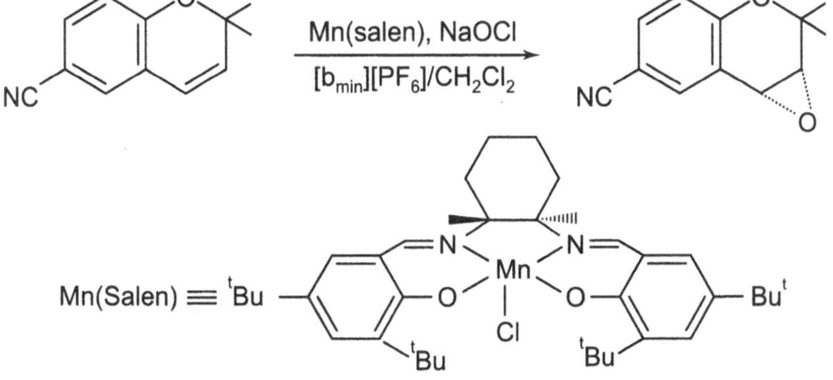

Mn(salen), NaOCl
$[b_{min}][PF_6]/CH_2Cl_2$

Mn(Salen) ≡ ^{t}Bu

A convenient procedure for the epoxidation of electrophilic alkenes in ionic liquids $[b_{min}][PF_6]$, $[b_{min}][PF_4]$ as solvent using aqueous solution of H_2O_2 in the presence of sodium hydroxide as basic catalyst has been reported[10].

Using the above procedure, following alkene could be converted into epoxides.

R = H or Me

An interesting epoxidation reagent is peroxycarbonic acid[10a] which is obtained *in situ* from H_2O_2 and CO_2. Thus, the epoxidation of cyclohexene in biphasic system (SC — CO_2/olefin) and aqueous H_2O_2 phase to yield the corresponding epoxides and diols. The yield of epoxide can be increased by the addition of DMF to increase the aqueous solubility of the olefins, suggesting that epoxidation occurs in aqueous phase.

(v) Hydroxylation of Alkenes: The dihydroxylation of alkenes to give glycols is usually carried out with O_sO_4 in presence of an oxidizing agent like BuOOH, $H_2O_2/\overline{O}H$ or $NaHSO_3$. The main problem is the high cost and toxicity of the contamination of the product with osmium catalyst. This problem is now overcome by using ionic liquids. The hydroxylation is carried out either in biphasic $[b_{min}][PF_6]/$ water or monophasic $[b_{min}][PF_6]/$tert. butanol system. Both procedures has been used for chiral substrates using the chiral ligands. This procedure permits the recycling and reuse of osmium ligand catalyst. The use of supercritical carbondioxide extraction helped in minimising the osmium leaching from the room temperature ionic liquid phase[11].

It is of interest to note that the above hydroxylation using O_sO_4 gives syn diols. The anti hydroxylation of alkenes can be achieved by acid catalysed ring opening of epoxide ring to give *trans* diols.

Cyclic alkenes, *e.g.*, cyclo octene can be oxidised to *cis*-1,2-cyclooctane diol in 50 per cent yield by aqueous $KMnO_4$ in presence of a PTC [benzyltriethyl-ammonium chloride (BTEAC)] and $NaOH$[12].

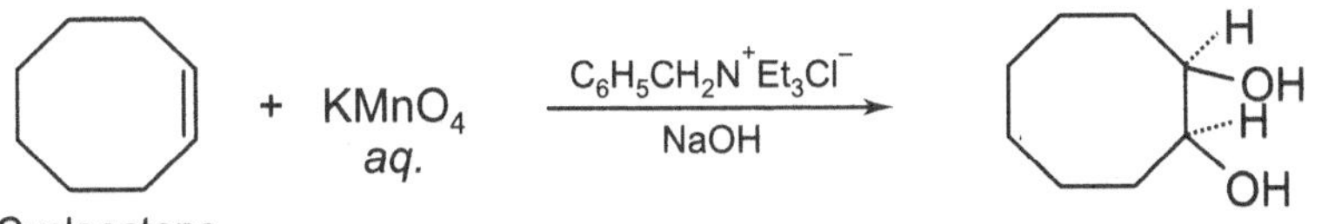

$$Cyclooctene + KMnO_4 \text{ aq.} \xrightarrow[NaOH]{C_6H_5CH_2N^+Et_3Cl^-}$$

cis-1,2-Cyclooctane diol
50%

The yield in above procedure is 50 per cent compared to about 7 per cent by classical technique.

Dihydroxylation of olefins can also be conveniently affected[12a] using polyethylene glycol (PEG-400) as solvent and osmium tetroxide as catalyst in presence of N-methylmorpholine (NMO). Thus, the hydroxylation of stilbene is represented below.

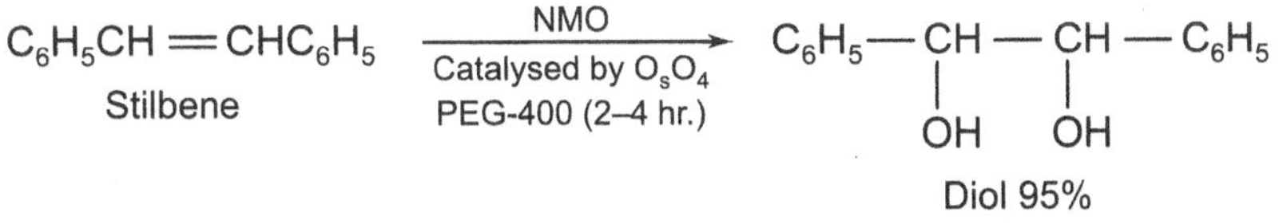

$$C_6H_5CH = CHC_6H_5 \xrightarrow[\substack{\text{Catalysed by } O_sO_4 \\ \text{PEG-400 (2–4 hr.)}}]{\text{NMO}} C_6H_5 - \underset{\underset{OH}{|}}{CH} - \underset{\underset{OH}{|}}{CH} - C_6H_5$$

Stilbene

Diol 95%

In the above reaction, the PEG-400 and O_sO_4 can be reused by extraction of the product diol using ether.

Asymmetric dihydroxylation of olefins has been carried out[12a] using catalytic amount of osmium tertraoxide and PEG-400. The ligand is efficiently recovered and recycled with good enantioselectivity. Some examples of asymmetric dihydroxylation are given below:

Styrene R═H
α-Methyl styrene R═CH₃

Diol R═H   94%
Diol R═CH₃   97%

Trans stilbene

1,2-Diphenyl-1,2-ethane diol
95%

Trans ethyl cinnamate R═H
4-Methoxy trans ethyl cinnamate R═OCH₃

92–93%

(*vi*) **Oxidative Cleavage of Alkenes:** Oxidative cleavage of $C = C$ is known to be brought about by treating with acidified KMnO$_4$ (or K$_2$Cr$_2$O$_7$) solution or aqueous solution of KMnO$_4$/NaIO$_4$. The initially formed of aldehydes are further oxidised to carboxylic acids.

$$CH_3CH_2CH = CH_2 \xrightarrow[H^+]{KMnO_4} CH_3CH_2\overset{\overset{O}{\|}}{C}-OH + CO_2$$

1-Butene   Propionic acid

It has been shown[13] that oxidation of 1-alkenes with aqueous KMnO$_4$ in presence of a PTC gives the carboxylic acid due to cleavage of $C = C$. In this case, a small amount of the next lower carboxylic acid is produced as the byproduct.

$$C_7H_{15}CH_2CH = CH_2 + KMnO_4 \xrightarrow[RT]{C_{16}H_{35}N^+(CH_3)_2CH_2C_6H_5Cl^-}$$

1-Decene   aq.
(in benzene)

$$\longrightarrow C_7H_{15}CH_2COOH + C_7H_{15}COOH + CO_2$$
$$\qquad\qquad 77\% \qquad\qquad\qquad 8\%$$

The utility of the above PTC technique is illustrated by the oxidation[14] of stilbene to benzoic acid in 95 per cent yield.

Stilben in benzene + KMnO$_4$ aq. → 1) Tetrabutyl ammonium bromide, stirring 2–3 hr. 2) NaHSO$_3$ → Benzoic acid 95%

In place of PTC, crown ethers can also be used in the above oxidations.

A typical example is the oxidation of α-pinene with KMnO$_4$ in presence of crown ether[15] to give pinonic acid in 90 per cent yield.

α-Pinene → KMnO$_4$, C$_6$H$_6$, Dicyclohexano[18] crown-6, 25°C → Pinonic acid 90%

A typical example of oxidative cleavage of alkenes[15a] is the synthesis of adipic acid from cyclohexene in polyethylene glycol based biphasic system using sodium tungstate and H$_2$O$_2$.

Cyclohexene + 4H$_2$O$_2$ → Na$_2$WO$_4$ / PEG/NaHSO$_4$ → Adipic acid + 4H$_2$O

(*vii*) **Ozonolysis of Alkenes:** The reaction of alkenes with Ozone (ozonolysis) to give carbonyl compound is a synthetically useful reaction[16]. Two examples are given below:

Alkene → 1) Ozonolysis, O$_3$ 2) Zn/CH$_3$COOH → Ketone + Aldehyde

$(CH_3)_2CHCH=CCH_2CH_3$ (with CH$_3$ substituent) → 1) O$_3$ 2) Zn/CH$_3$COOH → $(CH_3)_2CHCHO$ + $O=CCH_2CH_3$ (with CH$_3$ substituent)

2,4-Dimethyl-3-hexene → 2-Methyl propanal + 2-Butanone

In the above procedure, if H$_2$O$_2$ is used in place of Zn/CH$_3$COOH, the product obtained is a carboxylic acid instead of aldehyde.

$$\underset{R}{\overset{R}{>}}\!\!=\!\!\underset{H}{\overset{R}{<}} \xrightarrow[\text{2) H}_2\text{O}_2]{\text{1) O}_3} \underset{R}{\overset{R}{>}}\!\!=\!\!O \; + \; O\!\!=\!\!\underset{\text{OH}}{\overset{R}{<}}$$

Ketone   Carboxylic
acid

A commercially useful method is the synthesis of α, ω-alkane dicarboxylic acids[17] by ozonolysis of a cyclic alkene (*e.g.*, cyclooctene) in presence of an emulsifier (peroxylethylated lauryl alcohol) followed by treatment with aqueous alkaline H_2O_2.

$$\text{Cyclooctene} \xrightarrow[\substack{\text{2) H}_2\text{O}_2/\text{OH}^- \\ \text{Emulsifier}}]{\text{1) O}_3/\text{H}_2\text{O}/10°} H_2OC\!\!-\!\!\sim\!\!\sim\!\!-\!\!COOH$$

Cyclooctene   α-ω-alkanedicarboxylic
acid

(*viii*) **Addition of Water to Alkenes:** The addition of water to alkenes in presence of acid gives alcohols. This reaction is called hydration of alkene.

$$CH_2\!\!=\!\!CH_2 + H_2O \xrightarrow{H_2SO_4} CH_3CH_2OH$$

Ethene   Ethyl alcohol

In case of unsymmetrical alkenes, the addition follows antimarkovnikov rule. Thus,

$$CH_3CH\!\!=\!\!CH_2 + H_2O \underset{}{\overset{H^+}{\rightleftharpoons}} CH_3\underset{OH}{\overset{|}{CH}}\!\!-\!\!CH_3$$

Propene   Isopropyl alcohol

The antimarkovnikov addition also takes place by oxymercuration/ demercuration.

$$CH_3CH\!\!=\!\!CH_2 + Hg(OAc)_2 \overset{H_2O}{\rightleftharpoons} CH_3\underset{OH}{\overset{|}{CH}}\!\!-\!\!CH_2\!\!-\!\!Hg\!\!-\!\!OAc$$

Propene

$$\downarrow NaBH_4/OH^-$$

$$CH_3\underset{OH}{\overset{|}{CH}}\!\!-\!\!CH_3$$

Isopropyl alcohol

As already stated (see section 5.2.2), the above reaction gives much better yield on sonication.

Markovnikov addition of elements of water can be affected by hydroboration followed by oxidation of the formed trialkylborone with alkaline H_2O_2.

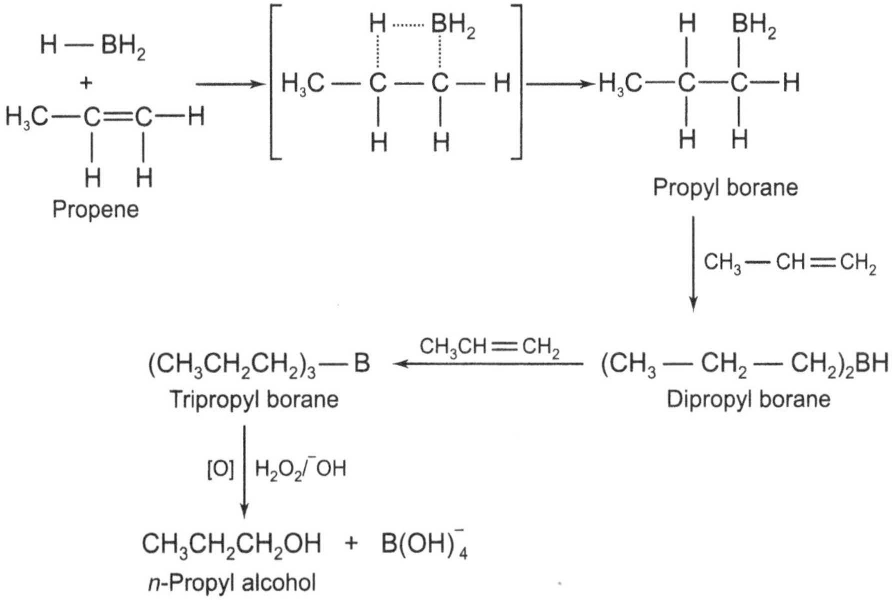

Propyl borane

Tripropyl borane (CH₃CH₂CH₂)₃—B ← CH₃CH=CH₂ — (CH₃—CH₂—CH₂)₂BH Dipropyl borane

$$(CH_3CH_2CH_2)_3-B \xleftarrow{CH_3CH=CH_2} (CH_3-CH_2-CH_2)_2BH$$

Tripropyl borane   Dipropyl borane

$$[O] \downarrow H_2O_2/^-OH$$

$$CH_3CH_2CH_2OH + B(OH)_4^-$$

n-Propyl alcohol

The hydroboration is considerably enhanced by low intensity ultrasound[16], especially in heterogeneous system. The reaction takes 1 hr. compared to 24 hrs. at 25° under the normal procedure.

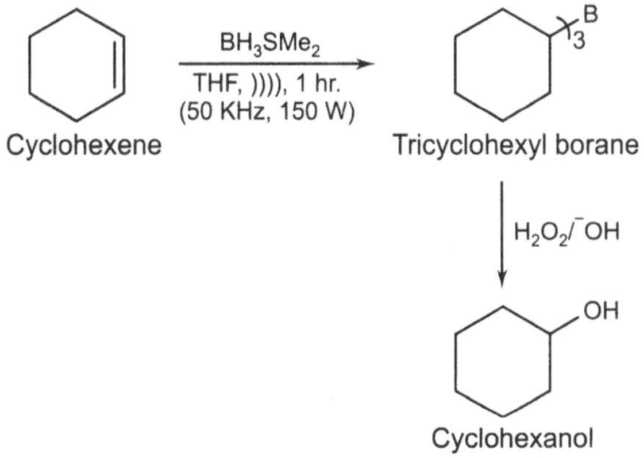

Cyclohexene $\xrightarrow[\substack{THF,)))), 1\ hr.\\(50\ KHz,\ 150\ W)}]{BH_3SMe_2}$ Tricyclohexyl borane

$$\downarrow H_2O_2/^-OH$$

Cyclohexanol

(*ix*) **Hydrosilation of Alkenes:** Hydrosilation of alkenes is achieved in presence of platinum-carbon catalyst in an ultrasonic cleaning bath at much lower temperature (30°) than is used without sonication.

$$\overset{}{=}\!\!< \ + \ R_3SiH \ \xrightarrow[\text{)))), 30°}]{\text{Pt-C, 1–2 hr.}} \ \overset{R_3Si \quad H}{\underset{}{\bigwedge\!\!\bigwedge}}$$

5.2.3 Alkynes

The alkynes contain $C \equiv C$ as the functional groups. However, the terminal alkynes contain besides $C \equiv C$, a $C - H$ bond also. Some typical reactions of $C \equiv C$ and $C - H$ are given below:

> **(i) Reduction of Alkynes:** In presence of Pt/H_2, the $C \equiv C$ is reduced to give alkane.

$$CH_3C \equiv CCH_3 \xrightarrow[\text{Pt}]{H_2} [CH_3CH = CHCH_3] \xrightarrow[\text{Pt}]{H_2} CH_3CH_2CH_2CH_3$$

2-Butyne   Butane

Nickel boride (also called P-2 catalyst, prepared by the reduction of nickel acetate with $NaBH_4$) causes *syn* addition of hydrogen to take place; the alkene that is formed with an internal triple bond has (Z) or *cis* configuration,

$$Ni\!\left(\overset{\overset{O}{\|}}{OCCH_3}\right)_{\!2} \xrightarrow[\text{C}_2\text{H}_5\text{OH}]{\text{NaBH}_4} Ni_2B$$

Nickel acetate   Nickel boride
(P-2)

$$CH_3CH_2C \equiv CCH_2CH_3 \xrightarrow[\text{Syn. addn.}]{H_2/Ni_2B(P\text{-}2)} \overset{CH_3CH_2}{\underset{H}{}}\!\!\diagup\!\!\overset{}{C} = C\!\!\diagdown\!\!\overset{CH_2CH_3}{\underset{H}{}}$$

3-Hexyne   (Z)-3-Hexene
(*cis*-3-Hexene) 97%

cis-Alkenes can also be obtained by reduction of alkynes with Lindlar's catalyst (Pd/CaCO$_3$; quinoline).

$$R - C \equiv C - R \xrightarrow[\substack{\text{Quinoline}\\ \text{(syn. addn.)}}]{\substack{\text{H}_2,\ \text{Pd-CaCO}_3\\ \text{(Lindlar's catalyst)}}} \overset{R}{\underset{H}{}}\!\!\diagdown\!\!\overset{}{C} = C\!\!\diagup\!\!\overset{R}{\underset{H}{}}$$

Alkyne   *cis*-Alkene

trans-alkenes are obtained by the reduction of alkynes with lithium or sodium metal in ammonia or ethyl amine at low temperature.

$$CH_3(CH_2)_2C \equiv C - (CH_2)_2CH_3 \xrightarrow[\text{2) NH}_4\text{Cl}]{\text{1) Li, C}_2\text{H}_5\text{OH, }-78°} \overset{CH_3(CH_2)_2}{\underset{H}{}}\!\!\diagdown\!\!\overset{}{C} = C\!\!\diagup\!\!\overset{H}{\underset{(CH_2)_2CH_3}{}}$$

4-Octyne   (E)-4-Octene
(*trans*-4-Octene)
52%

(*ii*) **Oxidative Cleavage of Alkynes:** Alkynes on treating with ozone or with basic potassium permanganate leads to cleavage at $C \equiv C$ to give carboxylic acid.

$$R - C \equiv C - R' \xrightarrow[\text{2) HOAc}]{\text{1) O}_3} RCO_2H + R'CO_2H$$

$$R - C \equiv C - R' \xrightarrow[\text{2) H}^+]{\text{1) KMnO}_4,\ \text{OH}^-} RCO_2H + R'CO_2H$$

For cleavage with alkaline $KMnO_4$, it is better to carry out the reaction in presence of a phase transfer catalyst.

(*iii*) **Addition of Water to Alkynes:** The addition of water to alkynes gives aldehydes or ketones depending on the structure of the alkyne.

Thus, the reaction of ethyne with dilute H_2SO_4 in presence of mercuric sulphate gives ethanal.

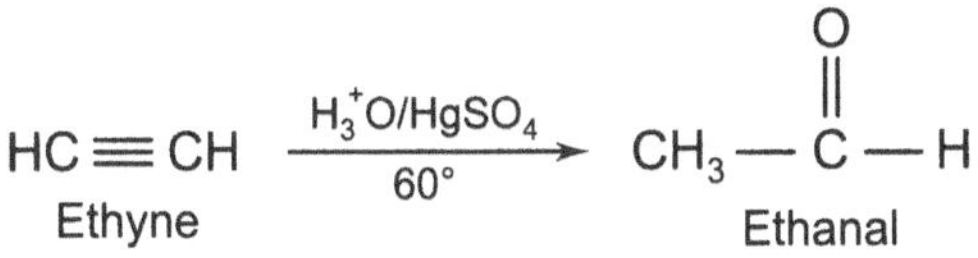

$$HC \equiv CH \xrightarrow[60°]{H_3^+O/HgSO_4} CH_3 - \overset{\overset{\displaystyle O}{\|}}{C} - H$$

Ethyne   Ethanal

In case of terminal alkynes, ketones are obtained.

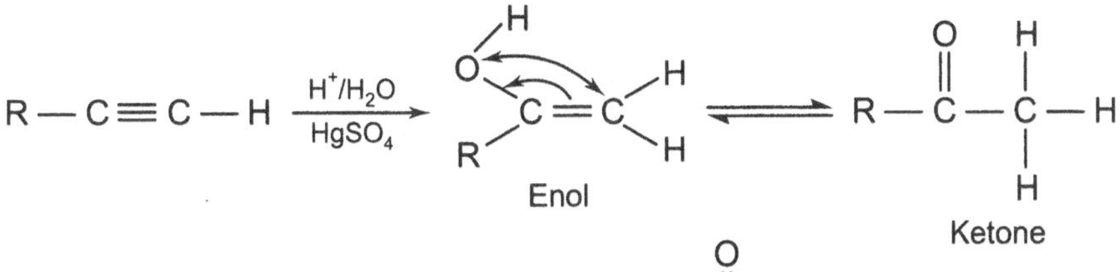

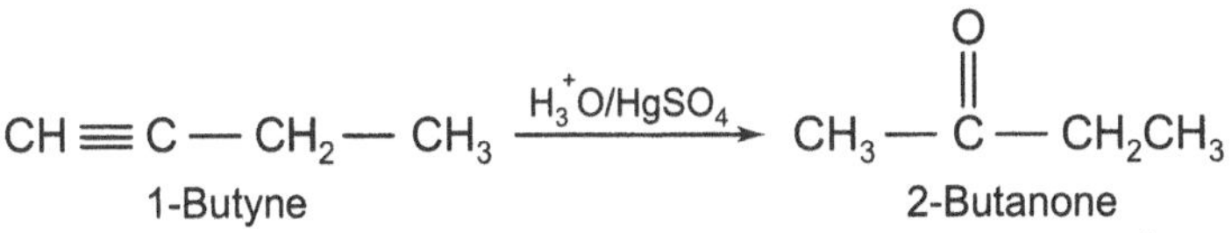

$$R - C \equiv C - H \xrightarrow[\text{HgSO}_4]{H^+/H_2O} \text{Enol} \rightleftharpoons R - \overset{\overset{\displaystyle O}{\|}}{C} - \overset{\overset{\displaystyle H}{|}}{\underset{\underset{\displaystyle H}{|}}{C}} - H$$

Enol   Ketone

$$CH \equiv C - CH_2 - CH_3 \xrightarrow{H_3^+O/HgSO_4} CH_3 - \overset{\overset{\displaystyle O}{\|}}{C} - CH_2CH_3$$

1-Butyne   2-Butanone

Internal alkynes give a mixture of two ketones

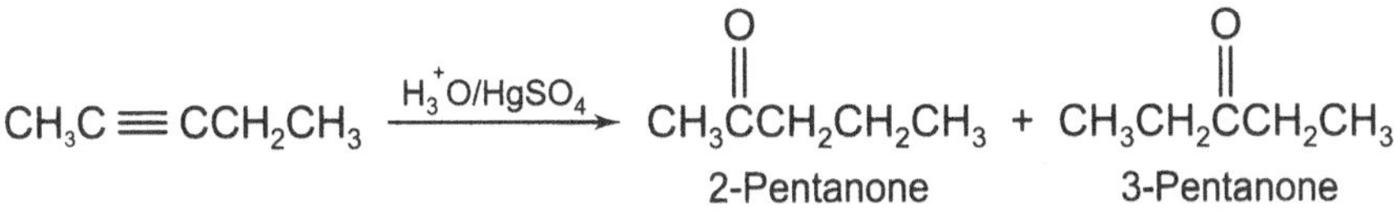

$$CH_3C \equiv CCH_2CH_3 \xrightarrow{H_3^+O/HgSO_4} CH_3\overset{\overset{\displaystyle O}{\|}}{C}CH_2CH_2CH_3 + CH_3CH_2\overset{\overset{\displaystyle O}{\|}}{C}CH_2CH_3$$

2-Pentanone   3-Pentanone

The above reactions can best be performed by sonication.

In all the above cases, the addition of water follows Markovnikov's addition rule. However, anti-markovnikov addition of water to alkynes can be affected by hydroboration.

$$CH_3CH_2CH_2C \equiv CH \xrightarrow[\text{THF}]{\text{BH}_3} \left[\overset{\overset{\displaystyle H}{|}}{\underset{}{CH_3CH_2CH_2C}} = \overset{\overset{\displaystyle \dot{B}-}{|}}{CH} \right]$$

1-Pentyne

$\downarrow$ H_2O_2/NaOH

$$CH_3CH_2CH_2CH_2CHO \rightleftharpoons CH_3CH_2CH_2\overset{\overset{\displaystyle H}{|}}{C} = \overset{\overset{\displaystyle OH}{|}}{CH}$$

Pentanal   Enol

$$CH_3CH_2C \equiv CCH_2CH_3 \xrightarrow[\text{2) } ^-OH/H_2O_2]{\text{1) } B_2H_6} CH_3 - CH_2 - \overset{\overset{\displaystyle O}{\|}}{C} - CH_2 - CH_2 - CH_3$$

3-Hexyne   3-Hexanone

$$CH_3C \equiv CCH_2CH_2CH_3 \xrightarrow[\text{2) } H_2O_2/\ ^-OH]{\text{1) } B_2H_6} CH_3 - \overset{\overset{\displaystyle O}{\|}}{C}CH_2CH_2CH_2CH_3$$

2-Hexyne   2-Hexanone

+

$$CH_3CH_2\overset{\overset{\displaystyle O}{\|}}{C}CH_2CH_2CH_3$$

3-Hexanone

The hydroboration step is best performed by sonication.

(*iv*) **Alkylation of Terminal Alkynes:** Terminal alkynes have a C — H bond; the hydrogen is called acetylenic hydrogen. This acidic hydrogen can be replaced by an alkyl group. This type of reaction called an alkylation is of considerable use in organic synthesis.

The acetylenic hydrogen is weakly acidic and can be abstracted with a strong base. Following are given the various steps involved in the synthesis of substituted alkynes.

$$R - C \equiv C - H \xrightarrow[(-NH_3)]{\text{NaNH}_2} R - C \equiv C^- Na^+ \xrightarrow[-\text{NaX}]{R'-X} R - C \equiv C - R'$$

Alkyne   Alkynide anion

R′ must be methyl or 1° and unbranched at the second carbon.

This is a convenient procedure for the formation of a carbon-carbon bond.

5.2.4 Aromatic Compounds

The aromatic compunds are basically derived from benzene. The functional group in these is the aromatic ring. The aromatic ring contains a six membered cyclic ring. All carbon atoms in an aromatic ring are sp^2 hybridized. This means that each carbon atom can form three sigma and one *pi* bond.

5.2.4.1 Electrophilic Substitutions

The aromatic ring is very stable and is not affected by acids or alkalies. However, it undergoes electrophilic substitution reactions like nitration, sulfonation, halogenation, alkylation and acylation.

In the last two reactions, *viz.* Friedel-Crafts alkylation and Friedel-Crafts acylation, it is better to use polymer supported aluminium chloride[19]. The recovered polymer supported aluminium chloride can be reused; this avoids the disposal problems.

The alkyl substituted benzene can be oxidised with $KMnO_4$ in presence of a phase transfer catalyst[20] or crown ether give benzoic acid in excellent yield.

CH₃

Toluene

alk. $KMnO_4$
PTC(Cetyltrimethyl ammonium chloride)
or [18] crown 6

COOH

Benzoic acid
90%

Biochemical oxidation of side chain proceeds without degradation or with only limited degradation. Thus, ethyl benzene, butyl benzene and dodecyl benzene give phenyl acetic acid[20a] on incubation with Nocardia strain 107–332 at 30°.

$C_6H_5CH_2CH_3$
Ethyl benzene

$C_6H_5CH_2CH_2CH_2CH_3$
Butyl benzene

$C_6H_5CH_2CH_2C_{10}H_{21}$
Dodecyl benzene

Oxidn.
Nocardia strain 107–332
30°

$C_6H_5CH_2COOH$
Phenyl acetic
acid

It is found that pentyl benzene (amyl benzene) on oxidation with *Cellulomonas galba* gives 5-phenylvaleric acid, which is further converted into trans-cinnamic acid[20b].

$C_6H_5(CH_2)_4CH_3$ $\xrightarrow[\text{C. galba}]{\text{Oxidn.}}$ $C_6H_5(CH_2)_4CO_2H$ $\longrightarrow$ $C_6H_5CH=CH-COOH$

Amyl benzene   0–12%   *trans*-Cinnamic acid
                        5-Phenylvaleric acid   88–100%

Enzymatic oxidation[20c] of benzene by micrococcus spheroids like organisms gives *trans, trans*-muconic acid.

Benzene

Micrococcus spheroids
like organism

COOH

COOH

trans, trans-Muconic
acid

5.2.4.2 Oxidation

Benzene is very stable under normal conditions and is unaffected by usual oxidising agents like KMnO$_4$, CrO$_3$ *etc*. However, under drastic conditions and in presence of a suitable catalyst, benzene ring is ruptured.

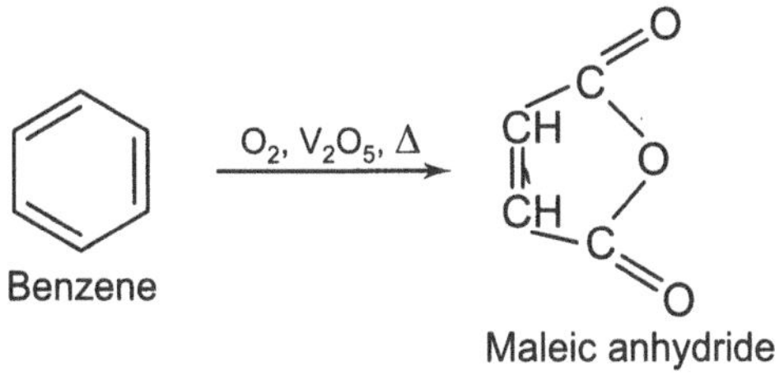

Benzene $\xrightarrow{O_2, V_2O_5, \Delta}$ Maleic anhydride

5.2.4.3 Hydroxylation of Aromatic Rings

Considerable work has been carried in the hydroxylation of aromatic rings. Thus, hydroxylation of benzene with *Pseudomonas putida*[21] in presence of oxygen gives *cis*-diol.

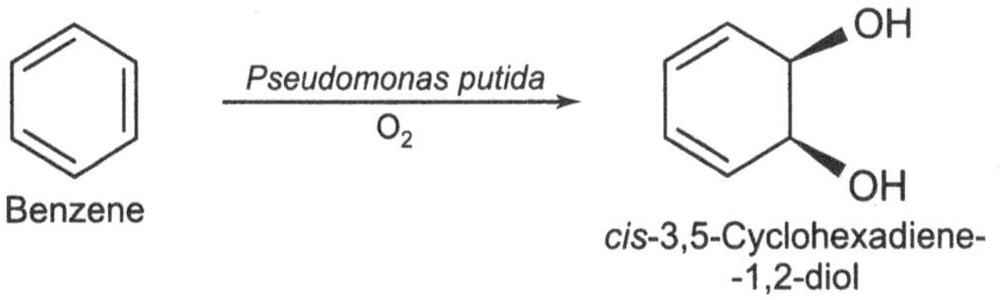

Benzene $\xrightarrow[O_2]{Pseudomonas\ putida}$ *cis*-3,5-Cyclohexadiene-
-1,2-diol

Benzene is converted into *trans, trans*-muconic acid by *Micrococus spheroids* like organisms[22].

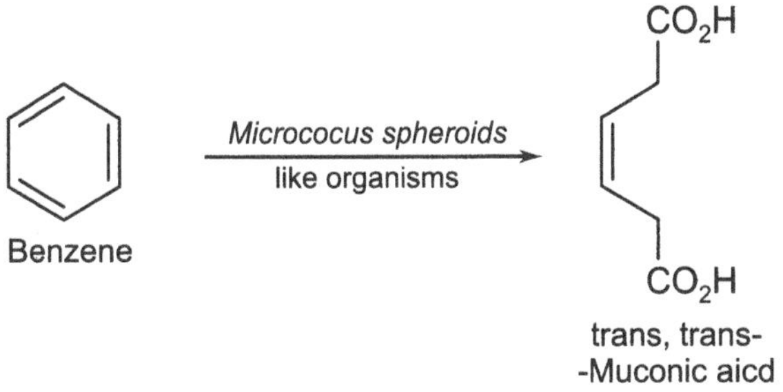

Benzene $\xrightarrow[\text{like organisms}]{Micrococus\ spheroids}$ trans, trans-
-Muconic aicd

In a similar way, toluene, halogen substituted benzenes and halogen substituted toluene gave the corresponding *cis*-diols[23].

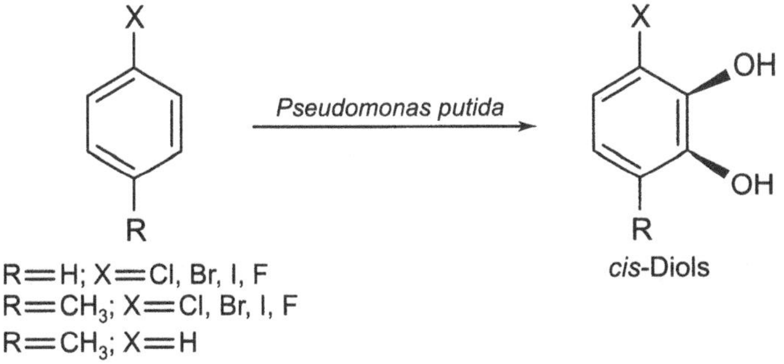

$\xrightarrow{Pseudomonas\ putida}$

R=H; X=Cl, Br, I, F
R=CH$_3$; X=Cl, Br, I, F
R=CH$_3$; X=H

cis-Diols

5.2.5 Alkyl Halides

In alkyl halides, the halogen is attached to an sp^3 hybridised carbon. However, in aryl halides, it is attached sp^2 hybridised carbon atom.

5.2.5.1 Conversion of C — Cl Group into Other Functional Groups *via* S$_N$2 Reactions

Alkyl halides, R — X (where R is methyl, primary or secondary alkyl halides) can be transformed into alcohols, ethers, nitriles, thioethers, quaternary ammonium salts, azides, esters, alkynes, thio alcohols, sulphonate and nitrites or nitro compounds.

Most of the above transformation can be performed under green conditions. Some of these are given as follows:

(*i*) Conversion of Alkyl Halides into Alcohols:

(*a*) The hydrolysis of benzyl chloride with water in a microwave oven gives benzyl alcohol[24] in 2 min. in 90 per cent yield.

$$C_6H_5CH_2Cl + H_2O \xrightarrow[\text{2 min.}]{\text{MW}} C_6H_5CH_2OH$$

Benzyl alcohol   Benzyl alcohol (97%)

(*b*) Hydrolysis of alkyl halides in near critical water (NCW) region in high temperature water. For near critical water (NCW) region especially designed microwave oven is used in which temperature is between 200–300° with higher pressure limits (80–100 bars)[25].

$$R-X \xrightarrow{\text{HTW}} ROH + HX$$

Alkyl halide

(*c*) The reaction of tert. butyl chloride with water in polyethylene glycol (PEG-300) gives tert. butyl alcohol[26].

$$(CH_3)_3CCl + H_2O \xrightarrow{\text{PEG-300}} (CH_3)_3COH$$

(*d*) Alkyl halides (*e.g.*, 1-bromooctane) is converted into the corresponding alcohol in 90–100 per cent yield by using $(n\text{-}C_4H_9)_4$ N^+OH^-, $C_6H_{13}(C_2H_5)_3$ N^+OH^- or $(C_8H_{17})_3$ C_2H_5 N^+OH^- as PTC[27]. The reaction is conducted at 80° for 48 hr.

$$C_8H_{17}Br \xrightarrow[\text{48 hr., 80°}]{\text{PTC}} C_8H_{17}OH$$

1-Bromo octane   1-Octanol

It has been found that the displacement of halogen by hydroxyl group is best accomplished[28] by the use of betanine, quaternary salts, $R_3N^+CH_2CO_2^-$.

(*ii*) Conversion of Alkyl Halides into Ethers:
Alkyl halides on reaction with aqueous alcohol using ultrasound[29] gives the corresponding ether.

$$(CH_3)_3CCl \xrightarrow[\text{))))}]{\text{ROH/H}_2\text{O}} (CH_3)_3C^+Cl^- \longrightarrow (CH_3)_3C-OR$$

Ether

Aryl ethers are obtained[30] by the reaction of phenols in conc. aqueous NaOH in presence of a PTC, *viz.*, $C_6H_5CH_2(C_4H_9)_3$ N^+Cl^- and an alkylating agent. The reaction is carried out in methylene chloride.

$$ArOH + RX \xrightarrow[\substack{NaOH \\ CH_2Cl_2}]{PTC} ArOR$$

The PTC method is also used for the alkylation of chelated hydroxyl group, *e.g.*, salicylaldehyde.

The PTC technique provides a simple and convenient method for conducting **Williamsons ether synthesis**[29a]. Using this procedure, larger alcohol (*e.g.*, $C_8H_{17}OH$) react with alkyl halide to give higher yield of ethers.

$$C_8H_{17}OH + C_4H_9Cl \xrightarrow[\text{NaOH Soln.}]{PTC} \underset{\text{Ether}}{C_8H_{15}OC_4H_9} + \underset{\text{Byproduct}}{C_8H_{17}OC_8H_{17}}$$

PTC is $(n\ C_4H_9)_4\ N^+HSO_4^-$.

In the above reaction, polyethylene glycol (PEG-2000) can also be used as PTC[29b].

(*iii*) **Conversion of Alkyl Halides into Nitriles:** Alkyl halides on reaction with sodium cyanide in presence of a PTC (hexadecyl tributyl phosphorium bromide, $C_{16}H_{33}P^+(C_4H_9)_3\ Br^-$ give the corresponding nitriles[31,32].

$$\underset{\text{1-Chlorooctane}}{CH_3(CH_2)_6CH_2Cl} + NaCN \xrightarrow[\text{stirring, }105°,\ 2\ hr.]{C_{16}H_{33}P^+\ (C_4H_9)_3Br^-} \underset{\substack{\text{1-Cyanooctane} \\ 97\%}}{CH_3(CH_2)_6CH_2CN}$$

$$\underset{\substack{\text{2-Phenyl ethyl} \\ \text{bromide}}}{C_6H_5CH_2CH_2Br} + \underset{aq.}{NaCN} \xrightarrow[90-95°,\ 3\ hr.]{[(C_2H_5)_3NCH_2C_6H_5]^+Cl^-} \underset{\substack{\text{3-Phenyl propionitrile} \\ 91\%}}{C_6H_5CH_2CH_2CN}$$

The PTC method is also useful to obtain benzoyl cyanides[33] from benzoyl chloride.

$$\underset{\substack{\text{Benzoyl} \\ \text{chloride}}}{\overset{\overset{\displaystyle O}{\|}}{C_6H_5C}-Cl} + \underset{aq.}{NaCN} \xrightarrow[\substack{MeCN,\ 50° \\ )))}]{Bu_4N^+X^-} \underset{\substack{\text{Benzoyl} \\ \text{cyanide} \\ 70-85\%}}{\overset{\overset{\displaystyle O}{\|}}{C_6H_5CCN}} + NaCl$$

Halides can be converted into cyanides by reaction with KCN. The reaction is conducted in toluene and catalysed by alumina on sonication[34].

$$\text{C}_6\text{H}_5\text{CH}_2\text{Br} + \text{KCN} \xrightarrow[\substack{\text{Toluene} \\ \text{))))}}]{\text{Al}_2\text{O}_3} \text{C}_6\text{H}_5\text{CH}_2\text{CN}$$

Benzyl   Benzyl cyanide
bromide   76%

Alkyl halides of the type RCH_2Br can also be converted into the corresponding cyanide by reaction with CN^- using PEG-400 as solvent[34a].

$$\text{RCH}_2\text{Br} + \text{CN}^- \xrightarrow{\text{PEG-400}} \text{RCH}_2\text{CN}$$

$$\text{R} = \text{C}_6\text{H}_5, \text{C}_3\text{H}_7 \text{ etc.}$$

(*iv*) **Conversion of Alkyl Halides into Thioethers:** Alkyl halides are converted into thioethers (dialkyl sulphides) by reacting with sodium sulphide using PTC technique[35].

$$\text{RX} + \text{Na}_2\text{S} \xrightarrow[70°, 40 \text{ hr.}]{\text{C}_{16}\text{H}_{35}\text{P}^+\text{Bu}_3\text{Br}^-} \text{R}-\text{S}-\text{R}$$

Alkyl   *aq.*
halide

(*v*) **Conversion of Alkyl Halides into Esters:** The reaction of alkyl halides with carboxylic acid sodium salt in presence of triethylamine gives esters[36,37]. In this case, the PTC is obtained *in situ* by the reaction of triethylamine and alkyl halide.

$$\text{Et}_3\text{N} + \text{RX} \longrightarrow \text{Et}_3\text{N}^+\text{RX}^-$$

Triethyl   Alkyl   PTC
amine   halide   (generated *in situ*)

$$\text{R}'\text{CO}_2\text{Na} + \text{RX} \xrightarrow{\text{Et}_3\text{N}} \text{R}'\text{CO}_2\text{R} + \text{NaX}$$

Carboxylic   Alkyl   Ester
acid sod. salt   halide
(aq. soln.)

In the above esterification, the alkyl halides must be reactive, *e.g.*, benzyl chloride, which rapidly forms the quaternary salts. Alternatively, quaternary phosphonium or ammonium salts can be directly used[38] in the esterification of carboxylic acids with alkyl halides. In place of PTC, crown ethers can also be used[39].

$$R = H,\ CH_3,\ CH_3CH_2,\ CH_3CH_2CH_2,\ C_6H_5,\ 2-CH_3C_6H_4-,$$
2, 4, 6-trimethyl benzoyl

The penacyl esters of fatty acids are obtained in quantitative yield using crown ethers as catalysts[40,41].

Low intensity ultrasound (cleaning bath) has also been used for the esterification of carboxylic acids with alkyl halides[42].

$$RCO_2H\ +\ R'X\ \xrightarrow[))))]{KOH,\ PEG}\ RCO_2R'$$

$$X = Cl,\ Br,\ I$$

(*vi*) **Conversion of Alkyl Halides into Alkynes:** For details, see section 5.2.3.

(*vii*) **Conversion of Alkyl Halides into Thioalcohols:** Alkyl halides are converted into thioalcohols via the formation of grignard reagent[43,44] which is prepared by the reaction of alkyl halide with magnesium in dry ether under sonication using laboratory cleaning bath. Under these conditions, magnesium is activated and the reaction goes to completion in a short time.

$$R-X\ +\ Mg\ \xrightarrow[))))]{ether}\ RMgX$$
$$(90\%)$$

$$RMgX\ +\ S\ \longrightarrow\ Mg\underset{X}{\overset{SR}{\big<}}\ \xrightarrow{H_3O^+}\ RSH\ +\ Mg(OH)X$$

(*viii*) **Conversion of Alkyl Halides into Alkyl Sulphonates:** Alkyl halides on treatment with sodium sulphite in presence of a PTC [$(C_2H_5)_4$ N^+Cl^-] give 98% yield of the corresponding sodium alkyl sulphonates[45].

$$RCl\ +\ Na_2SO_3\ \xrightarrow[100°,\ 2\ hr.]{(C_2H_5)_4N^+Cl^-}\ RSO_3Na$$
(aqueous)   Sodium alkyl
                                        sulphonate

R = alkyl or aryl

(*ix*) **Conversion of Alkyl Halids into Alkyl Nitrites and Nitroalkanes:** Alkyl halides can be converted into alkylnitrites and nitroalkanes by reaction with silver nitrate or sodiumnitrite in presence of a PTC[46].

$$RBr + NaNO_2 \xrightarrow{R_4N^+X^-} RNO_2 + NaBr$$

aq.   Nitroalkane

$$\xrightarrow{+ AgNO_3(aq.),\ R_4N^+X^-} R-O-N=O + RNO_2$$

Alkyl nitrate   Nitroalkane
                      minor

The above conversions can also be achieved by the use of crown ether.

(*x*) **Conversion of Alkyl Halides into Thiocyanates, Cyanates and *p*-Toluenesulfonates:** Alkyl bromides on reaction with sodium thionate (NaSCN), sodium cyanate (NaCNO) and sodium *p*-toluene sulphonate ($NaO_2SC_6H_4CH_3(p)$) in presence of a PTC (Quaternary ammonium nitrate) give the corresponding thiocyanate[32,47], cyanate[48] and *p*-toluenesulphonates[49], respectively.

$$RBr + NaX \xrightarrow{PTC} RX + NaBr$$

Alkyl
bromide

$$X = SCN,\ CNO,\ O_2SC_6H_4 - CH_3(p)$$

Alkyl thiocyanates can also be obtained from alkyl bromides on sonication with KSCN in presence of a quaternary ammonium salt.

$$\text{Br} \xrightarrow[))))]{\text{KCNS/H}_2\text{O/Bu}_4\text{N}^+\text{Br}^-,\ \text{RT, 6 hr.}} \text{SCN}$$

62%

A rapid and efficient synthesis of thiocyanates is promoted by microwaves[50].

$$C_6H_5CH_2Br \xrightarrow[MW]{KCNS/H_2O} C_6H_5CH_2SCN$$

Benzyl   Benzyl
bromide   thiocyanate

(*xi*) **Conversion of Alkyl Halides into Azides:** Alkyl halides can be converted into the corresponding azides by reaction with sodium azide in presence of a PTC[51,52].

$$R — Cl + NaN_3 \xrightarrow[\substack{DMSO,\ stirring \\ 1.5\ hr.,\ 40–50°}]{PTC} R — N_3$$

Alkyl   Alkyl azide
chloride   75%

Under mild conditions, the chloroaldehydes in a heterocyclic moiety react with sodium azide in DMSO and given high yields of the corresponding azides[53].

Alkyl, allylic and propargylic halides on sonication at room temperature with sodium azide give the corresponding azides[54].

$$R — X \xrightarrow[\substack{))))}]{NaN_3/H_2O,\ 60°} RN_3$$

R = Propargyl or allyl, yield 60–90%
R = Alkyl, yield 20%
R = Allyl, yield 86%

Alkyl halides can also be converted into the corresponding azides by reaction in microwave oven[50].

$$C_6H_5CH_2Br \xrightarrow[MW]{NaN_3/H_2O} C_6H_5CH_2N_3$$

Benzyl bromide   Benzyl azide

(*xii*) **Conversion of Alkyl Chlorides or Bromides into Alkyl Fluorides or Iodides:** Alkyl fluorides can be obtained from alkyl chloride or bromide by the displacement of chloride or bromide with fluorine in presence of PTC[55].

$$R = X + KF \xrightarrow{PTC} R — F + KX$$

X = Cl, Br   70–80%
R = CH_3(CH_2)_7

PTC is hexadecyl tributyl phosphonium bromide, $C_{16}H_{33}P^+(C_4H_9)_3\ Br^-$

In place of PTC, the reaction can also be carried out by using quaternary ammonium salts supported on an insoluble resin, such as Amberlyst A-26[56].

$$C_6H_5CH_2Cl + [resin^+]F^- \xrightarrow{24\ hr.} C_6H_5CH_2F$$

95%

The reaction of alkyl halides of the type RCH_2Br on reaction with I^- in PEG-400 give the corresponding iodide[34a].

$$RCH_2Br + I^- \xrightarrow{\text{PEG-400}} RCH_2I$$
$$R = C_6H_5, C_3H_7 \text{ etc.}$$

(*xiii*) **Conversion of Primary Alkyl Halides into Aldehydes:** Primary alkalyl halides on reaction with DMSO and $NaHCO_3$ give the corresponding aldehydes.

$$RCH_2X + \underset{\text{DMSO}}{(CH_3)_2S{=}O} + NaHCO_3 \longrightarrow RCHO + (CH_3)_2S + H_2O + NaX + CO_2$$

In this case, DMSO acts as a reagent and also as a solvent.

5.2.5.2 Conversion of C — Cl Group into Other Functional Groups via the Formation of Grignard Reagent

The alkyl and aryl halides can be transformed into a number of functional groups *via* the formation of grignard reagent. As already stated, (see section 2.1.29.1) the grignard reagents are prepared by the reaction of alkyl halide with magnesium in dry ether using sonication[43,44] (a laboratory cleaning bath can be used). The alkyl or aryl magnesium halide (the grignard reagent) can be be converted into alcohols (1°, 2° or 3°), carbonyl compounds (aldehydes) carboxylic acids, ethers, 1° amines, alkyl cyanides, hydrocarbons (alkanes alkenes, alkynes and cycloalkanes), deuterated hydrocarbons, thioalcohols, dithionic acid, sulfinic acid, alkyl iodides and other organometallic compounds.

Secondary alcohols can also be obtained by the reaction of grignard reagent with cyclic ketones like cyclohexanone or epoxides.

Tertiary alcohols can also be obtained by the reaction of girgnard reagent with acid chlorides, esters and acid anhydrides; in all these cases, ketones are obtained as intermediate products, which further react with grignard reagent to give tertiary alcohols.

$$RMgX + H-\underset{\substack{| \\ OC_2H_5 \\ \text{Ethyl} \\ \text{orthoformate}}}{\overset{OC_2H_5}{C}}-OC_2H_5 \longrightarrow \underset{\substack{\text{Acetal of} \\ \text{aldehyde}}}{RCH(OC_2H_5)_2} + Mg(OC_2H_5)I$$

$$\Big\downarrow H^+$$

$$\underset{\textbf{Aldehyde}}{RCHO + 2C_2H_5OH}$$

In this procedure, aldehyde is the main product.

$$RMgX + \underset{\text{Alkyl cyanide}}{R'C\equiv N} \longrightarrow RR'C=NMgX \xrightarrow{H_2O} \underset{\text{Ketimine}}{RR'C=NH} \xrightarrow{H^+} \underset{\text{Ketone}}{RCOR'}$$

$$RMgX + \underset{\substack{H_2N \\ \text{Amide}}}{\overset{R'}{>}}C=O \longrightarrow R-H + \left[R'-\underset{\substack{| \\ NHMgX \\ {}^{-}\ {}^{+}}}{\overset{\overset{O}{\|}}{C}} \right]$$

$$\Big\downarrow \underset{-\ +}{RMgX}$$

$$\underset{\substack{R' \\ \text{Ketone}}}{\overset{R}{>}}C=O \xleftarrow{H_3O^+} \underset{R'}{\overset{R}{\diagup}}\underset{\substack{| \\ NHMgX \\ {}^{-}\ {}^{+}}}{C}\underset{}{\diagdown}^{OMgX}$$

$$RMgX + \underset{\substack{\text{Dimethyl} \\ \text{formamide}}}{H\overset{\overset{O}{\|}}{C}NMe_2} \xrightarrow[\text{2) } H_3O^+]{\text{1) THF, 0--20°}} \underset{\textbf{Aldehyde}}{RCHO}$$

$$RMgX + \underset{\substack{\text{Dimethyl} \\ \text{acetamide}}}{CH_3\overset{\overset{O}{\|}}{C}NMe_2} \xrightarrow[\text{2) } H_3O^+]{\text{1) THF, 0--20°}} \underset{\textbf{Ketone}}{RCOCH_3}$$

$$C_6H_5COCH_2Cl + RMgBr \xrightarrow[\text{2) ice/HCl}]{\text{1) ether, } -30°} C_6H_5COCH_2R$$

α-Haloketones **Ketones**

$$RMgX + \underset{O}{\overset{O}{\underset{\|}{\overset{\|}{C}}}} \longrightarrow R - \underset{O}{\overset{O}{\underset{\|}{\overset{\|}{C}}}} - \bar{O}\overset{+}{M}gX \xrightarrow[H_2O]{HCl} RCOOH + Mg(X)Cl$$

Carbon dioxide

Carboxylic acid

$$R\overset{\frown}{CH_2}MgX + CH_3OCH_2\overset{\frown}{}Cl \longrightarrow RCH_2CH_2 - OCH_3$$

Methoxy methyl chloride

Ethers

$$RMgI + H_2N\overset{\frown}{}Cl \longrightarrow RNH_2 + Mg(Cl)I$$

Chloramine

1° Amines

$$RMgI + NC\overset{\frown}{}CN \longrightarrow RC\equiv N + Mg(CN)I$$

Cyanogen

Alkyl cyanides

$$RMgI \xrightarrow[\text{ammonia or amine}]{H_2O, \text{ alcohol}} RH$$

Hydrocarbons

$$RMgI + CH_3CH_2Br \xrightarrow{CoCl_2} RCH_2CH_3 + Mg(Br)I$$

Hydrocarbons

$$R - MgX + CH_2=CHCH_2Br \longrightarrow CH_2=CHCH_2R + Mg(Br)I$$

Unsaturated alkyl halides

Alkenes

$$R-MgX + CH_3C\equiv C-H \longrightarrow CH_3C\equiv CMgI \xrightarrow[CH_3I]{S_N2} CH_3C\equiv C-CH_3 + MgI_2$$

Propyne

Propynyl magnesiumiodide

Alkynes (2-Butyne)

Cyclobutyl bromide   →(Mg, Et₂O)→   Cyclobutyl magnesium bromide   →(H₂O)→   **Cycloalkane** (Cyclobutane)   + Mg(X)OH

$$RMgX \xrightarrow{D_2O} RD$$

**Deuterated
hydrocarbons**

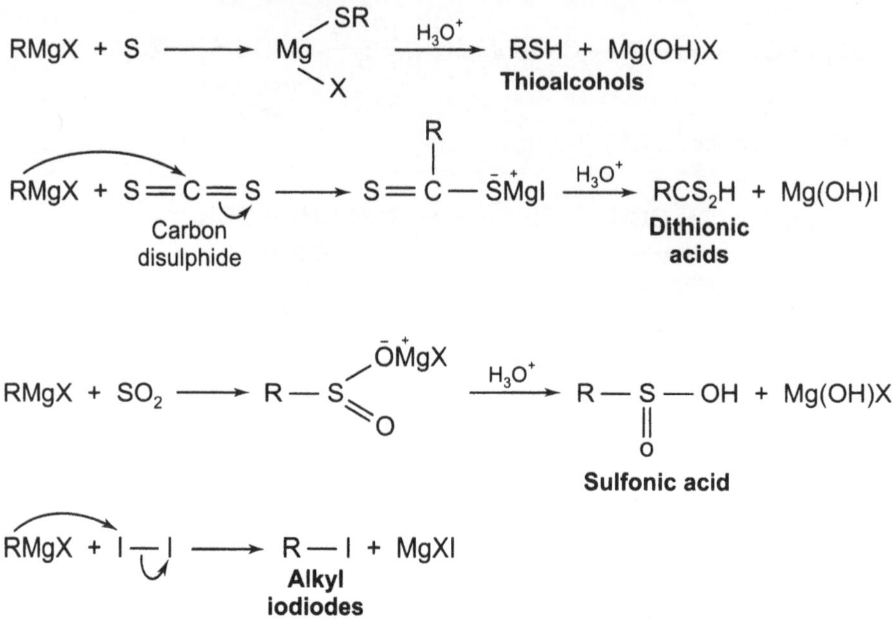

$$RMgX + S \longrightarrow Mg\begin{smallmatrix} SR \\ \\ X \end{smallmatrix} \xrightarrow{H_3O^+} RSH + Mg(OH)X$$

Thioalcohols

$$RMgX + S{=}C{=}S \longrightarrow S{=}\overset{\displaystyle R}{\underset{\displaystyle |}{C}}{-}\bar{S}\overset{+}{M}gI \xrightarrow{H_3O^+} RCS_2H + Mg(OH)I$$

Carbon
disulphide

**Dithionic
acids**

$$RMgX + SO_2 \longrightarrow R{-}\overset{\displaystyle \bar{O}\overset{+}{M}gX}{\underset{\displaystyle O}{S}} \xrightarrow{H_3O^+} R{-}\underset{\displaystyle \overset{\|}{O}}{S}{-}OH + Mg(OH)X$$

Sulfonic acid

$$RMgX + I{-}I \longrightarrow R{-}I + MgXI$$

**Alkyl
iodiodes**

$$(CH_3)_3CCH_2Cl \xrightarrow{Mg} (CH_3)_3CCH_2MgCl \xrightarrow{I_2} (CH_3)_3CCH_2I$$

Neopentyl
chloride

**Neopentyl
iodide**

$$4\ RMgBr + PbCl_4 \longrightarrow R_4Pb + 4\ Mg(Br)Cl$$

Tetra alkyl
lead

$$2\ RMgBr + HgCl_2 \longrightarrow R_2Hg + 2\ Mg(Cl)I$$

Dialkyl
mercury

$$4\ RMgBr + SiCl_4 \longrightarrow R_4Si + 4\ Mg(Cl)I$$

Tetra alkyl
silane

$$3\ RMgBr + PCl_4 \longrightarrow R_3P + 3\ Mg(Cl)Br$$

Trialkyl
phosphine

$$2\ RMgBr + CdCl_2 \longrightarrow R_2Cd + 2\ Mg(Cl)Br$$

Dialkyl
cadmium

5.2.5.3 Some Miscellaneous Transformations of Alkyl Halides

(*i*) **Synthesis of Aldehydes:** Alkyl or aryl halides on sonication with lithium, dimethyl formamide in THF gave 76–80% yield of the corresponding aldehydes[57].

$$\text{RX} \xrightarrow[\substack{\text{RT, 10–40 min.} \\ ))))}]{\text{Li/DMF/THF}} \left[\underset{\underset{\text{R}}{} \quad \underset{\text{NMe}_2}{}}{\overset{\overset{\text{LiO}}{} \quad \overset{\text{H}}{}}{\times}} \right] \longrightarrow \underset{\text{76–80\%}}{\text{RCHO}}$$

R=alkyl, aryl, benzyl

(*ii*) **Synthesis of Alkanes with Odd Number of Carbon Atoms:** This is affected by the well known **Corey-Posner-Whitesides-House Synthesis**[58]. In this procedure, an alkyl halide is converted into lithium dialkyl cuprate by sonication of alkyl halide with lithium and cuprous iodide in diethyl ether.

$$\underset{\substack{\text{Alkyl} \\ \text{halide}}}{\text{R} - \text{X}} + 2\,\text{Li} \xrightarrow[))))]{\text{diethyl ether}} \underset{\substack{\text{Alkyl} \\ \text{lithium}}}{\text{RLi}} + \text{LiX}$$

$$\underset{\substack{\text{Alkyl} \\ \text{lithium}}}{2\,\text{RLi}} + \underset{\substack{\text{Cuprous} \\ \text{iodide}}}{\text{CuI}} \longrightarrow \underset{\substack{\text{Lithium} \\ \text{dialkylcuprate}}}{\text{R}_2\text{CuLi}}$$

The formed lithium dialkylcuprate is reacted with the second alkyl halide (R′ – X). Coupling between one alkyl group of the lithium dialkyl cuprate and the alkyl group of the alkyl halide, R′X takes place.

$$\underset{\substack{\text{Lithium} \\ \text{dialkyl cuprate}}}{\text{R}_2\text{CuLi}} + \underset{\substack{\text{Alkyl} \\ \text{halide}}}{\text{R}' - \text{X}} \longrightarrow \underset{\text{Alkane}}{\text{R} - \text{R}'} + \text{RCu} + \text{LiX}$$

The last step gives a good yield of the alkane, if the alkyl halide, R′—X is a methyl bromide, primary alkyl halide or a secondary cycloalkyl halide. The alkyl groups of lithium dialkylcuprate, R_2CuLi, may be methyl, 1°, 2° or 3°. The two alkyl groups being coupled can be similar or different. Various coupling reactions are summarised as shown ahead.

$$R_2CuLi$$
Lithium dialkyl
cuprate

CH$_3$X
or
R'CH$_2$X

R — CH$_3$
or
R'CH$_2$ — R

(*iii*) **Biochemical Oxidations of Alkyl Halides:** The oxidation of primary alkyl halides to carboxylic acids is accoumplished on incubation of long chain aliphatic halides (fluorides, chlorides, bromides and iodids) with the yeast, Torulopsis gropengiesseri[57a]. In this case, the oxidation starts at the terminal methyl groups to give of ω halogenoalkanoic acid; these are finally converted into α, ω. dicarboxylic acid. In addition dicarboxylic acids with two less carbons are also obtained.

$$CH_3(CH_2)_6CH_2X \xrightarrow{\text{Torulopsis gropengiesseri}} HOOC(CH_2)_6CH_2COOH + HOOC(CH_2)_{11}CuH$$

X = F	9%	4.5%
Cl	18%	29%
Br	17%	29%
I	15%	30%

Note: Data taken from reference 57a.

(*iv*) **Conversion of Aryl Halides into Carboxylic Acids:** Aryl halides on reaction with lithium metal followed by reaction with carbon dioxide and subsequent acdification give the corresponding carboxylic acids, one example is given below:

p-Bromotoluene

1) Li metal, hexane
2) CO$_2$
3) H$_3$O$^+$

p-Toluic acid

Alternatively, the aryl halides are converted into the corresponding grignard reagent, which on treatment with CO$_2$ and subsequently acidification give the corresponding carboxylic acid.

(For details, see section 2.1.29.2)

5.2.6 Alcohols

In alcohols, R—OH, the hydroxyl group is attached to an sp^3 hybridised carbon. However, in case of phenols (in which the OH group is attached to an aromatic ring), the hydroxyl group is attached to an sp^2 hybridised carbon.

5.2.6.1 Conversion of Hydroxyl Group into Other Functional Group *via* S$_N$2 Reaction

Unlike chloro group in alkyl halides, the hydroxyl group is not a good leaving group and so alcohols do not ordinarily undergo S$_N$2 reaction. This difficulty is overcome by converting alcohols into the corresponding tosylates or mesylates. The alkyl sulfonates are frequently used as substrate for nucleophilic substitution reactions. This is because sulfonate ions are excellent leaving groups. The mesylates and tosylates of alcohols are prepared as given below:

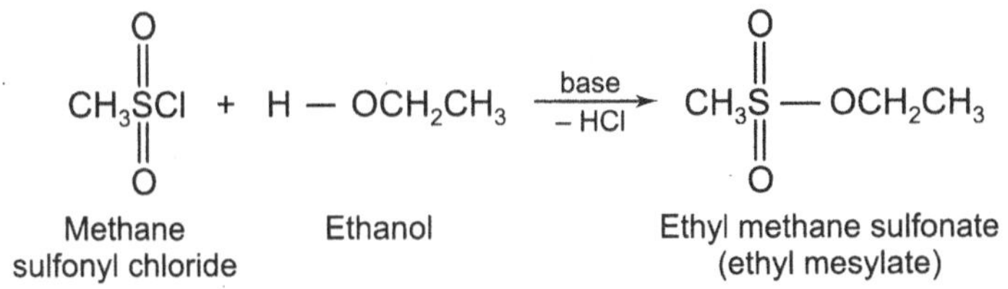

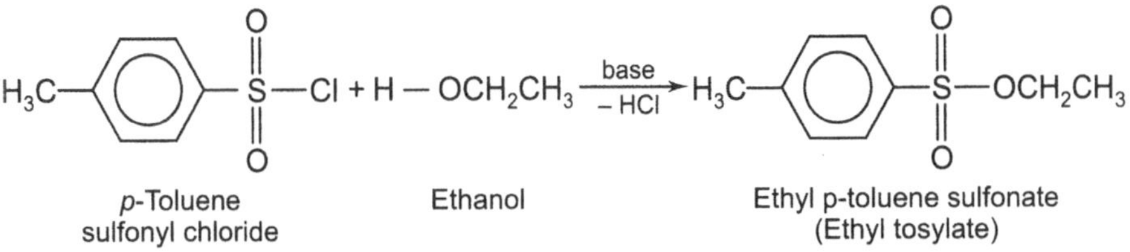

$$
\underset{\substack{\text{Methane} \\ \text{sulfonyl chloride}}}{CH_3\overset{\overset{O}{\|}}{\underset{\underset{O}{\|}}{S}}Cl} + \underset{\text{Ethanol}}{H-OCH_2CH_3} \xrightarrow[-HCl]{\text{base}} \underset{\substack{\text{Ethyl methane sulfonate} \\ \text{(ethyl mesylate)}}}{CH_3\overset{\overset{O}{\|}}{\underset{\underset{O}{\|}}{S}}-OCH_2CH_3}
$$

$$
\underset{\substack{\text{p-Toluene} \\ \text{sulfonyl chloride}}}{H_3C-\bigcirc\!\!\!-\overset{\overset{O}{\|}}{\underset{\underset{O}{\|}}{S}}-Cl} + \underset{\text{Ethanol}}{H-OCH_2CH_3} \xrightarrow[-HCl]{\text{base}} \underset{\substack{\text{Ethyl p-toluene sulfonate} \\ \text{(Ethyl tosylate)}}}{H_3C-\bigcirc\!\!\!-\overset{\overset{O}{\|}}{\underset{\underset{O}{\|}}{S}}-OCH_2CH_3}
$$

The nucleophilic substitution reaction takes place as shown below.

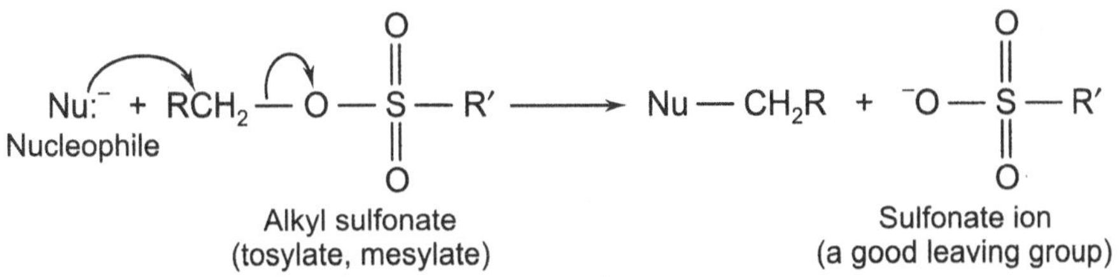

$$
\underset{\text{Nucleophile}}{Nu:^-} + RCH_2-O-\overset{\overset{O}{\|}}{\underset{\underset{O}{\|}}{S}}-R' \longrightarrow Nu-CH_2R + {}^-O-\overset{\overset{O}{\|}}{\underset{\underset{O}{\|}}{S}}-R'
$$

$$
\underset{\substack{\text{Alkyl sulfonate} \\ \text{(tosylate, mesylate)}}}{} \qquad\qquad \underset{\substack{\text{Sulfonate ion} \\ \text{(a good leaving group)}}}{}
$$

The alkyl sulfonates (tosylates etc.) undergo all the nucleophilic substitution reactions like alkyl halides. Thus, in this way the hydroxyl group can be replaced by alkoxide ($^-$OR') to form ether, by C≡N (to give nitriles), by $^-$SR' (to give thioethers), by NR_3' to give quaternary ammonium salt, by $-O-\overset{\overset{\displaystyle O}{\|}}{C}-R'$ to give esters by $C \equiv C - R'$ to give alkynes, by SH^- to give thiolcohols, with SO_3^2 to give sulphonates etc.

(For more details see section 5.2.5.1).

5.2.6.2 Conversion of Alcohols to Alkyl Halides (Cleavage of C — O Bond of Alcohols)

Alcohols can be converted to alkyl halides by reacting with hydrogen halide (HX), thionyl chloride ($SOCl_2$), Phosphorus trichloride (PCl_3) or phosphorus pentachloride (PCl_5).

$$R - CH_2OH \; + \; HX \longrightarrow RCH_2X \; + \; H_2O$$
$$\text{Alcohol} \qquad\qquad\qquad \text{Alkyl halide}$$
$$X = Cl, Br, I$$

In this case, the reactivity of HX is $3° > 2° > 1° > CH_3$. Halogen halide reactivity is of the order

$HI > HBr > HCl > HF$

In the above reaction, better yields are obtained in presence of $ZnCl_2$ (catalyst).

$$R - CH_2CH_2OH \; + \; Cl-\overset{\overset{\displaystyle O}{\|}}{S}-Cl \longrightarrow RCH_2CH_2Cl \; + \; SO_2 \; + \; HCl$$
$$\text{Thionyl} \qquad\qquad \text{Alkyl halide}$$
$$\text{chloride}$$

The above reaction is normally carried out in presence of pyridine to neutralise the formed HCl.

$$3\,RCH_2OH \; + \; PCl_3 \longrightarrow 3\,RCH_2Cl \; + \; H_3PO_3$$
$$RCH_2OH \; + \; PCl_5 \longrightarrow RCH_2Cl \; + \; POCl_3 \; + HCl$$

Phosphorus tribromide is commonly used for converting alcohols into alkyl bromides. PBr_3 is generally prepared *in situ* by reading phosphorus with liquid bromine.

$$2\,P \; + \; 3\,Br_2 \longrightarrow 2\,PBr_3$$

Tertiary alcohols can be converted with the corresponding chlorides by treatment with conc. HCl at 25°.

In all the above reactions, there is cleavage at the C — O bond of alcohol.

Alcohols on reaction with dichlorocarbene generated *in situ* by the reaction of chloroform with aqueous NaOH in presence is a PTC, $C_6H_5CH_2N^+Et_3Cl^-$) gave good yield of the chlorides[59].

$$ROH + CHCl_3 + NaOH \xrightarrow{C_6H_5CH_2N^+Et_3Cl^-} RCl + NaCl + H_2O$$

In case of steroidal alcohols, the OH is replaced with Cl with retension of configuration[60].

5.2.6.3 Dehydration of Alcohols to Alkenes

Alcohols on refluxing with a catalyst (conc. H_2SO_4) or by passing vapours of alcohol over heated Al_2O_3 give alkenes. The order of dehydration of various alcohols is $3° > 2° > 1°$.

$$CH_3CH_2OH \xrightarrow[180°]{Conc.\ H_2SO_4} CH_2{=}CH_2$$

1° Alcohol   Ethene

In case of 2-butanol, the dehydration follows **Saytzeff's rule** (the more highly substituted alkene is the major product).

$$\underset{\text{2-Butanol}}{CH_3{-}\overset{\displaystyle OH}{\underset{|}{CH}}{-}CH_2{-}CH_3} \xrightarrow[\Delta]{H_3PO_4} \underset{\substack{\text{2-Butene} \\ \text{(major)}}}{CH_3{-}CH{=}CH{-}CH_3} + \underset{\substack{\text{1-Butene} \\ \text{(minor)}}}{CH_2{=}CHCH_2CH_3}$$

5.2.6.4 Conversion of Alcohols into Ethers

Alcohols on heating with conc H_2SO_4 at 140° undergo intermolecular dehydration to give alcohols.

$$R{-}OH + HO{-}R \xrightarrow[(-H_2O)\ 140°]{H^+} R{-}O{-}R$$

In case the temperature is kept at 180°, alkene is the major product.

An important route for unsymmetrical ethers is a nucleophilic substitution reaction known as the **Williamsons Synthesis**. It consists in an S_N2 reaction of sodium alkoxide with an alkyl halide, alkyl sulfonate or alkyl sulfate.

$$R{-}\overset{..}{\underset{..}{O}}{:}Na^+ + R'{-}X \longrightarrow R{-}\overset{..}{\underset{..}{O}}{:}{-}R' + Na^+ + {:}X^-$$

Sodium   Alkyl halide,   Ether
(potassium)   alkyl sulfonate
alkoxide   or dialkyl sulfate

The Williamsons synthesis can be best conducted in presence of a PTC[29a]. (See also section 3.1.16.1).

5.2.6.5 Esterfication of Alcohols on Reaction with Carboxylic Acids Esters

(For details see carboxylic acid section 5.2.12)

5.2.6.6 Oxidation of Alcohols

Normally, primary alcohols on oxidation give aldehydes under controlled conditions. Under drastic conditions and by using strong oxidising agents, the formed aldehydes may be further oxidised to the corresponding carboxylic acid.

Secondary alcohols on oxidation give ketones. However, under drastic conditions, the formed ketones may be further oxidised into a mixture of carboxylic acids.

Following procedures are used for the oxidation of 1° alcohols to aldehydes.

(*i*) **Chromate Oxidation:** It has been found that quaternary bichromate salts in organic media are of low oxidizing power, but have high selectivity. A commercial resin, Amberlyst. A-26 can be converted into $HCrO_4^-$ form by stirring. The chloride form of the resin (35 g) into a solution of CrO_3 (15 g in water (100 mL). The resin so obtained (3.8 mol of CrO_3 per g of resin) is stable at room temperature for several weeks. This resin has been used for the oxidation of alcohol[61] in solvents like hexane, benzene, chloroform, THF and gives the corresponding aldehyde or ketones in high yield. In these oxidations, carboxylic acids are not obtained.

(*ii*) **Oxidation with Polymeric Thioanisolyl Resin:** The polystyrene methyl sulfide on reaction with chlorine in presence of triethylamine gives an S-chloro sulfonium chloride resin, which acts as a selective oxidation for alcohols[62].

Polystyrene
Methylsulphide

$HO(CH_2)_nOH$

$HO(CH_2)_{n-1}CHO$ (50.2%)

+

$OHC(CH_2)_{n-2}CHO$ (2.2%)

(*iii*) **Oxidation with Palladium Acetate and O$_2$ in Ionic Liquid as Solvent:** Benzyl alcohols could be oxidised selectively to benzaldehyde in dry ionic liquids[63] using Pd(OAc)$_2$ as a catalyst and O$_2$. Benzyl alcohols can also be oxidised in ionic liquid containing a cyclic hexalkylguanidium cation with sodium hypochlorite as an oxidant[64] at room temperature.

Hexalkyl guanidium
cation

(*iv*) **Oxidation of Benzylic Alcohols to Aldehydes with KMnO$_4$ in Ionic Liquids:** The oxidation of benzyl alcohols with KMnO$_4$ in 1-butyl-3-methylimidazolium tetrafluoroborate, [b$_{min}$] [BF$_4$] ionic liquid at room temperature gave the corresponding aldehydes in 90 per cent yield[65].

Benzyl alcohols

Benzaldehydes
90%

(*v*) **Oxidation with Manganese Dioxide:** Sonication of cinnamyl alcohol, geraniol or 1-phenylethanol in presence of MnO$_2$ gives the corresponding aldehydes. Best results are obtained by using a less volatile solvent like octane[65a].

$$C_6H_5CH = CHCH_2OH \xrightarrow[))))]{MnO_2/octane} C_6H_5CH = CHCHO$$

Cinnamyl alcohol   Cinnamaldehyde

Geraniol   Geranial

$$C_6H_5CH_2CH_2OH \xrightarrow[))))]{MnO_2/octane} C_6H_5CH_2CHO$$

1-Phenylethanol   Phenyl acetaldehyde (*vi*) **Biochemical Oxidation of Primary Alcohols into Carboxylic Acids:** A typical example is the biooxidation of the benzyl ether of Kojic acid into the benzyl ether of comenic acid by Arthrobacter ureafaciens in a phosphate buffer at pH7.2.

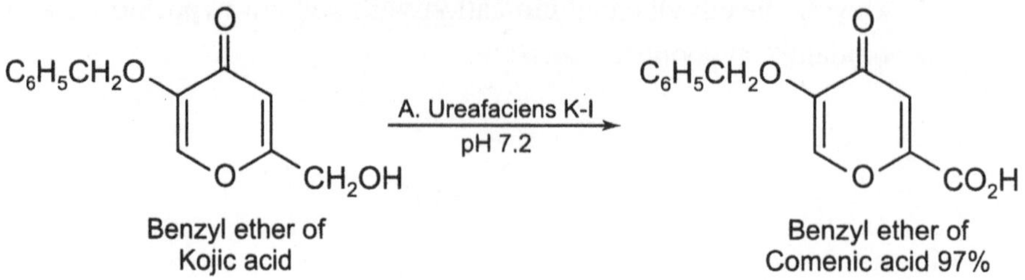

Benzyl ether of
Kojic acid

A. Ureafaciens K-I
pH 7.2

Benzyl ether of
Comenic acid 97%

Following procedures are used for the oxidation of secondary alcohols to ketones:

(*i*) **Oxidation with KMnO₄:** The oxidation of secondary alcohols to ketones is affected by $KMnO_4$ in hexane in presence of ultrasound[66] (in an ultrasonic bath).

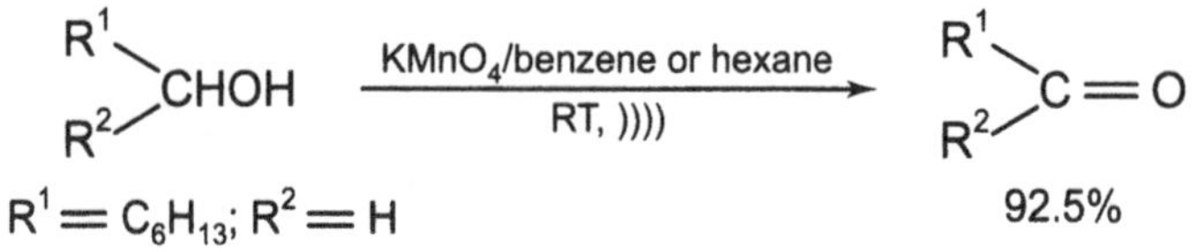

$R^1 = C_6H_{13}; R^2 = H$ 92.5%

KMnO₄/benzene or hexane
RT,))))

(*ii*) Oxidation in solid state using clayfen[67], MnO_2-silica[67], chromium trioxide supported on wet alumina[68] and iodobenzene diacetate (IBD) 'doped' Alumina[69] on Microwave irradiation.

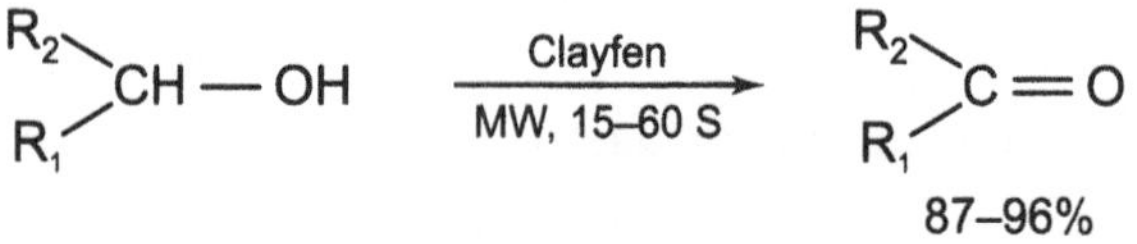

Clayfen
MW, 15–60 S

87–96%

Clayfen is montmorillonite K10 clay supported on iron (III) nitrate.

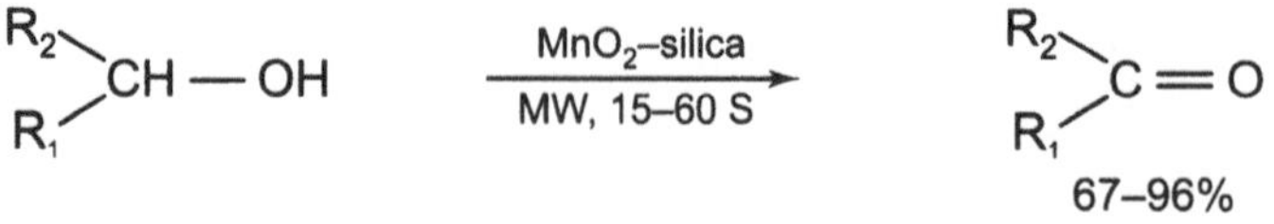

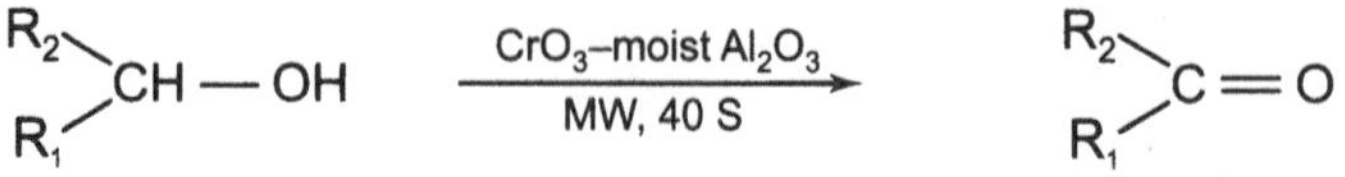

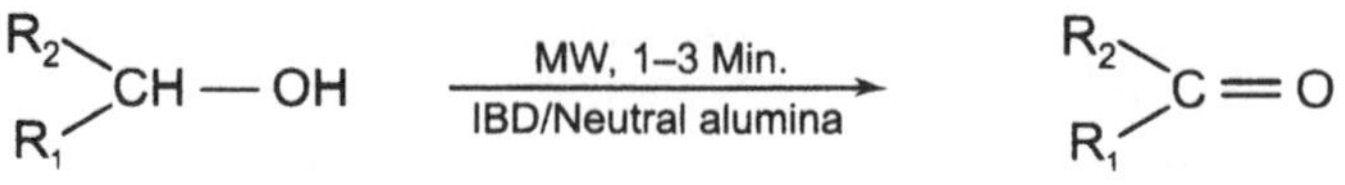

MnO₂–silica
MW, 15–60 S

67–96%

CrO₃–moist Al₂O₃
MW, 40 S

MW, 1–3 Min.
IBD/Neutral alumina

In all the above procedures, there is no overoxidation to carboxylic acids.

(***iii***) **Enzymatic Oxidation:** Secondary alcohols an oxidation with enzymes give ketones. Thus, (±)-trans-3-methylcyclohexanol and (±)-cis-2-methylcyclopentanol are dehydrogenated with horse liver alcohol dehydrogenase into (–)-S-3-methylcyclohexanone (50% yield; ee 100%) and to (+) – (S)-methylcyclopentanone (55% yield ee 96%).

The oxidation of secondary alcohol in presence of primary alcohol is defficult by chemical procedures. Such oxidation, can, however be affected by **Acetobacter suboxydans**[71]. Thus, adonit is converted into adinose.

L-Adonit → (Acetobacter suboxydans, 18°, 17 days) → L-Adinose (95%)

Secondary alcoholic groups in steroids are biochemically oxidised to keto group, one such example is given below[72].

Dehydroepiandrosterone → (Corynebactrium simplex ATCC 6946) → 4-Androstene-3, 17-dione 55%

Oxidation of Tertiary Alcohols

Tertiary alcohols are resistant to oxidation. However, some tertiary alcohols are converted into tertiary hydroperoxides[73] on treatment with hydrogen peroxide in sulphuric acid. One such example is given ahead.

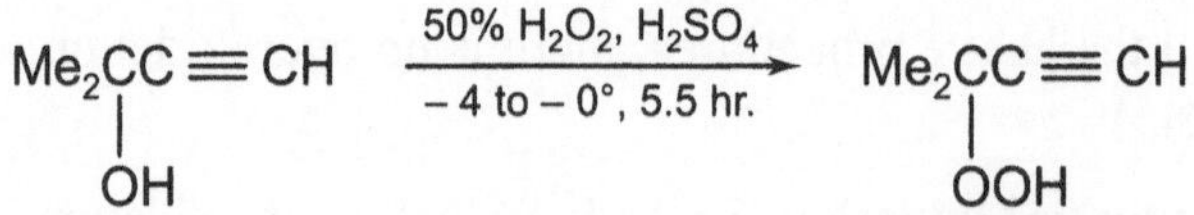

$$Me_2CC\equiv CH \xrightarrow[-4\ to\ -0°,\ 5.5\ hr.]{50\%\ H_2O_2,\ H_2SO_4} Me_2CC\equiv CH$$

with OH below the left structure and OOH below the right structure.

98%
(73% pure)

5.2.7 Phenols

These undergo the following reactions:

(*i*) Acetylation/benzoylation

$$C_6H_5OH + RCOCl \xrightarrow[Alkali]{-HCl} C_6H_5-OCOR$$

$$R=CH_3\ or\ C_6H_5$$

(*ii*) **Electrophilic Aromatic Substitution:** Phenols undergo nitration, bromination, sulphonation, Friedel-Crafts alkylation and acylation, Reimer-Tiemen reaction and Kolb's carboxylation. All these have been discussed earlier (see C — C bond formation).

(*iii*) **Hydroxylation:** Oxidation of phenol with alkaline persulphate (**Elbs hydroxylation**)[74] gives hydroxy phenols. The OH group enters the para or ortho position with respect to the phenolic hydroxyl group.

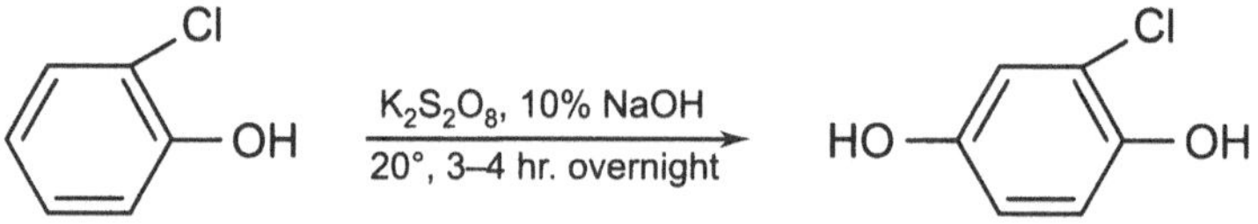

o-Chlorophenol

$\xrightarrow[20°,\ 3–4\ hr.\ overnight]{K_2S_2O_8,\ 10\%\ NaOH}$

o-Chlorohydroquinone
62%

Another procedure for hydroxylation of phenols is by enzymatic oxidation using mushroom polyphenol oxidase in chloroform. In this case, the hydroxy phenols are further oxidised to quinones, which can be reconverted into dihydroxy compounds by ascorbic acid.[75]

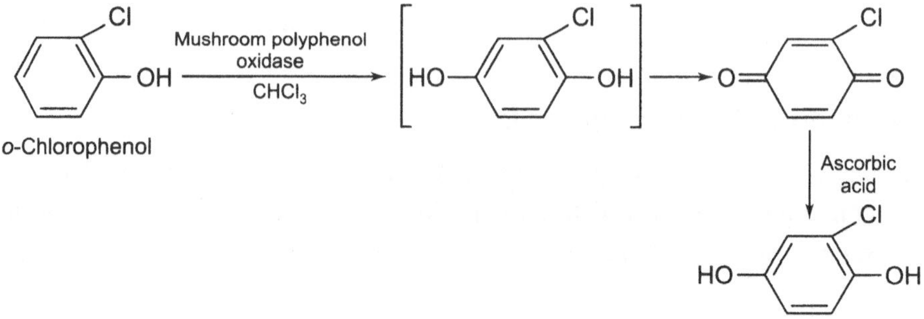

o-Chlorophenol

Mushroom polyphenol oxidase
CHCl₃

Ascorbic acid

(*iv*) **Hydrogenolysis:** It is used for the removal of hydroxyl groups from aromatic compounds (phenols). In this procedure, the hydroxyl group is first converted into the tosyl ester and then hydrogenolysed with Raney Nickel and hydrogen in alcohol.

Conversion to ortho- and para- quinones can also be affected in the presence of horse radish peroxidase[76].

5.2.8 Ethers

Ethers are virtually unreactive under usual condition. However, on reaction with strong acids (like HI, HBr, H_2SO_4), the carbon-oxygen bond breaks. Thus, diethyl ether on treatment with hot conc. HBr gives ethyl bromide.

$$CH_3CH_2OCH_2CH_3 + 2\ HBr \longrightarrow 2\ CH_3CH_2Br + H_2O$$
Diethyl ether

An important reaction of ethers is their oxidation with benzyl triethylammonium permanganate[77] to give esters.

$$C_3H_7CH_2OC_4H_9 \xrightarrow[CH_2Cl_2]{PhCH_2NEt_3MnO_4} C_3H_7CO_2C_4H_9$$
Dibutyl ether   Butyl propionate
90%

However, oxidation with RuO_4 in CCl_4 at room temperature gives the ester in quantitative yield[78].

5.2.9 Amines

Amines are regarded as derivatives of ammonia. These can be primary, secondary or tertiary depending on the number of alkyl groups attached to nitrogen.

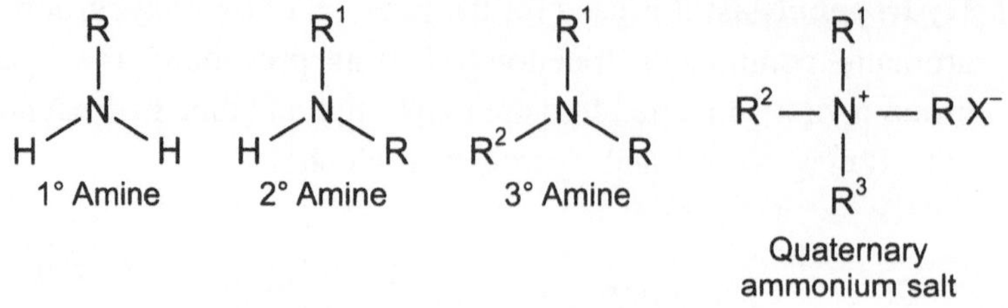

R, R¹, R², R³ structures labeled:
1° Amine   2° Amine   3° Amine   Quaternary ammonium salt

In aryl amines, nitrogen is directly attached to an aromatic ring. These can also be 1°, 2° or 3° aromatic amines.

The amines particulary primary amines contain $-NH_2$ group as the functional group.

5.2.9.1 Primary Amines

The $-NH_2$ group can be transformed into a number of functional groups, some of these are given below.

  (*i*) **Conversion of NH_2 into NO_2:** This can be affected by refluxing the amine with H_2O_2, Ac_2O, H_2SO_4, $CHCl_3$[79,80].

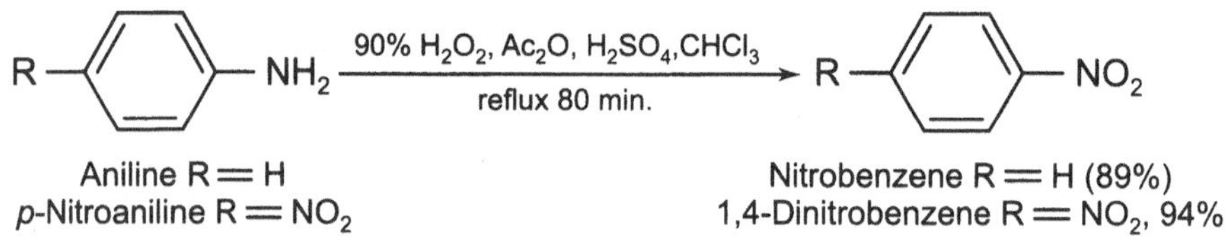

$$R-\text{benzene}-NH_2 \xrightarrow[\text{reflux 80 min.}]{90\% \ H_2O_2, \ Ac_2O, \ H_2SO_4, CHCl_3} R-\text{benzene}-NO_2$$

Aniline R = H
p-Nitroaniline R = NO_2

Nitrobenzene R = H (89%)
1,4-Dinitrobenzene R = NO_2, 94%

The conversion of NH_2 into NO_2 can also be affected bichemically. Thus, *p*-aminobenzoic acid on treatment with *Streptomyces thioluteus* at pH 7 and 30° give *p*-nitrobenzoic acid in 90 per cent yield[81].

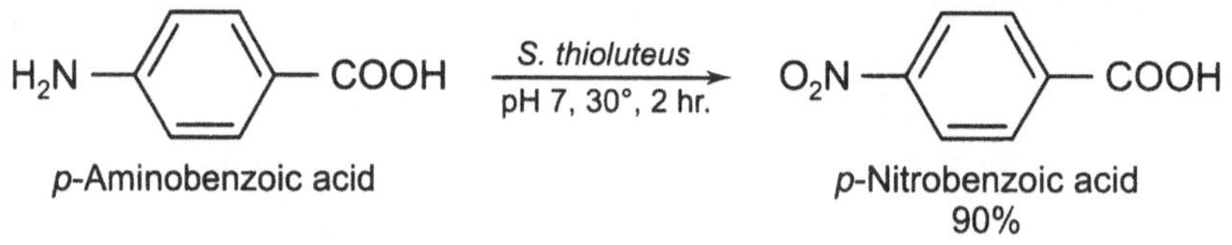

$$H_2N-\text{benzene}-COOH \xrightarrow[\text{pH 7, 30°, 2 hr.}]{S. \ thioluteus} O_2N-\text{benzene}-COOH$$

p-Aminobenzoic acid

p-Nitrobenzoic acid
90%

  (*ii*) **Conversion of NH_2 to $C \equiv N$:** Primary amines can be dehydrogenated to cyanides by a number of reagents as shown below:

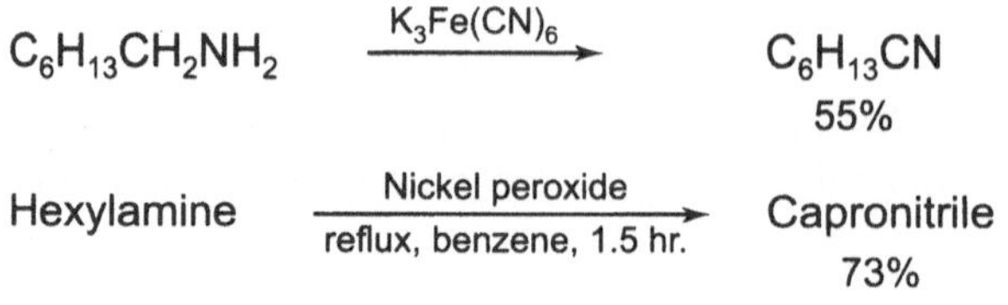

$$C_6H_{13}CH_2NH_2 \xrightarrow{K_3Fe(CN)_6} C_6H_{13}CN$$
                                                                    55%

$$\text{Hexylamine} \xrightarrow[\text{reflux, benzene, 1.5 hr.}]{\text{Nickel peroxide}} \text{Capronitrile}$$
                                                                    73%

$$n\text{-}C_7H_{15}CH_2NH_2 \xrightarrow[\substack{Bu_4NHSO_4EtOAc, \\ RT,\ 35\ min.}]{NaOCl} n\text{-}C_7H_{15}CN$$

1-Octylamine   1-Cyanoheptane
60%

$$C_6H_5CH_2NH_2 \xrightarrow{AgO,\ H_2O_2,\ 25\text{–}70°} C_6H_5CN$$

Benzyl amine   Benzonitrite

Biochemical oxidation with Hygrophorus conicus converts tryptamine into 2-indolone-3-acetic acid[86].

Tryptamine   2-Indolone-3-acetic acid
43%

(*iii*) **Conversion of NH$_2$- to Isonitriles – N ≡ C:** Primary amines react with dichlorocarbene generated by the PTC method[87] to give isonitriles.

$$RNH_2 + CHCl_3 + NaOH \xrightarrow[aq.]{C_6H_5CH_2N^+Et_3Cl^-} R\text{—}N\equiv C$$
$$40\text{–}60\%$$

The above one step procedure is convenient than the two steps process involving conversion of an amine to its formamide derivative followed by dehydration to isonitrile[88,89].

Using this process, tert. butylamine is converted into butyl isocyanide[90].

$$(CH_3)_3C\text{—}NH_2 + CHCl_3 + NaOH \xrightarrow{C_6H_5CH_2N^+Et_3Cl^-} (CH_3)_3CN\equiv C$$

tert.-Butyl amine   *tert.*-Butyl isocyanide
73%

This procedure has been used for the preparation of large number of alkyl isocyanides[87].

(*iv*) **Conversion of –NH$_2$ Group to Aldehydes and Ketones:** Aliphatic 1° amines on treatment with sodium thiosulphate/AgNO$_3$ in presence of aqueous sodium hydroxide gives aldehydes[91]. In this case, the elimination of H from N of the amino group and from the C of the adjacent methylene or methyl group converts primary amines into imino compound, which in the presence of water gets converted into aldehydes.

$$C_6H_5CH_2NH_2 \xrightarrow[\text{NaOH, H}_2\text{O, 0}\rightarrow\text{20°, 1–3 hr.}]{\text{Na}_2\text{S}_2\text{O}_8/\text{AgNO}_3} C_6H_5CHO$$

Benzylamine   Benzaldehyde
96%

Primary amino group bonded to a secondary carbon (as in the case of cyclohexylamine) on dehydrogenation give ketones[92].

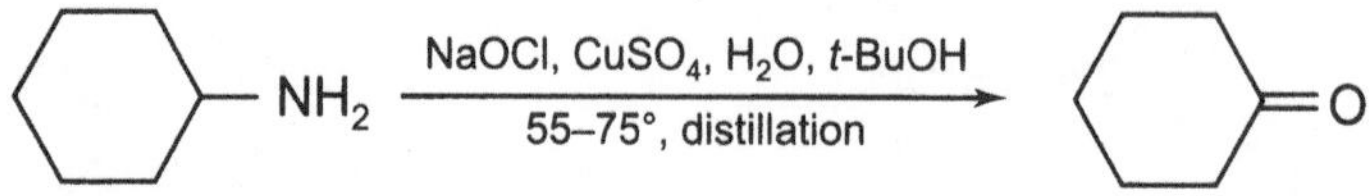

Cyclohexylamine $\xrightarrow[\text{55–75°, distillation}]{\text{NaOCl, CuSO}_4\text{, H}_2\text{O, }t\text{-BuOH}}$ Cyclohexanone

(*v*) **Conversion of NH₂ Group (in Aromatic Amines) into Various Functional Groups *via* the Formation of Diazo Compounds:** Aromatic amines on treatment with sodium nitite solution in presence of dilute HCl at 0–5°, give the corresponding diazonium compounds, which can be converted into phenol, chlorides, bromides, cyanides, iodides, fluorides and aromatic compounds (elimination of NH₂ group) as shown below.

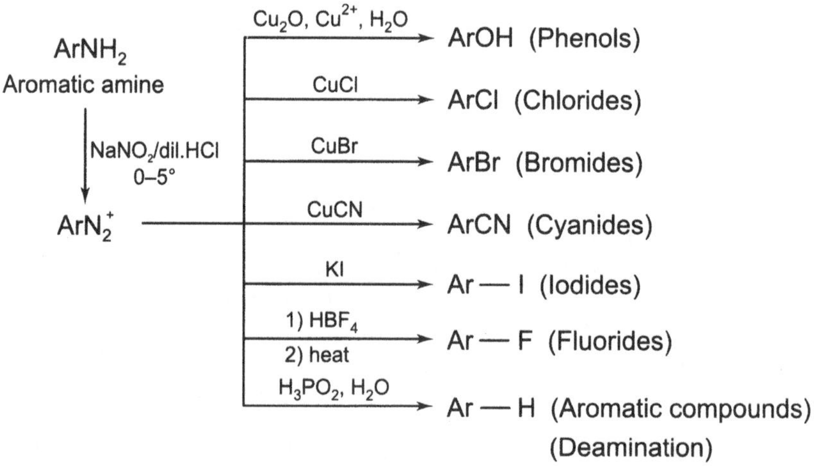

5.2.9.2 Secondary Amines

(*i*) **Conversion into Formamides:** Secondary amine reacts with dichlorocarbene (generated *in situ* by the PTC method[87]) to give N, N-disubstituted formamids[93,94].

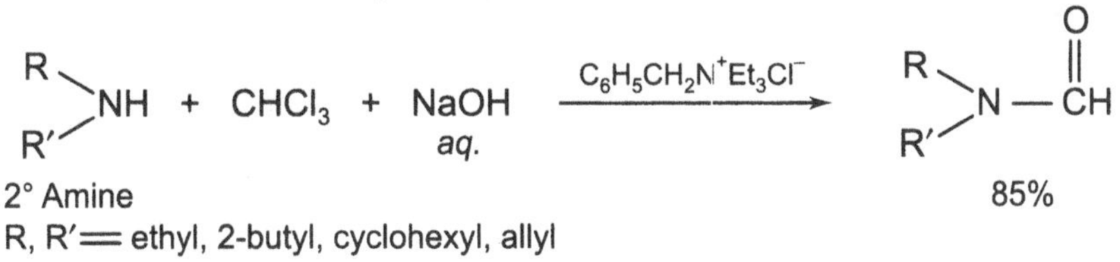

2° Amine
85%
R, R'= ethyl, 2-butyl, cyclohexyl, allyl

(*ii*) **Conversion into Imines:** Secondary amines having hydrogens on the α-carbon are dehydrogenated to the imines by manganese dioxide[95] or mercuric acetate[96].

$$C_6H_5CH_2NHC_6H_5 \xrightarrow[81°]{MnO_2, C_6H_6} C_6H_5CH = NC_6H_5$$

Benzyl phenylamine
84%

Piperidines
R = ^{t}Bu
R = Pr

75%
42%

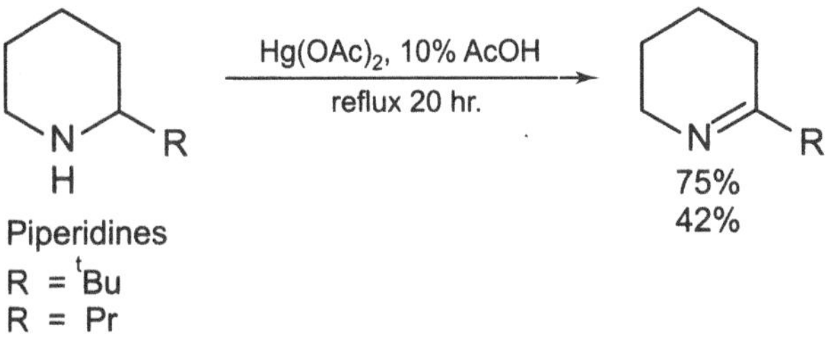

5.2.9.3 Tertiary Amines

The tertiary amines are oxidised to amine oxides by peroxy compounds like hydrogen peroxide and peroxyacetic acid. Aromatic tertiary amines also give N-oxides[97] by treatment with 50% H_2O_2 as shown below.

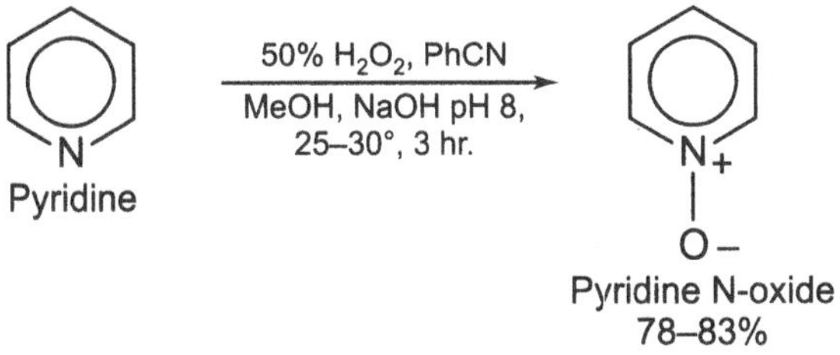

Pyridine

Pyridine N-oxide
78–83%

Tertiary amines, unlike secondary amine on treatment with dichlorocarbene (generated *in situ*) do not give formamides. Thus, bridgehead tertiary amine results in elimination of carbon bridge. One such example is given below[98].

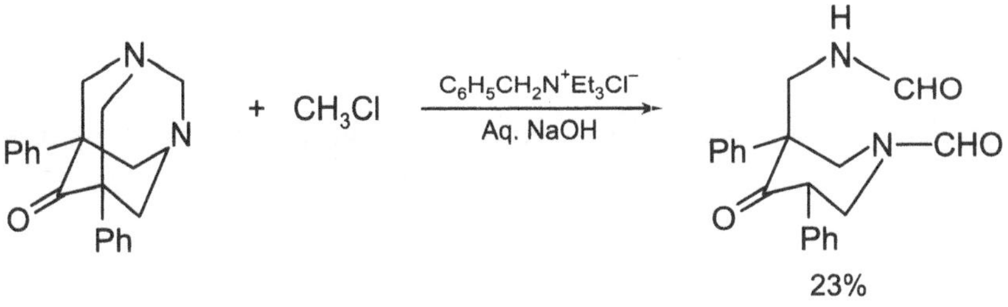

23%

5.2.10 Aldehydes

The functional group in aldehydes is $-\overset{\overset{\displaystyle O}{\|}}{C}-H$, commonly known as aldehyde group. Following are given the conversion of aldehyde group into other functional groups.

(i) Oxidation to Carboxylic Acids:

(a) Aldehydes on oxidation with aqueous $KMnO_4$ in presence of a PTC give quantitative yield of the corresponding carboxylic acid.

$$R-CHO + aq.\ KMnO_4 \xrightarrow{PTC} \underset{100\%}{RCHO}$$

PTC used in tetra butylammonium bromide.

(b) Aldehydes which do not have α-hydrogen on **Cannizzaro reaction** give carboxylic acids along with alcohol.

$$RCHO \xrightarrow{KOH} RCOOK + RCH_2OH$$

(c) Aromatic aldehydes which have an hydroxyl group in ortho or para position on **Dakin oxidation**[100] are converted into the corresponding hydroxy compounds.

alk. H_2O_2
^-OH

Salicylic aldehyde   Catechol

(d) **Baeyer-Villiger Oxidation**[101] of aldehydes with peracids followed by hydrolysis give carboxylic acids.

$$RCHO \xrightarrow{R'CO_3H} ROCOH \xrightarrow{hydrolysis} R-\underset{\underset{\displaystyle OH}{|}}{C}=O$$

In place of peracid, it is more convenient to use polymer supported per acid[102].

(e) The *ortho* or *para* hydroxy aldehydes can also be oxidised in solid state[103] by heating with **urea-hydrogen peroxide adduct (UPH)**. For more details see section 3.1.9.

$$\text{HO}\text{—C}_6\text{H}_3\text{—CHO} \xrightarrow[85°, 75 \text{ min.}]{\text{UPH}} \text{HO}\text{—C}_6\text{H}_3\text{—OH}$$

82%

Using this procedure a large number of aldehydes could be synthesised.

(*ii*) **Oxidation of Aldehydes to Peroxy Acids:** This is accomplished[103a] by passing oxygen under irradiation through a solution of benzaldehyde in acetone.

$$\underset{\text{Benzaldehyde}}{C_6H_5CHO} \xrightarrow{O_2,\ hv,\ Me_2CO} \underset{\text{Perbenzoic acid}}{C_6H_5CO_3H}$$

(*iii*) **Biochemical Oxidation:** Oxidation of phenyl acetaldehyde with cyclohexanone oxygenase[103b] produces phenyl acetic acid in 65 per cent yield in addition to small amounts of benzyl formate and benzyl alcohol.

$$\underset{\substack{\text{Phenyl} \\ \text{acetaldehyde}}}{C_6H_5CH_2CHO} \xrightarrow[\text{Oxygenase}]{\text{Cyclohexanone}} \underset{\substack{\text{Phenyl} \\ \text{acetic acid} \\ 65\%}}{C_6H_5CH_2CO_2H} + \underset{\substack{\text{Benzyl} \\ \text{formate} \\ 12\%}}{C_6H_5CH_2OCHO} + \underset{\substack{\text{Benzyl} \\ \text{alcohol} \\ 23\%}}{C_6H_5CH_2OH}$$

(*iv*) **Conversion of Aldehydes into Nitriles:** A single step conversion of aryl halides into the corresponding nitriles[104,105] is achieved by heating with hydroxylamine hydrochloride supported on montmorillonite K10 clay in a microwave oven.

$$\xrightarrow[\text{Microwave 1–1.5 min.}]{\text{K10 clay NH}_2\text{OH . HCl}}$$

99–95%

$R_1 = H$; $R_2 = OH$, Br, Me, NO_2
$R_1 = R_2 = OMe$

(*v*) **Reduction of Aldehydes:**

   (*a*) Aldehydes can be reduced using $NaBH_4$ give alcohols (— CHO ⟶ CH_2OH). Some examples are given below.

$$C_6H_5CHO \xrightarrow{\text{NaBH}_4/\text{MeOH}} C_6H_5CH_2OH$$

Benzaldehyde   Benzyl alcohol

$$O_2NCH_2CH_2CH_2CHO \xrightarrow{\text{NaBH}_4/\text{EtOH}} O_2NCH_2CH_2CH_2CH_2OH$$

4-Nitrobutyraldehyde   4-Nitrobutanol

OHC —⟨○⟩— CO₂Et $\xrightarrow{\text{NaBH}_4/\text{EtOH}}$ HOH₂C —⟨○⟩— CO₂Et

NO₂ ⟨○⟩— CHO $\xrightarrow{\text{NaBH}_4/\text{EtOH}}$ NO₂ ⟨○⟩— CH₂OH

m-Nitrobenzaldehyde   *m*-Nitrobenzyl alcohol

$$NCCH_2CH_2CHO \xrightarrow{\text{NaBH}_4} NCCH_2CH_2CH_2OH$$

3-Cyanopropionaldehyde   4-Hydroxybutyronitrile

Sodium borohydride has also been used to reduce CHO $\longrightarrow$ CH$_2$OH in solid state[105a]. The procedure consisting is heating aldehydes witll NaBH$_4$ –Al$_2$O$_3$ in microwave oven for 0.5–2 min.

R —⟨○⟩— CHO $\xrightarrow[\text{MW, 0.5–2 min.}]{\text{NaBH}_4\text{- Al}_2\text{O}_3}$ R —⟨○⟩— CH₂OH

60–90%

(*b*) Sodium/alcohol also reduces aldehydes to alcohols.

$$R - CHO \xrightarrow[\text{Reflux}]{\text{Na/C}_2\text{H}_5\text{OH}} RCH_2OH$$

This is known as **Bouveault Blanc reduction.**

(*c*) Formic acid and ethyl magnesium bromide convert aldehydes to alcohols[107].

$$C_9H_{19}CHO \xrightarrow[\text{2) H}_2\text{O}]{\text{1) HCOOH, EtMgBr}} C_9H_{19}CH_2OH$$

n-Decanal   Decanol

(*d*) Aldehydes on reduction using RuHCl (CO) (PPh$_3$)$_3$ in presence of formic acid under microwave irradiation give[108] the corresponding alcohol.

$$C_6H_5CHO \ + \ HCO_2H \ \xrightarrow[\text{RuHCl(CO)(PPh}_3\text{)}_3]{\text{MW, 7 min.}} \ C_6H_5CH_2OH$$

| Benzaldehyde | Formic acid | | Benzyl alcohol |

Benzaldehyde   Formic   Benzyl
                      acid   alcohol

(*e*) Aldehydes (not containing α-hydrogen) on irradiation in microwave with Ba(OH)$_2$. 8H$_2$O in presence of *para* formaldehyde give alcohols as the major product.

$$RCHO \ + \ (CH_2O)_n \ \xrightarrow[\text{Ba(OH)}_2,\ 8\ H_2O]{\text{MW}} \ RCH_2OH \ + \ RCOOH$$

                                          Major   Minor
                                      (80–99%)   (1–20%)

This is known as **Solid State Cannizzaro reaction[109]**.

(*vi*) **Reaction with Grignard Reagent:** By this process aldehydes can be converted into 1° alcohols (by reacting with formaldehyde) and 2° alcohols (by reacting with aldehydes other than formaldehyde).

(*vii*) **Aromatic Aldehydes undergo Perkin Reaction and Benzoin Condensation**

$$C_6H_5CHO \ + \ (CH_3CO)_2O \ \xrightarrow[\text{Perkin reaction}]{\text{CH}_3\text{COONa}} \ C_6H_5CH\!=\!CHCOOH$$

Benzaldehyde   Acetic anhydride   Cinnamic acid

$$C_6H_5CHO \ \xrightarrow[\text{Benzoin condensation}]{\text{KCN}} \ C_6H_5\!-\!\overset{\displaystyle OH}{\underset{\displaystyle |}{CH}}\!-\!\overset{\displaystyle O}{\overset{\displaystyle ||}{C}}\!-\!C_6H_5$$

Benzaldehyde   Benzoin (83%)

Benzoin condensation can also be achieved by reacting an aldehyde with a catalyst (3-benzyl-4-methyl thiazolium chloride) in presence of KOH.

$$CH_3CHO \ \xrightarrow[\text{KOH}]{\text{3-Benzyl-4-methyl thiazolium chloride}} \ CH_3\!-\!\overset{\displaystyle}{\underset{\displaystyle OH}{\underset{\displaystyle |}{CH}}}\!-\!\overset{\displaystyle O}{\overset{\displaystyle ||}{C}}\!-\!CH_3$$

                                                                Acetoin
                                                                (100%)

Benzoin condensation can also be brought about[111] by coenzyme 'Thiamine'.

(*viii*) **Enzymatic Reduction of – CHO into CH_2OH:** Aldehydes can be enzymatically reduced to CH_2OH. Two examples are given below.

$$\underset{trans\ \text{Crotonaldehyde}}{CH_3CH{=}CH{-}CHO} \xrightarrow{\text{Beauveria sulfurescens}^{110a}} \underset{\substack{\text{2-Buten-1-ol}\\ 89\%}}{CH_3CH{=}CH{-}CH_2OH}$$

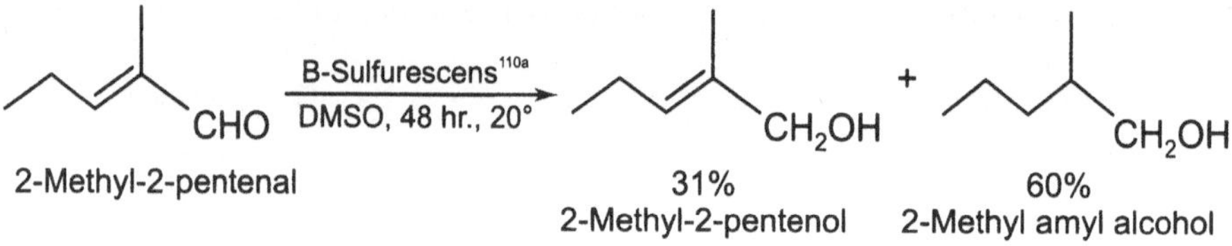
B-Sulfurescens[110a]
DMSO, 48 hr., 20°
CHO
CH2OH
+
CH2OH
2-Methyl-2-pentenal
31%
2-Methyl-2-pentenol
60%
2-Methyl amyl alcohol

(*ix*) **Conversion of Aldehydes into Ketones:** Aldehydes can be converted into ketones by the sequence of reactions as shown below.

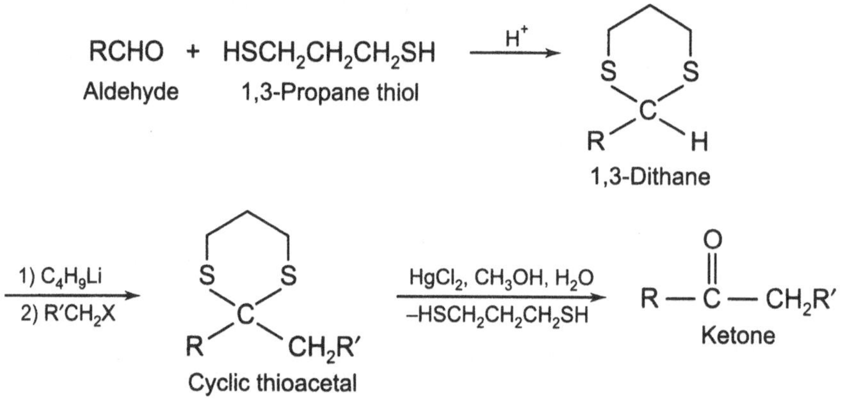
RCHO + HSCH2CH2CH2SH
Aldehyde
1,3-Propane thiol
H+
S S
C
R H
1,3-Dithane
1) C4H9Li
2) R'CH2X
S S
C
R CH2R'
Cyclic thioacetal
HgCl2, CH3OH, H2O
–HSCH2CH2CH2SH
O
R — C — CH2R'
Ketone

5.2.11 Ketones

As in case of aldehydes, the functional group in ketones is the keto group $\rangle C{=}O$. Following are given the conversion of keto group into other functional groups.

(*i*) **Oxidation of Methyl Ketones for Carboxylic Acids:** This is a convenient method of converting — $COCH_3$ into — COOH. In this procedure[112], methyl ketones are treated with potassium hypochlorite in presence it a PTC (Bu_4 N^+X^-). One example is given ahead.

$$C_6H_5CH = \underset{\underset{O}{\overset{CH_3}{|}}}{C} - \underset{\underset{\parallel}{|}}{C} - CH_3 \xrightarrow[\text{Bu}_4\text{N}^+\text{X}^-, \text{ EtOAc}]{\text{KOCl}} C_6H_5CH = \underset{\overset{CH_3}{|}}{C} - COOH$$

70–80%

(*ii*) **Conversion of —COCH₃ into Esters:** This is achieved by the well known **Baeyer-Villeger Oxidation**[113]. The procedure consists is treating the ketone with peracids in organic solvents.

COCH₃ group on benzene ring	OCOCH₃ group on benzene ring
R	R
Acetophenone R═H	Phenyl acetate R═H
R═Cl, OMe	70–90%

PhCOOOH, CHCl₃ / 25°

Some Baeyer-Villiger oxidation of ketones with *m*-chloroperbenzoic acid proceed much faster at room temperature in the solid state[114]. some examples are given below.

Br— ⟨benzene ring⟩ —COCH₃ + *m*-CPBA $\xrightarrow[\text{Solid state}]{\text{RT, 5 days}}$ Br —⟨benzene ring⟩— OCOCH₃

p-Bromoacetophenone

p-Bromophenyl acetate
64%

PhCOCH₂Ph + *m*-CPBA $\xrightarrow[\text{Solid state}]{\text{RT, 24 hr.}}$ PhCOOCH₂Ph

97%

PhCOPh + *m*-CPBA $\xrightarrow[\text{Solid state}]{\text{RT, 24 hr.}}$ PhCOOPh

85%

(*iii*) **Conversion of $>$CO Group into $>$CH—OH Group:** Ketones on reduction with NaBH₄/MeOH to yield the corresponding alcohols.

R — COCH₃ $\xrightarrow{\text{NaBH}_4/\text{MeOH}}$ R — CHOHCH₃

⟨cyclohexanone⟩ ═O $\xrightarrow{\text{NaBH}_4/\text{MeOH}}$ ⟨cyclohexyl⟩—OH

Keto group can also be reduced the OH group by using alumina supported $NaBH_4$ using microwaves[105a].

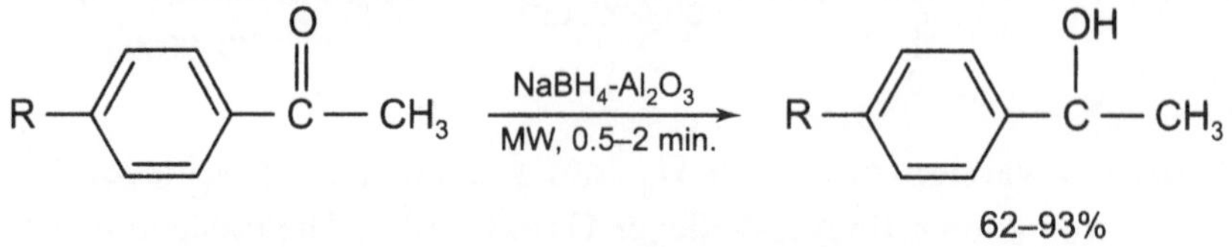

62–93%

(See also enzymatic reduction given below).

(iv) Reductive Amination of Carbonyl Groups: Carbonyl compounds on reaction with primary amines in presence of catalytic amount of clay using microwave give 95–98 per cent of the corresponding imine[115,116], which on reduction with montmorillonite K10 clay supported $NaBH_4$ in presence of microwave irradiation give the corresponding amination products[117].

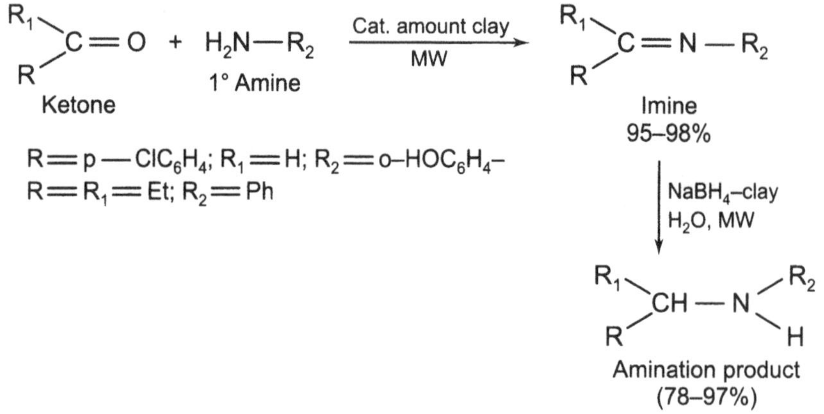

$R = p\!-\!ClC_6H_4;\ R_1\!=\!H;\ R_2\!=\!o\!-\!HOC_6H_4\!-$
$R = R_1\!=\!Et;\ R_2\!=\!Ph$

(v) Conversion of C=O into CH₂ Group: This conversion in normally effected by **Clemmensons reduction** using Zn/Hg/HCl. This reduction proceeds much faster by sonication[118]. Thus, camphor on sonication in THF in presence of a metal (Li, Na or K) yields a mixtures of endo and exo borneol[119].

(vi) Enzymatic Reduction of $>$C=O Group: Ketones can be enantioselectively reduced using enzymes. Some examples are given below.

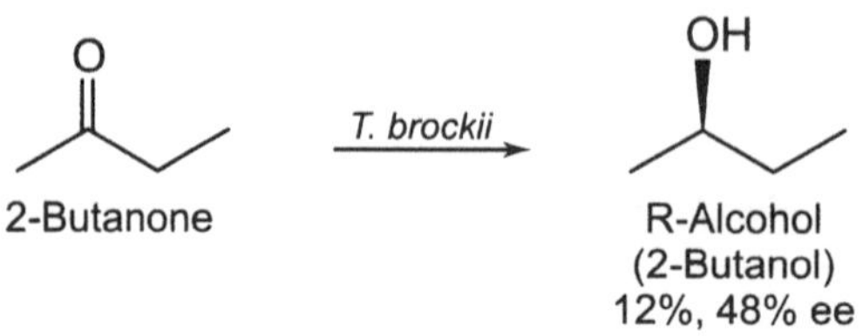

...(Ref. 119a)

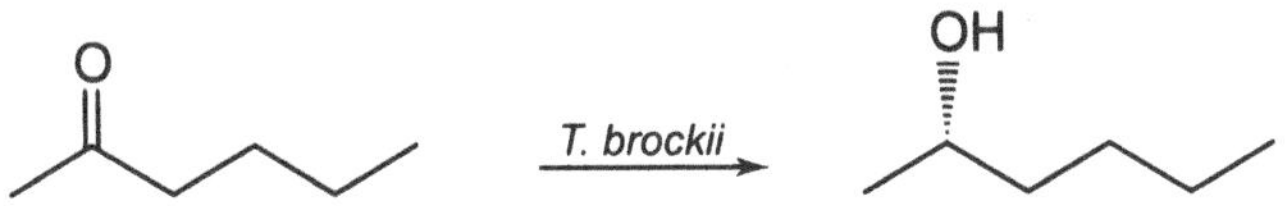

2-Hexanone → S-Alcohol (2-Hexanol) 85%, 96% ee — *T. brockii* ...(Ref. 119a)

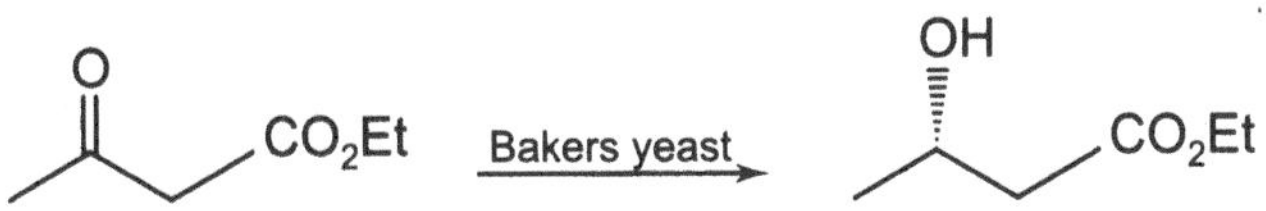

Ethyl acetoacetate → S-Alcohol 67% — Bakers yeast ...(Ref. 119b)

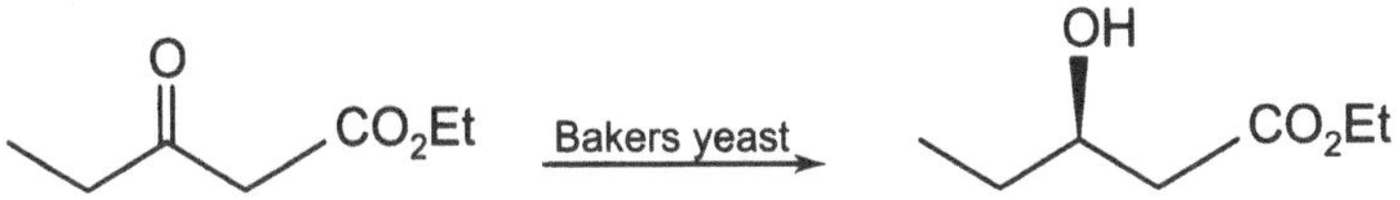

Ethyl β-ketovalerate → R-Alcohol — Bakers yeast ...(Ref. 119c)

5.2.12 Carboxylic Acids

The functional group in carboxylic acids is the carboxyl group (— COOH). The carboxyl group can be converted into following functional groups:

(*i*) **Conversion of COOH into COOR (Esters):** The conversion of COOH into — COOR is known as esterification and is usually performed by heating a carboxylic acid with an alcohol in presence of acid catalyst.

(*a*) **Using PTC:** Carboxylic acids can be esterified with alkyl halides in the presence of triethyl amine. For details see section 5.2.5.1(v).

Crown ethers have also been used for esterification[120]. One example is shown below.

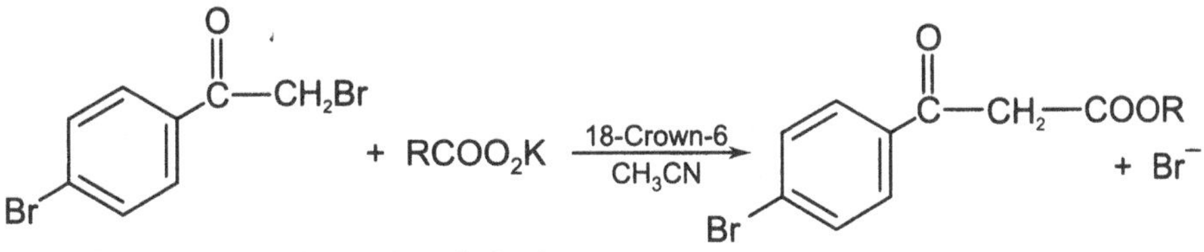

p-Bromophenacyl bromide + RCOO$_2$K (Pot. Salt of carboxylic acid) $\xrightarrow[\text{CH}_3\text{CN}]{\text{18-Crown-6}}$ p-Bromophenacyl esters + Br$^-$

R = H, CH$_3$, CH$_3$CH$_2$, CH$_3$CH$_2$CH$_2$, C$_6$H$_5$, 2-CH$_3$C$_6$H$_5$, 2, 4, 6-Trimethyl benzoyl (*b*) **Using microwaves:** Carboxylic acids and alcohols on microwave irradiation[121] and catalytic amount of H_2SO_4 yield esters.

$$C_6H_5COOH \ + \ ROH \ \xrightarrow[\text{Conc. } H_2SO_4(0.1 \text{ mL})]{\text{MW, 5–10 min.}} \ \underset{\text{70–80\%}}{C_6H_5COOR}$$

$$R = CH_3, \ C_2H_5, \ C_3H_7, \ C_4H_9$$

(*c*) **Using ultrasound:** A simple procedure for the esterification of a variety of carboxylic acids with different alcohols at ambient temperature is by using ultrasound[122].

$$RCOOH \ + \ R'OH \ \xrightarrow[))))]{H_2SO_4, \ RT} \ RCOOR'$$

For the above esterification low intensity ultrasound[123] (cleaning bath) has been used.

The ester group can be easily converted into — CH_2OH by reduction with $NaBH_4$ using PEG-400 as solvent[124].

(*ii*) **Decarboxylation:** Carboxylic acids can be decarboxylated under microwave irradiation[125] using quinoline as solvent. A typical example is given below.

Indole-2-carboxylic acid   Indole
                                                                            99%

The above dicarboxylation can also be affected in quantitative yield at 255° in near critical region[126].

The decarboxylation is also possible by heating carboxylic acid in MW owen in presence of methylimidazole and aqueous $NaHCO_3$ (as catalyst) in polyethylene glycol[127].

(*iii*) **Conversion of COOH group into CH₂OH:** Carboxylic acids on reduction with lithium aluminium hydride give the corresponding alcohols. Some examples are given below.

$$R(CH_2)_nCOOH \xrightarrow[\text{2) H}^+]{\text{1) LiAlH}_4, \text{ Ether}} R(CH_2)_nCH_2OH$$

Carboxylic acid   Alcohol

$$CH_3CH=CH-COOH \xrightarrow[\text{2) H}^+]{\text{1) LiAlH}_4, \text{ Ether}} CH_3CH=CH-CH_2OH$$

Crotonic acid   Crotyl alcohol

Furoic acid   Furyl alcohol
87%

(*iv*) **Conversion of COOH group into CHO:** Carboxylic acids can be converted into the corresponding aldehydes via the formation of acid chloride followed by **Rosenmund reduction**[128].

$$RCH_2COOH \xrightarrow{\text{SOCl}_2} RCOCl \xrightarrow[\text{Rosenmund reduction}]{\text{H}_2, \text{ Pd/BaCO}_3} RCH_2CHO$$

Carboxylic   Acid   reduction   Aldehyde
acid   chloride

(*v*) **Conversion of COOH group into COCH₃:** The reaction of carboxylic acid with methyl lithium gives the corresponding ketone.

$$R-COOH + CH_3Li \longrightarrow RCOCH_3$$

A special feature of this reaction is that original stereochemistry of the carboxylic acid is maintained in the formed ketone.

trans-2-phenycyclobutane   *trans*-2-phenycyclobutyl
carboxylic acid   methyl ketone

5.2.13 Esters

The functional group in esters is — COOR′. The ester group can be converted into the following functional groups.

(*i*) **Conversion of — COOR into COOH:** The conversion of — COOR into COOH is known as **saponification**. This is conducted using microwave activation under solid-liquid PTC conditions[125] without any solvent.

$$R-\underset{\underset{Ester}{}}{\overset{\overset{O}{\|}}{C}}-OR' \quad \xrightarrow[\text{2) HCl}]{\text{1) KOH-Aliquat, MW, 4–10 min.}} \quad R-\underset{\underset{Carboxylic\ acid}{}}{\overset{\overset{O}{\|}}{C}}-OH$$

This procedure is useful for saponification of even hindered esters. In place of PTC, crown ethers[129] can also be used.

In toluene; R = CH$_3$,
t-Bu, neopentyl

1) K$^+$ Dicyclohexyl [18] crown-6
2) H$^+$

+ $^-$OH

(*ii*) **Conversion of — COOR into — CH$_2$OH:** As in the case of — COOH, the ester group, — COOR can also be reduced by LiAlH$_4$ (LAH).

$$\underset{RCH_2COOR'}{CH_2OH\ group} \quad \xrightarrow[\text{2) H}^+]{\text{1) LAH}} \quad RCH_2OH$$

$$\underset{Ethyl\ acetoacetate}{CH_3COCH_2COOC_2H_5} \quad \xrightarrow[\text{2) H}^+]{\text{1) LAH, Ether}} \quad \underset{\underset{\underset{Butane\text{-}1,3\text{-}diol}{}}{OH}}{CH_3CHCH_2CH_2OH}$$

$$\underset{Ethyl\ crotonate}{CH_3CH=CH-COOC_2H_5} \quad \xrightarrow[\text{2) H}^+]{\text{1) LAH, Ether}} \quad \underset{Crotyl\ alcohol}{CH_3CH=CH-CH_2OH}$$

(*iii*) **Conversion of — COOR into — COOR′:** This process is known as **transesterification**. In fact, one alcohol is capable of displacing another alcohol (in ester) form a new ester.

$$\underset{Ester}{R-\overset{\overset{O}{\|}}{C}-OR'} + \underset{Alcohol}{R''-OH} \quad \underset{}{\overset{H^+\ or\ OR^-}{\rightleftharpoons}} \quad \underset{New\ ester}{R-\overset{\overset{O}{\|}}{C}-OR''} + R'O-H$$

Trans esterification is catalysed by acid (H$_2$SO$_4$ or dry HCl) or base (usually alkoxides). Trans esterification is an equilibrium reaction.

To shift the equilibrium to the right, it is necessary to use a large excess of the alcohol whose we wish to make ester. Alternatively, one of the reaction products is removed from the reaction mixture; this second approach is better (if feasible).

Trans esterification can be achieved using a suitable enzyme[134] using SC — CO_2. One such example is given below.

N-Acetyl-1-phenylalanine
chloroethyl ester

Subitism carlsbert
2.5 vol. % ethanol
SC — CO_2
45°, 150 bar

N-Acetyl-1-phenylalanine
ethyl ester, 100%

Trans esterification can also be carried out using candida lipase B(CaLB) either as free enzyme (SP 525) or in an immobilized form (Novozym 435) in ionic liquid $[b_{min}][BF_4]$ or $[b_{min}][PF_6]$[135].

$$R^1CO_2Et + R^2OH \xrightarrow[\text{or } [b_{min}][BF_4], 40°]{\text{Cal. B}[b_{min}][PF_6]} R^1CO_2R^2 + EtOH$$

Using this procedure, trans esterification of N-acetyl-L-phenylalanine ethyl ester with propanol was affected[136].

N-Acetyl-L-Phenyl
alanine ethyl ester

α-Chymotypsin
$[b_{min}][PF_6]$, PrOH

Another example of trans esterification is given below[135].

Ethyl butanoate

+ nBuOH

CALB(lipase)
$[b_{min}][PF_6]$

Butyl butanoate
95%

(*iv*) **Conversion of — COOR into CONH₂:** Esters on reaction with liquor amonia gives the corresponding amide.

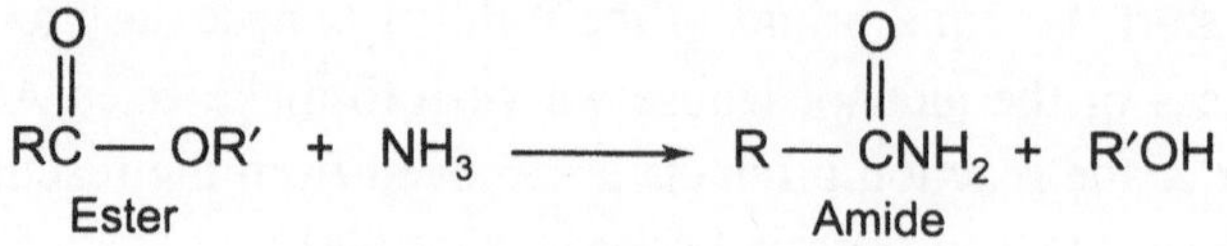

$$\underset{\text{Ester}}{RC\overset{\displaystyle O}{\overset{\|}{—}}OR'} + NH_3 \longrightarrow R\overset{\displaystyle O}{\overset{\|}{—}}CNH_2 + R'OH$$
$$\underset{\text{Amide}}{}$$

(*v*) **Conversion of Esters into Tertiary Alcohols:** This can be affected by the treatment of esters with grignard reagent.

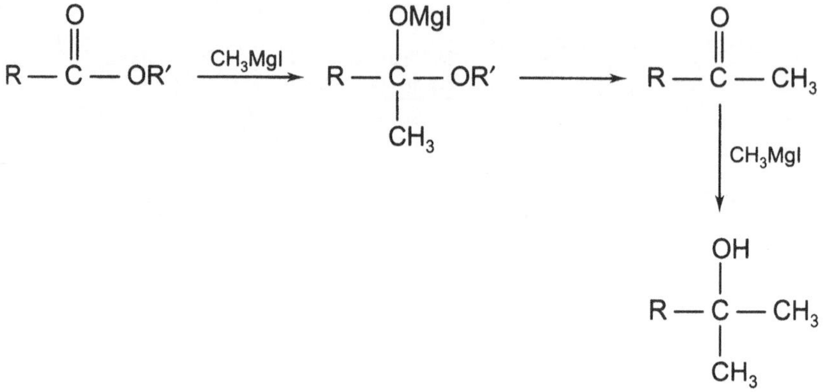

5.2.14 Amides

The functional group in amides is — $CONH_2$. It can be converted into following functional groups:

(*i*) **Conversion of — CONH$_2$ Group into — COOH Group:** Amides on acid hydrolysis under microwave irradiation[130,131] give the corresponding carboxylic acids.

$$\underset{\text{Benzamide}}{C_6H_5CONH_2} + 20\% \ H_2SO_4 \xrightarrow[\text{H}^+]{\text{MW, 10 min.}} \underset{\substack{\text{Benzoic acid}\\99\%}}{C_6H_5COOH}$$

$$\underset{\text{N-Phenylbenzamide}}{C_6H_5CONH_2C_6H_5} + 20\% \ H_2SO_4 \xrightarrow[\text{H}^+]{\text{MW, 10 min.}} \underset{\substack{\text{Benzoic acid}\\74\%}}{C_6H_5COOH}$$

Amides can also be hydrolysed to carboxylic acids by high temperature water (HTW)[132].

(*ii*) **Conversion of CONH$_2$ into CN:** Normally amides on heating with P_2O_5 give nitriles. A better procedure[133] is to treat the amide with dichlorocarbene (generated *in situ*).

$$\underset{\text{Benzamide}}{C_6H_5CONH_2} + CHCl_3 + aq.\ NaOH \xrightarrow{\underset{\text{ammonium chloride}}{\text{Benzyl triethyl}}} \underset{\underset{84\%}{\text{Benzonitrile}}}{C_6H_5CN}$$

(*iii*) **Conversion of CONH$_2$ into NH$_2$:** This conversion is affected by the well-known **Hoffman rearangement**[137]. In this method, the amide is treated with Br$_2$ and KOH.

$$\underset{\text{Propanamide}}{CH_3CH_2CONH_2} + Br_2 + 4\ KOH \longrightarrow \underset{\text{Ethyl amine}}{CH_3CH_2NH_2} + 2\ KBr + K_2CO_3 + 2\ H_2O$$

(*iv*) **Conversion of CONH$_2$ into COCH$_3$:** Amides as treatment with grignard reagent give ketones.

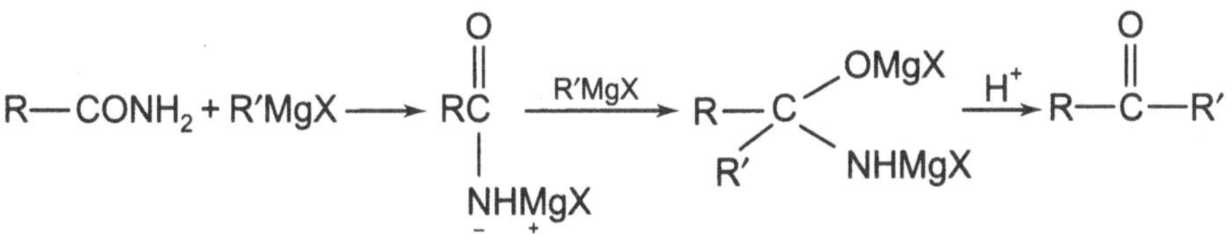

$$R-CONH_2 + R'MgX \longrightarrow \underset{\overset{|}{NHMgX}}{\overset{\overset{O}{\|}}{RC}} \xrightarrow{R'MgX} R-\underset{R'}{\overset{OMgX}{C}}\overset{}{\diagdown}NHMgX \xrightarrow{H^+} \overset{\overset{O}{\|}}{R-C-R'}$$

5.2.15 Nitriles

Nitriles contain $-\,C\equiv N$ as the functional group. The $-\,C\equiv N$ can be converted into other functional group as given below:

(*i*) **Conversion of $-\,C\equiv N$ into $-COOH$:** Nitriles on basic hydrolysis give carboxylic acids. The yield is greatly improved by sonication[137].

$$ArCN \xrightarrow[))))]{HO^-/H_2O} ArCOOH$$

(*ii*) **Conversion of $-\,C\equiv N$ into CH$_2$NH$_2$:** Nitriles on reduction with H$_2$/PtO$_2$ in alcohol give the reduced product. The yield is considerably improved by sonication.

$$RC\equiv N \xrightarrow[\underset{))))}{C_2H_5OH}]{H_2,\ PtO_2} [RCH=NH] \longrightarrow RCH_2NH_2$$

The above transformation can also be affected with LiAlH$_4$

$$RC\equiv N \xrightarrow[\text{2) }H_3O^+]{\text{1) LiAcH}_4} RCH_2NH_2$$

(*iii*) **Conversion of Nitriles into Ketones:** Nitriles in reaction with a Grignard reagent followed by hydrolysis give ketones.

$$RC \equiv N \ + \ R'MgX \ \longrightarrow \ RR'C = NMgX$$

$$\downarrow H_2O$$

$$RCOR' \xleftarrow{\ H_3O^+\ } [RR'C = NH]$$

$$\text{Ketamine}$$

(*iv*) **Conversion of Nitriles into N-substituted Amides:** Nitriles on reaction with alcohols in presence of acid give N-substituted amides. This reaction is known as **Ritter Reaction**[139].

$$RC \equiv N \ + \ R'OH \ \xrightarrow{\ H^+\ } \ R - \overset{\overset{\displaystyle O}{\|}}{C} - NHR'$$

REFERENCES

1. I. Tabushi, Z. Yoshida and N. Takahashi, J. Am. Chem. Soc., 1970, **92**, 6670.

2. S.H. Goh, K.C. Chan, T.S. Kam and H.L. Chong, Aust. J. Chem., 1975, **28**, 381.

2*a*. Z. Cohen, H. Varkony, E. Keinan and Y. Mazur, Org. Synth. Coll. Vol., 1988, **6**, 43.

3. S.L. Regan and A. Singh, J. Org. Chem., 1982, **47**, 3361.

4. M. Makosza and M. Wawrzniewicz, Tetrahedron Lett., 1969, 4659.

5. O. Repic and S. Vogt, Tetrahedron Lett., 1982, **23**, 2729.

6. J. Einhorn, C. Einhorn and J.L. Luchi, J. Org. Chem., V.K. Ahluwalia and Renu Aggarwal, Organic Synthesis, Special Techniques, Narosa Publishing House, 2006, Page 144.

7. J.M.J. Frechet and K.E. Haque, Macromolecules, 1975, **8**, 130; C.R. Harrison and P. Hodge, J. Chem., Soc. Chem., Commun., 1974, 1009.

8. C.E. Song and E.J. Roh, Chem. Commun., 2000, 837.

9. L. Gallion and F. Bedioui, Chem., Commun., 2001, 1458.

10. O. Bortolini, V. Conte, C. Chiappe, G. Fantin, M. Fogagnolo and S. Maietti, Green Chem., 2002, **4**, 94.

10*a*. I. Komoto and S. Kobayashi, Org. Letters, 2004, **4**, 1115.

11. N. Jain, A. Kumar, S. Chauhan and S.M.S. Chauhan, Tetrahedron 2005, **61**, 1015–1060.

12. W.B. Weber and J.P. Shephard, Tetrahedron Lett., 1972, 4907.

12a. S. Chandrasekhar, Ch. Narsihmulu, S.S.Sultana and N.R. Reddy, Chem., Commun., 2003, 1716.

13. A.P. Krapcho, J.R. Larsen and J.M. Eldridge, J. Org. Chem., 1971, **42**, 3749.

14. A.W. Herriott and D. Picker, Tetrahedron Lett., 1974, 1511.

15. D. Sam and H.E. Simmons, J. Am. Chem., Soc., 1972, **94**, 4024.

15a. Y. Deng, I. Ma, K. Wang and J. Chen, Green Chem., 1999, **1**, 275.

16. H.C. Brown and U.S. Racheria, Tetrahedron Lett., 1985, **26**, 2187.

17. H.H. Han and P. Boudjouk, Organometallics 1983, **2**, 769.

18. B.H. Han and P. Boudjouk, Tetrahedron Lett., 1981, **22**, 2757.

19. D.C. Neckers, D.A. Kooistra and G.W. Green, J. Am. Chem. Soc., 1972, 9284.

20. A.W. Herriott and D. Picker, Tetrahedron Lett., 1974, 1511.

20a. J.B. Davis and R.L. Raymond, Appl. Microbiol, 1961, **9**, 383.

20b. J.D. Douros, Jr, and J.W. Frankenfeld, Appl. Microbiol, 1968, **16**, 320.

20c. A. Kleinzeller and Z. Fencl, Chem., Listy, 1952, **46**, 300; Chem., Abstr., 1953, **47**, 4290.

21. L.M. Shirley and S.C. Taylor, J. Chem., Soc. Chem. Commun., 1983, 954.

22. A. Kleinzeller and Z. Fenel, Chem., Listy, 1952, **46**, 300; Chem., Abstr., 1953, **47**, 4290.

23. D.T. Gibson, J.R. Koch, C.L. Schuld and R.E. Kallio, Biochemistry, 1968, **7**, 3797; D.T. Gibson, M. Hansley, H. Yoshioka and T.J. Mabry, Biochemistry, 1970, **9**, 1626.

24. R.N. Gedye, W. Rank and K.C. Westaway, Can. J. Chem., 1991, **69**, 706.

25. N. Akiya and P.E. Savage, Chem., Rev., 2002, **107**, 2725 and the references cited there in.

26. N.F. Leininger, R. Clontz, J.L. Gainer and D.V. Kirwan in Clean Solvents, Alternative Media for Chemical Reactions and Processing, ed. M.A. Abraham and L. Moens, ACS symposium series 819, American Chemical Society, Washington DC, 2002, P. 208.

27. P.W. Herriott and D. Picker, Tetrahedron Lett., 1972, 4521.

28. Charles M. Stakes and Charles L. Liotta, Phase Transfer Catalysts, Principles and Techniques, Academic Press Inc. Ny, 1918, p. 127.

29. H. Priebe, Acta. Chem., Scand. Ser. B, 1987, **1341**, 640.

29a. J. Jarrouse and C.R. Hebd, Scances Acad. Sci. Ser. C. 1951, **232**, 1424; H.H. Freeman and R.A. Dubois, Tetrahedron Lett., 1975, 3251.

29b. B. Apribat, Y. Le Bigot and A. Gaset, Synth. Commun., 1994, **24**, 2091; Tetrahedron, 1996, **52**, 2119.

30. A Mckillop, J.C. Fiaud and R.P. Hug, Tetrahedron, 1974, **30**, 1379; A.P. Basall and J.F. Collins, Tetrahedron Lett., 1975, 3489.

31. C.M. Starks, J. Am. Chem., Soc., 1971, **93**, 195.

32. N. Sugimoto, T. Fujita, N. Shigematsu and A. Ayada, Chem., Pharm. Bull., 1962, **10**, 427.

33. K.E. Koening and W.P. Weber, Tetrahedron Lett., 1974, 2275; H. Habner, Ann. Chem., Pharm., 1861, **120**, 330.

34. T. Ando, S. Sumi, T. Kawate, J. Ichihara and T. Haatusa, J. Chem., Soc. Chem., Commun., 1984, 439.

34a. E. Santaniello, A. Manzocchi and P. Sozzani, Tetrahedron Lett., 1979, **20**, 4581.

35. H. Dou, P. Hassanby, G. Vernin and J. Metzger, Helv. Chem., Acta, 1978, **61**, 3143.

36. H.E. Hennis, L.R. Thompson and J.P. Long, Ind. Eng. Chem. Prod. Res. Dev., 1968, **7**, 96.

37. H.E. Hennis, J.P. Esterly and L.R. Thompson Ind. Eng. Chem. Prod. Res. Dev., 1967, **6**, 193.

38. R. Holmberg and S. Hansen, Tetrahedron Lett., 1975, 2307.

39. H.D. Durst, Tetrahedron Lett., 1974, 2421.

40. H.D. Durst, M. Milano, E.J. Kihta, S.A. Connelly and E. Grushka, Anal. Chem., 1975, **47**, 1797.

41. C.J. Pedersen, J. Am. Chem. Soc., 1967, **89**, 2485, 7017; 1970, **92**, 386, 391.

42. R.S. Davidson, A. Safdar, J.D. Spencer and D.W. Lewis, Ultrasonics, 1987, **25**, 35.

43. J.M. Khurana and P.K. Sahoo, Syn. Commun, 1992, 1691.

44. J. Jamakawi, S. Sumi, T. Ando and J. Hanafusa, Chemistry Lett., 1983, 379.

45. R. Lantzsch, A. Markhold and K.F. Lehmint (To Bayer A.G.), German Patent, 2, 545, 644 (1977); C.A., 1977, **87**, 133781.

46. C. Kimura, K. Murari, Y. Ishikawa and K. Kashiwaya, Sekiyu Gakkari Shi, 1973, **19**, 49; C.A., 1977, **87**, 133781.

47. M. Kiroha, T. Nakamura and T. Ozeki (Nippon Soda Co. Ltd.), Japanese Patent, 7606928; C.A. 1976, 164159.

48. D.G. Brady (to Phillips Petrolium Co.), US Patent, 3, 376, 750 (1972).

49. G.E. Vinnstra and B. Zwansberg, Synthesis, 1975, 519.
50. Y. Ju, D. Kumar and R.S. Varma, J. Org. Chem., 2006, **71**, 6697.
51. Nakajima, oda and Inouge, Tetrahedron Lett., 1978, 3107.
52. V.K. Ahluwalia and Bindu Goyal, Unpublished results.
53. J. Bicher, K. Pluta, N. Kraka, K. Brondum, N.J. Christensen and M.V. Vinalder, Synthesis, 1989, 530.
54. H. Priebe, Acta. Chem. Scand. Ser B. 1984, **38**, 895.
55. D. Landini, F. Montanan and F. Rolla, Synthesis, 1974, 428.
56. G. Cainelli and F. Manescalchi, Synthesis, 1976, 472.
57. C. Petrier, A.L. Gemal and J.–L. Luche, Tetrahedron Lett., 1982, **23**, 3361.
57a. D.F. Jones and R. Howe, J. Chem. Soc. C. 1968, 2801 and 2816.
58. G. H. Posner, Organic Reactions, Wiley, New York, 1975, **22**, p. 253 400.
59. I. Tabushi, Z. Yoshida and N. Takahashi, J. Am. Chem. Soc., 1971, **93**, 1820.
60. R. Ikan, A. Markus and Z. Goldschmidt, Isr. J. Chem. 1973, **11**, 591.
61. G. Gainelli, G. Cardillo, M. Orena and S. Sandri, J. Am. Chem. Soc., 1976, **98**, 6737.
62. G.A. Crossby, N.M. Weinshenker and H.S. Un, J. Am. Chem. Soc., 1975, **97**, 2232.
63. K. R. Seddon and A. Stark, Green. Chem., 2002, **4**, 119.
64. H. Xie, S. Zhang, H. Duan, Tetrahedron Lett., 2004, **45**, 2013.
65. A. Kumar, N. Jain and S.M.S. Chauhan, Synthetic Communication, 2004, **34**, 2835.
65a. K. Kimura, M. Fujita and T. Ando, Chemistry Lett., 1988, 137.
65b. I. Ichmoto, K. Fujii, F. Sekido, S. Nonomuna and C. Tatsumi, Agric. Biol. Chem., 1965, **29**, 99.
66. J. Yamakawi, S. Sumi, T. Ando and J. Hanafusu, Chemistry Lett., 1983, 379.
67. R.S. Varma, R.K. Saini and R. Dahiya, Tetrahedron Lett., 1997, **38**, 7823.
68. R.S. Varma and R.K. Saini, Tetrahedron Lett., 1998, **39**, 1481.
69. R. Ballini, Y. Bosica and M. Parrini, Tetrahedron Lett., 1998, **39**, 7963.
70. J. Grunwald, B. Wirz, M.P. Scollar and A.M. Klibanov, J.Am. Chem. Soc., 1986, **108**, 6732.
71. T. Reichstein, Helv. Chem. Acta, 1934, 17, 996.
72. W. Charney, A. Nobile, C. Federbush, D. Sutter, P.L. Perlman, H.L.

Herzog, C.C. Payne, M.E. Tully, M.J. Gentles and E.B. Hershberg, Tetrahedron, 1962, **18**, 591.

73. N.A. Milas and O.L. Mageli, J. Am. Chem., Soc., 1952, **74**, 1471.

74. W. Baker and N.C. Brown, J. Chem., Soc., 1948, 2303.

75. R.Z. Kazandjian and A.M. Klibanov, J. Am. Chem. Soc., 1985, **107**, 5448.

76. B.C. Saunders and B.P. Stark, Tetrahedron, 1967, **23**, 1867.

77. H.J. Schmidt and H.J. Schäfer, Angew. Chem. Int. Ed. Engl, 1979, **18**, 68.

78. H.J. Schmidt and H.J. Schäfer, Angew. Chem. Int. Ed. Engl, 1981, **20**, 109.

79. W.D. Emmons, J. Am. Chem. Soc., 1957, **79**, 5528.

80. W.D. Emmons, J. Am. Chem. Soc., 1954, **76**, 3470.

81. S. Kawai, T. Oshima and F. Egami, Biochem. Biophys. Acta, 1965, **97**, 391.

82. G.I. Nikishin, E.I. Troyanskii and V.A. Joffe, Izv. Akad. Nauk SSSR, 1982, 2758; Chem. Abstr. 1983, **98**, 142911.

83. K. Nakagawa, H. Onoue and K. Minami, Chem. Commun, 1966, 730.

84. G.A. Lee and H.H. Freedman, Tetradron Lett., 1976, 1641.

85. T.G. Clarke, N.A. Hampson, J.B. Lee, J.R. Morley and B. Scanlon, Tetradron Lett, 1968, 5685.

86. A. Burgher and L.R. Modlin, J. Am. Chem. Soc., 1940, **62**, 1079.

87. W.B. Weber and G.W. Gukil, Tetrahedron Lett., 1972, 1637.

88. L. Fieser and M. Fieser, Reagents for Organic Synthesis, Wiley, N.Y., 1967, p. 405.

89. K. Kugi, M. Migr, F. Lipinski, F. Bodesheim and F. Rosendahl, Org. Synth., 1961, **41**, 13.

90. W.B. Weber, G.W. Gukil and I. Ugi, Angew. Chem. Int. Ed. Engl., 1972, **11**, 530.

91. R.G.R. Bacon and D. Stewart, J. Chem., Soc. C 1966, 1384.

92. G.A. Lee and H.H. Freedman, Tetrahedron Lett., 1976, 1641.

93. J. Graefe, I. Forehlick and M. Muehlstiedt, 1974, **14**, 434.

94. M.Makosza and A. Kacprowicz, Rocz. Chem., 1975, **49**, 1627.

95. K. Kariyone and H. Yazawa, Tetrahedron Lett., 1970, 2885.

96. J.M. Coxan, E. Dansted and M.P. Harstsorn, Org. Synth. Collective Volume, 1988, **6**, 946.

97. G.B. Payne, P.H. Deming and P.H. Williams J. Org. Chem., 1961, **26**, 659.

98. T. Sasaki, S. Eguchi, T. Kiriyama and Y. Sakito, J. Org. Chem., 1973, **38**, 1648.

99. S. Cannizzaro, Ann., 1853, 88, 129; C.G. Swain, A.L. Powell, W.A. Shippard and C.R. Morgan, J. Am. Chem. Soc., 1979, **101**, 3576.

100. H.D. Dakin, OS, 1941, **1**, 149; J.E. Letter, Chem., Rev., 1949, **45**, 385.

101. A.V. Baeyer and V. Villiger, Ber., 1899, **32**, 3625.

102. J.M.J. Frechet, K.E. Haque, Macromoleucles, 1975, **8**, 130.

103. V.K. Ahluwalia and R.S. Varma, Alternate Energy Processes in Chemical Synthesis, Narosa Publishing House, 2008, Page 13.1 and the references cited therein.

103a. W.P. Jorissen and P.A.A. Van Der Beek, Rec. Travi Chem. Pays-Bas, 1926, **45**, 245.

103b. B. Branchaud and C.T. Walsh, J. Am. Chem. Soc., 1985, **107**, 2153.

104. R.S. Varma and K.P. Naicker, Molecules on line, 1998, **2**, 94.

105. R.S. Varma, K.P. Naicker, D. Kumar, R. Dahiya and P.J. Liesen, Microwave Power Engineering Energy, 1999, **34**, 113.

105a. R.S. Varma and R.K. Saini, Tetrahedron Lett., 1997, **38**, 4337.

106. G. Bouveault and G. Blanc, Compt. Red., 1903, **136**, 1676; H.O. House, Modern Synthetic Reactions, W.A. Benjamin, California, p. 150.

107. J.H. Bablar and B.J. Inergo, Tet. Lett., 1981, **22**, 621.

108. E.M. Gorden, D.C. Gaba, K.A. Jebber and D.M. Zacharias, Organometallics, 1993, **12**, 5020.

109. R.S. Varma, G.W. Kabalka, L.T. Evans and R.M. Pagni, Synth. Commun., 1985, **15**, 279.

110. C.M. Starks and C. Liotta, Phase Transfer Catalysts, Principles and Techniques, Academic Press, Inc. Ny, 1978, p. 343.

110a. M. Desrut, A. Kergomard, M.F. Rerard and H. Veschmbre, Tetrahedron, 1981, **37**, 3825.

111. R. Breslow, J. Am. Chem., Soc., 1958, **80**, 3719.

112. G.A. Lee and H.H. Freedman, Tetrahedron Lett. 1976, 1641.

113. N.A. Milas and S. Sussman, J.Am. Chem. Soc., 1936, 58, 1302; R. Daniels and J.L. Fischer, J. Org. Chem., 1963, **28**, 320.

114. K. Tanaka and F. Toda, Chem. Rev., 2000, **100**, 1028–29.

115. R.S. Varma, R. Dahiya and S. Kumar, Tetrahedron Lett., 1997, **38**, 2039.

116. R.S. Varma and R. Dahiya, Syn. Lett., 1997, 1245.

117. R.S. Varma and R. Dahiya, Tetrahedron, 1998, **54**, 6293.

118. W.P. Reeves, J.A. Murry, D.W. Willoughby and W.J. Friedrick, Synth. Commun., 1988, **18**, 1961.

119. J.W. Huffman, W. Liao and R.H. Wallace, Tetrahedron Lett., 1987, **28**, 3315.

119*a*. E. Kienan, E.K. Hafeli, K.K. Seth and R. Lamed, J. Am. Chem. Soc., 1986, **108**, 162.

119*b*. B. Zhou, A.S. Gopalan, F. Van Middlesworth, W.–R. Shieh and C.J. Sih, J. Am. Chem., Soc., 1983, **105**, 5925.

120. H.D. Durst, Tetrahedron Lett., 1974, 2421.

121. R. Gedye, F. Smith, K. Westaway, H. Ali, L. Baldisera, L. Laberge and J. Rousell, Tetrahedron Lett., 1986, **27**, 279.

122. J.M. Khurana, P.K. Sahoo and G.C. Maikap, Synth. Commun., 1990, 2267.

123. R.S. Davidson, A. Safdar, J.D. Spencer and D.W. Lewis, Ultrasonics, 1987, **25**, 35.

124. E. Santaniello, P. Ferraboschi and P. Sozzani, J. Org. Chem., 1981, **46**, 4584.

125. A. Loupy. P. Pigeon, M. Ramdani and P. Jacqualt, Synth. Commun., 1994, **24**, 159.

126. J. An. L. Bagnelli, T. Cablewski, C.R. Straus and R.W. Trainer, J. Org. Chem., 1997, **62**, 2505.

127. D. Kumar, V.B. Reddy, B.D. Mishra, R.K. Rana, M.N. Nadagouda and R.S. Varma, Tetrahedron, 2007, **63**, 3093.

128. Boehm, Schumann and Hasen, Arch. Pharm, 1933, 271, 490; Hershberg and Cason, Org. Synthesis, 1941, **21**, 84.

129. C.J. Pedersen and H.K. Friensdorff, Angew. Chem. Int. Ed. Engl, 1972, 11, 16; C.J. Pedersen, J. Am. Chem. Soc., 1967, **89**, 2485, 7017; 1970, **92**, 386, 391.

130. R.N. Gedye, F.E. Smith, K.C. Westaway, Can. J. Chemistry, 1988, **66**, 17.

131. R.N. Gedye, W. Rank, K.C. Westaway, Can. J. Chem., 1991, **69**, 706.

132. A. Akiya and P.E. Savage, Chem. Rev., 2002, **107**, 2725 and the references cited therein.

133. T. Sarai, K. Ishiguro, K. Kawashina and K. Morita, Tetrahedron Lett., 1973, 2121.

134. P. Pasta, G. Mazzola, G. Carrea and S. Riva, Biotech. Lett., 1989, **11**, 643.

135. R.M. Lau, F. Van Rantwijk, K.R. Seddon and R.A. Sheldon, Org. Lett., 2002, **2**, 4189.

136. J. A. Laszo and D.L. Compton, Biotechnol. Bioeng. 2001, **75**, 181.

137. E.S. Wallis and J.F. Lane, OR, 1946, III, 267; A.W. Hofmann, Ber., 1881, **14**, 2725.

138. J. Elguero, P. Goya, J. Lissavetzky and A.M. Valdeomillos, C.R. Acad. Scier. Paris, 1984, **298**, 877.

139. J.J. Ritter and P.P. Mineri, J. Am. Chem. Soc., 1948, **70**, 4045; J.J. Ritter and J. Kalish, J. Am. Chem. Soc., 1948, **70**, 4048.

Protecting Groups

6.1 INTRODUCTION

In substrates, if there are two functional groups, both of which react with the same reagent, in such cases one of the functional group has to be protected and reaction carried out. After the reaction, the protected group is deprotected. For example, if it is necessary to prepare a Grignard reagent from an alkyl halide that also contains a hydroxyl group, the Grignard reagent can still be prepared if the hydroxy group is first protected by conversion to a functional group that does not react with the Grignard reagent, *e.g.*, a silyl ether. The Grignard reaction can be conducted with the alkyl halide in which the hydroxyl group has been protected. After the preparation of the Grignard reagent, the original hydroxyl group can be liberated by cleavage of the silyl ether with fluoride ion.

Another example is that of a substrate containing a keto group and a ester.

In an attempt to reduce an ester by lithium aluminium hydride, the keto group which can be more easily be reduced) has to be protected, by protecting the keto group, by converting it to a cyclic acetal.

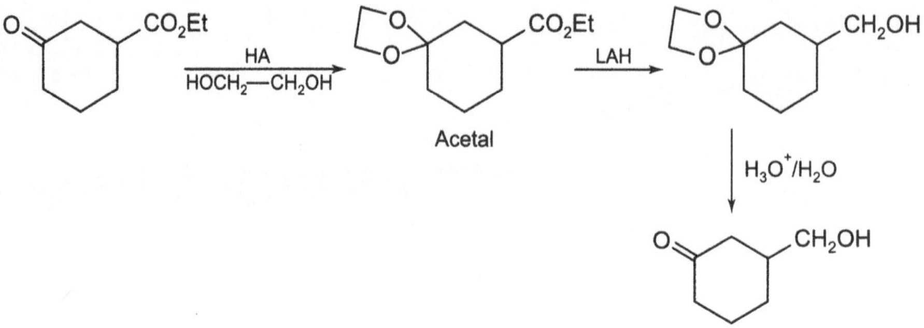

Acetal

The blocking group which is used to block a functional group is called a protecting group (acetal in the above illustration). This process is called **protection**.

The synthesis, in which protection and deprotection is required adds two synthetic steps to an overall synthesis. This process generates wastes and should be avoided as far as possible.

Following are given the protecting groups which are used to protect various functional groups. The functional groups included are alcohols, diols, aldehydes and ketones, carboxyl group and amines.

6.2 PROTECTION OF FUNCTIONAL GROUPS

A number of reviews[1,2] are available which give information about the protection of various functional groups. In this section, some common protecting groups for various functional groups along with procedure for deprotection are given.

6.2.1 Protection of Alcohols

(*a*) **Methyl ethers:** The simplest protecting group is the methyl ether ($-OCH_3$), which can be prepared by any of the following procedures.

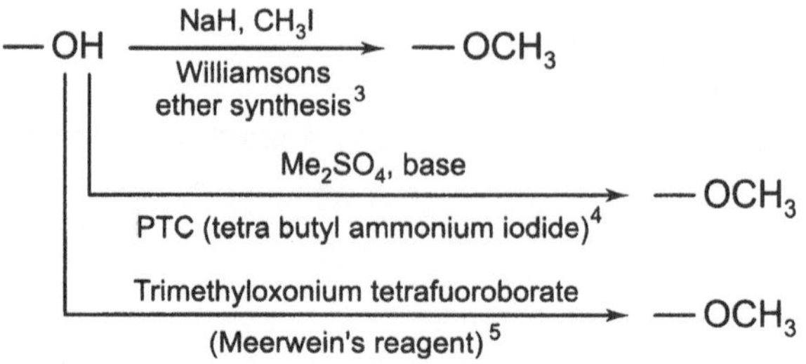

The methyl ethers can cleaved by heating with conc. HI or with trimethyl silyl iodide in chloroform at ambient temperatures.

(*b*) **Benzyl ethers (— O — CH₂Ph):** The alcohol on treatment with benzyl bromide or chloride in presence of a base (NaOH) gives the corresponding benzyl ether[6]. The benzyl ethers can be deprotected by hydrogenolysis (H₂/Pd catalyst).

$$—OH \xrightarrow{\text{NaOH, C}_6\text{H}_5\text{CH}_2\text{X}} —OCH_2C_6H_5$$

The phenolic OH can also be benzylated by heating with benzyl chloride in acetone is presence of K₂CO₃.

A convenient method of deprotection of benzyl ethers under solvent free condition is described (R.S. Varma, A.K. Chattrjee, M. Varma, Tetrahedron Lett., 1993, **34**, 4603). In this procedure, the benzyl ether is adsorbed on acidic alumina (by shaking an methylene chloride solution of benzyl ether with acidic alumina and subsequently the solvent evoporated) and irradiated in a beaker with microwaves using a MW oven for 3–10 min. The required product is isolated by extraction with hexane-ether and then with methanol-dichloromethane (4 : 1). Methanol is removed (vacuo) to give the hydroxy compound in about 95 per cent yield.

(*c*) ***t*-Butyl ether [— O —C(CH₃)₃]:** The reaction of alcohol with isobutylene (2-methyl-2-propene) in presence of acid catalyst gives[7] *t*-butyl ether.

$$—OH + CH_3—CH{=}CH_2 \xrightarrow{\text{H}^+} —OC(CH_3)_3$$
$$\qquad\qquad\;\; |$$
$$\qquad\qquad\;\; CH_3$$

The *t*-butyl group can be removed by strong aqueous acid or by treatment with anhydrous trifluoroacetic acid (TFA)[7] or with trimethyl silyl iodide[8].

(*d*) **Methoxy methyl methyl ethers (— O — CH₂OCH₃):** It is formed[9] by the reaction of the alcohol with a base (*e.g.*, NaH in THF) and chloromethyl methyl ether (ClCH₂OCH₃).

$$—OH + ClCH_2OCH_3 \xrightarrow{\text{NaH/THF}} —O—CH_2OCH_3$$

This group can be removed[10] by treatment with 50 per cent aqueous acetic acid with sulfuric acid as catalyst.

(e) **Methoxyethoxy methyl ethers (— O — CH$_2$OCH$_2$CH$_2$OCH$_3$):**
The alcohol on treatment with a base (NaH in THF or DME) and reacting with methoxyethoxychloromethyl ether gives methoxyethoxymethyl ether[11].

$$—OH + ClCH_2OCH_2CH_2OCH_3 \xrightarrow{\text{NaH/THF}} — O— CH_2OCH_2CH_2OCH_3$$

This group can be cleaved in very strong aqueous acid (pH < 1) or by titanium tetrachloride in dichloromethane[11].

(f) **Tetrahydropyranyl ethers (— O — 2C — C$_5$H$_9$O, THP):** The alcohol on reaction with dihydropyran in presence of *p*-toluene sulphonic acid (PTS) gives the THP derivative[12].

R — OH +   Dihydro pyran $\xrightarrow{\text{PTS}}$ THP deriv.

The THP group can be removed by treatment with aqueous acid[13].

(g) **Trimethyl silyl ethers (— O — Si(CH$_3$)$_3$, o– TMS):** Treatment of alcohol with chlorotrimethyl silane in presence of triethylamine or pyridine gives[14] the trimethyl silyl ether.

$$— OH + Me_3SiCl \xrightarrow{\text{Et}_3\text{N}} — O — Si(CH_3)_3$$

This group can be deprotected by fluoride ion (such as tetrabutyl ammonium fluoride)[15].

(h) **Acetates (— OCOCH$_3$):** Acetates are obtained by treating alcohols with acetic anhydride or acetyl chloride in presence of triethyl amine or pyridine.

$$— OH + CH_3COCl \xrightarrow{\text{Et}_3\text{N}} — OCOCH_3$$

Deprotection can be affected by hydrolysis with acids or bases (saponification). Also by using microwaves, the acetates can be deacetylated by absorbing on to neutral alumina and then heating the absorbed material in a MW Oven (R.S. Varma, M. Varma and A.K. Chatterjee, J. Chem. Soc. Perkin Trans-1, 1993, 999).

6.2.2 Protection of Diols
Cyclic Ketals

1,2-Diols on reaction with acetone in presence of acid catalyst give cyclic ketals[16].

| 2,3-Butanediol | Cyclic ketal (1,3-dioxolane) (acetonides) |

Cyclic ketals can also be obtained[17] by the reaction of diol with 2-methoxy-1-propene in presence of anhydrous HBr.

Cyclic ketals can be cleaved with HCl[18], acetic acid[19] or *p*-toluene sulfonic acid in methanol[20].

In place of acetone, other aldehydes or ketones can also be used to prepare a wide array of ketals or acetals.

6.2.3 Protection of Aldehydes and Ketones

    (*a*) **Ketals and acetals:** The reaction of carbonyl compound (aldehydes or ketones) with methanol or ethanol under anhydrous conditions in the presence of an acid catalyst and (dry HCl gas, sulfuric acid[21] or *p*-toluene sulfonic acid). The addition of molecular sieves to adsorb water improves the yield.

Ketone, R = R′ = alkyl   Methanol R″ = CH₃   Ketal or acetal
Aldehyde R = alkyl, R′=H   Ethanol R″=C₂H₅

    The ketals and acetals can be converted back to the ketone or aldehyde by treatment with aqueous acid (*e.g.*, trifluoroacetic acid[22]) or oxalic acid[23]. Alternatively, the ketal or acetal is treated with acetone along with *p*-toluenesulfonic acid; by this procedure acetone generates a new ketal (2,2-dimethoxy propane), liberating the originally protected carbonyl derivative[24].

    A common protecting group for ketones or aldehydes is to react with 1, 2-ethane diol (ethylene diol) and *p*-toluene sulfonic acid,

BF_3–OEt_2 or oxalic acid to yield the corresponding 1,3-dioxylane derivative.

Reaction of the carbonyl group with 1, 3-propane diol (propylene glycol) under similar conditions generates the corresponding 1, 3-dioxanes.

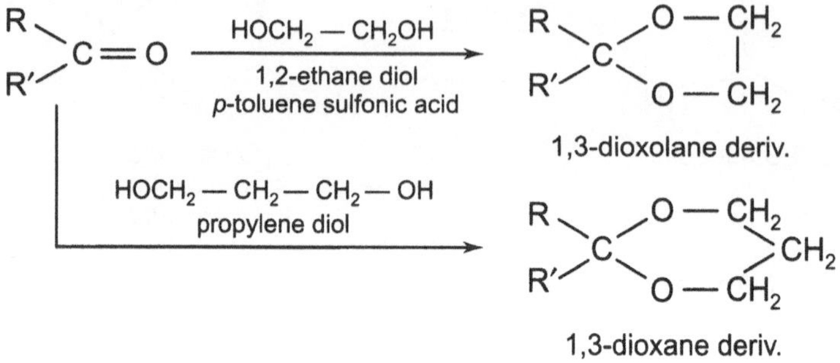

1,3-dioxolane deriv.

1,3-dioxane deriv.

A convenient method for the cleavage of dioxane or dioxolane derivative is to treat with aqueous acid (HCl in THF)[25] or aqueous acetic acid[26].

(*b*) **Dithioketals and dithioacetals:** Treatment of a ketone or a aldehyde with a thiol (mercaplan) under acidic conditions (HCl or BF_3) gives[27] the corresponding dithioketals and dithioacetals, respectively.

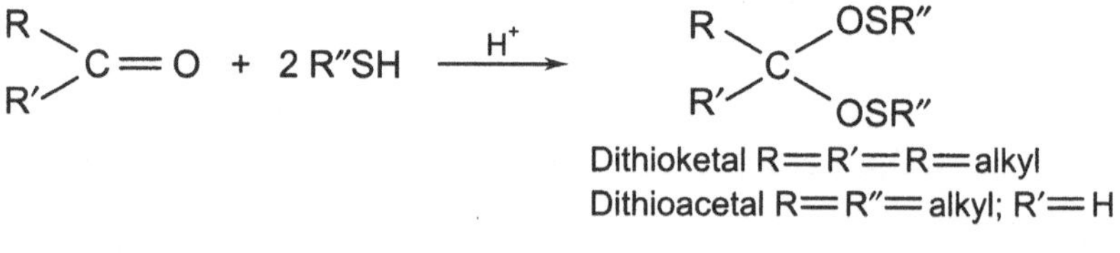

Dithioketal R=R'=R=alkyl
Dithioacetal R=R"=alkyl; R'=H

Use of 1, 2-ethanedithiol or 1, 3-propanedithiol in place of a thiol in the above reaction gives the corresponding 1, 3-dithiolanes and 1, 3-dithianes, respectively.

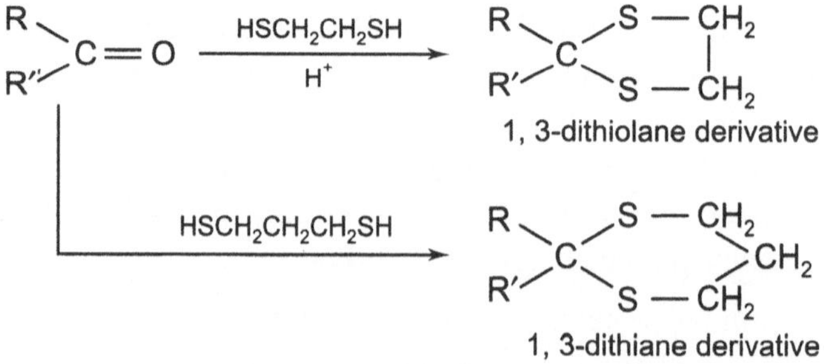

1, 3-dithiolane derivative

1, 3-dithiane derivative

The dithiolane or dithiane derivative can be converted back to the original carbonyl group by treatment with aqueous mercuric chloride under aqueous conditions[28].

A convenient method of dethioacetalization of thioacetals and thioketals of aldehydes and ketones is by heating them in solid state in a MW oven in presence of Clayfen for 40 seconds (R.S. Varma and A.K. Chatterjee, J. Chem., Soc. Perkin Trans I, 1993, 999).

6.2.4 Protection of Carboxyl Group

Protection of carboxyl group in α-amino acids is an important step in the synthesis of peptides. The usual means of carboxy protection is esterification. The parent carboxylic acid can be regenerated from esters by acyl-oxygen or alkyl-oxygen fission. Following esters are used for the protection of carboxyl group.

(*a*) **Methyl and ethyl esters:** Amino acids on reaction with hot methanolic hydrogen chloride give the corresponding methyl ester hydrochlorides. An alternative procedure is the treatment of the amino acid with thionyl chloride and methanol. The hydrochlorides on neutralisation give the corresponding free bases.

$$H_2NCH_2COOH \xrightarrow{\ CH_3OH\ +\ HCl(g)\ } Cl^-H_3^+N - CH_2COOMe$$

$$\downarrow \text{alkali}$$

$$H_2N - CH_2COOMe$$

For deprotection, saponification in a satisfactory procedure.

(*b*) **Benzyl esters:** Amino acid benzyl esters are prepared by reaction with benzyl alcohol in presence of *p*-toluene sulfonic acid, the formed water is removed in a Dean and Stark apparatus.

$$H_2N-CH-COOH + C_6H_5CH_2OH \xrightarrow[\Delta,\ -H_2O]{\substack{p\text{-toluene}\\ \text{sulfonic acid}}} TOS^-H_3N^+CHRCO_2CH_2C_6H_5$$
$$\overset{|}{R}$$

The corresponding free bases are unstable. The benzyl ester groups can be cleaved by saponification or by hydrogenolysis. These can also be cleaved by HBr/AcOH, HF and by catalytic hydrogenolysis, but not by TFA. The benzyl esters can also be deprotected as in the case of benzyl ethers of Phenols (see section 6.2.1).

A convenient green procedure for deprotection of benzyl esters by adsorbing on acidic alumina (by dissolving the benzyl ester in CH_2Cl_2, mixing with acidic alumina and evoparation of the solvent) and irradiation in a microwave oven in a beaker which is placed in an alumina bath (heat sink) inside the microwave oven (130–148°). The debenzylation is complete in 3–10 min. Finally, the debenzylated product is extracted with hexane-ether (4 : 1) and then with MeOH–CH_2Cl_2 (4 : 1). Methanol is removed under vocuo to give about 95 per cent yield of the carboxylic acid. The hexane-ether extract removes only the byproduct (benzyl alcohol) (R.S. Varma, A.K. Chatterjee and M. Varma, Tetrahedron Lett., 1993, **34**, 4603).

(*c*) *t*-**Butyl esters:** These are prepared from the amino acid by treatment with methylene chloride saturated with $Me_2C = CH_2$ in presence of catalytic amount of sulphuric acid.

$$\underset{\substack{\text{N–Protected amino}\\\text{acid}}}{Z\ NHCH(R)\ CO_2H} + \underset{\substack{\text{in } CH_2Cl_2}}{Me_2C = CH_2} \xrightarrow[\substack{20°,\ 3\ days}]{\substack{\text{solid soln.ol}\\ H^+}} Z\ NH\ CH(R)CO_2Bu^t$$

$$\downarrow H_2/Pd(C)/EtOH$$

$$H_2NCH(R)CO_2Bu^t$$

The *t*-butyl esters can also be cleaved by treatment with TFA.

(*d*) **Phenacyl esters:** These are prepared by the reaction of — COOH group with phenacyl bromide ($BrCH_2COPh$)

$$— COOH + BrCH_2COPh \longrightarrow \underset{\text{Phenacyl ester}}{— COOCH_2COPh}$$

Deprotection can be affected by Zn/HOAc.

6.2.5 Protection of Amines

(*a*) **Benzyl group (— CH_2Ph):** Benzyl group is a very common protecting group for NH_2 group. Thus, treatment of an amine with benzyl chloride or benzyl bromide in presence of potassium carbonate or hydroxide gives the *N*-benzyl compound[29].

$$— NH_2 + \underset{X = Cl\ or\ Br}{C_6H_5CH_2X} \xrightarrow{K_2CO_3\ or\ ^-OH} — NHCH_2Ph$$

The N — C bond is cleaved by hydrogenolysis by catalytic hydrogenation or by sodium in liquid ammonia[30].

(*b*) **Trimethyl silyl group (—SiMe₃):** Treatment of amine with chlorotrimethyl silane in presence of pyridine or triethylamine gives the *N*-trimethyl silyl compound[31].

$$— NH_2 + ClSiMe_3 \xrightarrow[\substack{anhyd. \\ conditions}]{Et_3N} — NHSiMe_3$$

The trimethyl silyl group can be deprotected under aqueous conditions.

(*c*) **Acyl group (— COCH₃):** The amine on treatment with acetic anhydride or acetyl chloride in presence of triethylamine or pyridine give[32] N-acetyl compound.

$$— NH_2 + CH_3COCl \xrightarrow{Et_3N} — NHCOCH_3$$

This group can be deprotected by treatment with aqueous acid[33] or by treatment with triethoxonium tetrafluoroborate $(Et_3O^+BF_4^-)$[34].

Use of trifluoro acetic anhydride in place of acetyl chloride gives the corresponding trifluoroacetamide[32,35].

$$— NH_2 + (CF_3CO)_2O \xrightarrow{Et_3N} — NHCOCF_3$$

This group is usually removed by potassium carbonate in aqueous methanol[30] or by reduction with sodiumborohydride[37].

(*d*) **N-Carbamate protecting groups:** These groups were developed for the protection of amino acids in peptide synthesis. A number of reagents are available, but the most oftenly used are benzyl chloroformate $\left(C_6H_5CH_2O - \overset{\overset{\displaystyle O}{\|}}{C} - Cl\right)$[38] and di-tert-butyl carbonate $(CH_3)_3 CO - \overset{\overset{\displaystyle O}{\|}}{C} - OC(CH_3)_3$[39]. Both these reagents react with amino group to form derivatives.

The benzyloxy carbonyl group (abbreviated Z) can be removed[40,41] by catalytic hydrogenation or by treating with cold HBr/HOAc.

The tert-butyloxycarbonyl group (abbreviated BOC) can be removed by treatment with HCl[42] or CF_3CO_2H in acetic acid[43].

$$\bigcirc\!\!\!\!-CH_2OC(=O)-Cl + H_2N\!-\!R \xrightarrow[25°]{OH^-} \bigcirc\!\!\!\!-CH_2OC(=O)-NH\!-\!R + Cl^-$$

Benzyl chloroformate   Benzyloxycarbonyl or Z group

$$(CH_3)_3COC(=O)\!-\!OC(CH_3)_2 + H_2N\!-\!R \xrightarrow[25°]{base} (CH_3)_3COC(=O)\!-\!NHR + (CH_3)_3COH$$

Di-tert-butylcarbonate   *tert*-butyloxy carbonyl group or BOC group

REFERENCES

1. T.W. Greene, Protective Group in Organic Synthesis, Wiley, New York, 1980.

2. T.W. Greene, Protective Groups in Organic Synthesis, Wiley, New York, 1991, Vol -II.

3. A.W. Willamson, J. Chem. Soc., 1954, **4**, 229; O.C. Dermer, Chem. Rev., 1934, **14**, 409.

4. A. Merz, Angew. Chem, Int. Ed. Engl, 1973, **12**, 846.

5. H. Meerwein, G. Hinz, P. Hofmann, E. Kroning and E. Pfeil, J. Prakt. Chem., 1937, **147**, 257.

6. C. Czernecki, C. Georgoulis and C. Provelenghiou Tetrahedron Lett., 1976, 3535; T. Iwashige and H. Salki, Chem. Pharm. Bull., 1967, **15**, 1803.

7. H.C. Beyerman and J.S. Bontekoe, Proc. Chem. Soc., 1961, 249; H.C. Beyerman and G.J. Heiszwolf, J. Chem. Soc., 1963, 755.

8. M.E. Jung and M.A. Lyster, J. Org. Chem., 1977, **42**, 3761.

9. A.F. Kluge, K.G. Untch and J.H. Fried, J.Am. Chem. Soc., 1972, **94**, 7827.

10. F.B. Laforge. J. Am. Chem. Soc., 1933, **55**, 3040.

11. E.J. Corey, J.–L. Gras and P. Ulrich, Tetrahedron Lett., 1976, 809.

12. K.F. Bernady, M.B. Floyd, J.F. Poletto and J.J. Weiss, J. Org. Chem., 1979, **44**, 1438; M. Miyashita, A. Yoshikoshi and P.A. Grieeo, ibid. 1977, **42**, 3772.

13. E. J. Corey, H. Niwa and J. Knolle, J. Am. Chem. Soc., 1978, **100**, 1942.

14. E.J. Corey and B.B. Snider, J. Am. Chem. Soc., 1972, **94**, 2549; C.C. sweely, R. Bently, M. Makita and W.W. Wells, ibid, 1963, **85**, 2497.

15. Cited in Organic Synthesis, Michael B. Smith Mc Graw-Hill International Edition, 1994, page 646.

16. O.Th. Schmidt, Methods Carbohydr. Chem., 1963, **2**, 318; A.N. deBelder, Adv. Carbohydr. Chem., 1965, 20, 219.

17. E.J.Corey, S. Kim, S. Yoo, K. Nicolau, L.S. Melvin, D.J. Brunelle, J.R. Falck, E.J. Trybulski, R. Lett and P.W. Sheldrake, J. Am. Chem. Soc., 1978, **100**, 4620.

18. I.J. Bolton, R.G. Harrison, B. Lythgoe and R.S. Manwaring, J. Chem. Soc., C. 1971, 2944.

19. M.L. Lewbart and J.J.Schneider, J. Org. Chem., 1969, **34**, 3505.

20. A. Ichihara, M. Ubukata and S.Sakamura, Tetrahedron Lett., 1977, 3473; J. Kimura and O. Mitsonobu, Bull. Chem. Soc. Jpn., 1978, **51**, 1903.

21. A.F.B. Cameron, J.S. Hunt, J.F. Oughton, P.A. Wilkinson and B.M. Wilson, J. Chem. Soc., 1953, 3864.

22. R.A. Ellison, E.R. lukenbach, and C.–W. Chiu, Tetrahedron Lett., 1975, 499.

23. F. Huet, A. Lechevallier, M. Pellet and J.M. Conia, Synthesis, 1978, 63.

24. E.W. Colvin, R.A. Raphael and J.S. Roberts J. Chem. Soc., Chem. Commun., 1971, 858.

25. P.A. Grieco, M. Nishizawa, T. Oguri, S.D. Burke and N. Marinovic, J.Am. Chem. Soc., 1977, **99**, 5773; P.A. Grieco, Y. Yokoyama, G. P. Withers, F.J. Okuniewics and C.–L.J. Wang, J. Org. Chem., 1978, **43**, 4178.

26. J.H. Babler, N.C. Malek and M. Coghlan, J. Org. Chem., 1978, **43**, 1821.

27. H. Zinner, Chem., Ber., 1950, **83**, 275; E. Fujita, Y. Nagao and K. Kanelo, Chem. Pharm. Bull., 1978, **26**, 3743.

28. J. English, Jr. and P.H. Griswold, Jr, J.Am. Chem. Soc., 1945, **67**, 2039.

29. L. Velluz, G. Amiard and R. Heymes, Bull. Soc. Chim. Fr., 1954, 1012.

30. V. du Vigneaud and O.K. Behrens, J. Biol. Chem., 1937, **117**, 27.

31. J. Pratt, W. D. Massey, F.H. Pinkerton and S.F. Thames, J. Org. Chem., 1975, **40**, 1090.

32. A.G.M. Barrett and J.C.A. Lana, J. Chem. Soc., Chem. Commun., 1978, 471.

33. G.A. Dilbeck, L. Field, A.A. Gallo and R.J. Gargiulo, J. Org. Chem., 1978, **43**, 4593.

34. S. Hanessian, Tetrahedron Lett., 1967, 1549.

35. J. F. Green, G. N. Jham, J.L. Neumeyer and P. Vouros, J. Pharm, Sci., 1980, **69**, 936.

36. M.A. Schwartz, B.F. Rose and B. Vishnuvajjala, J. Am. Chem. Soc., 1973, **95**, 612.

37. F. Weygand and E. Frauendorfer, Chem. Ber., 1970, **103**, 2437.

38. M. Bergman and L. Zervas, Ber., 1932, **65**, 1192.

39. D.S. Tarbell, G. Yamamoto and B.M. Pope, Proc. Natl. Acad. Sci., 1972, **69**, 730.

40. M. Itoh, D. Hagiwara and T. Kamiya, Bull. Chem. Soc. Jpn., 1977, **50**, 718.

41. J. Meienhofer and K. Kuromizu, Tetrahedron Lett., 1974, 3259.

42. G.L. Stahl, R.Walter and C.W. Smith, J. Org. Chem., 1978, **43**, 2285.

43. B.F. Lundt, N.L. Johansen, A.Volund and J. Markussen, Int. J. Pept. Protein Res., 1973, **12**, 258.

❑❑❑

Enantioselective Synthesis

7.1 INTRODUCTION

In a substrate, if there are two different functional groups and the reagent reacts preferentially with one rather than the other, such a reaction is termed **chemoselective**. A typical example is the reduction of *p*-benzoyl benzaldehyde with tetrabutylammonium triacetoxy borohydride to give the corresponding alcohol; in this case, the keto group remains unaffected.

p-Benzoyl benzaldehyde   p-Benzoyl benzylalcohol

Some other examples of chemoselective reactions are given below:

$$O_2N\ CH_2CH_2CH_2CHO \xrightarrow{NaBH_4/EtOH} O_2N\ CH_2CH_2CH_2CH_2OH$$

4-Nitrobutanal   4-Nitro-1-butanol

m-Nitrobenzaldehyde   m-Nitrobenzyl alcohol

Vitamin A   Retinal

On the other hand, if in a reaction an unequal mixture of enantiomers is obtained, the reaction is called **enantioselective**. If it produces only one enantiomer of the two possibilities, the reaction is **enantiospecific**. An example is the reduction of the diketone (2, 4, -hexadione) with Bakers yeast gives 2-(S) hydroxy-4-hexanone with 99% ee S

2,4-Hexadione   2(S)-Hydroxy-4-hexanone
                                               98% ee

The term % ee means % enantiomeric excess. A 0% ee means a 50:50 mixture (racemic mixture), 50% ee means a 75:25 mixture, and 90% ee means a 95:5 mixture. In the above examples, the predominance of the S enantiomer makes this reaction highly enantioselective.

What is the advantage of enantioselective synthesis? This is of special relevance in pharmaceutical industry. A typical example is that of **ibuprofin**, an anti inflamatory drug. It is found that only the (S) isomer is effective; the (R) isomer has no anti inflamatory action, even though it (R isomer) is slowly converted into the (S) isomer in the body.

(S) Ibuprofin

Other examples include methyldopa (an anti-hypertensive drug) and penicillamine (a drug for primary arthritis). In both these cases, the (S) isomer is exclusively effective (the R isomer of penicillamine is highly toxic).

Methyldopa   Penicillamine

An interesting example is that of the drug thalidomide. It was used for several years before 1963 to alleviate the symptoms of morning sickness in pregnant women. It was only in 1963 it was discovered that thalidomide was the cause of horrible birth defects in many children born subsequent to the use of the drug by the pregnant mothers.

Thalidomide

This drug was withdrawn from market due to its association with fetal abnormalities.

Subsequently, evidence was found indicating that one of the thalidomide enantiomers (The right-handed molecule) has the intended effect of curing morning sickness, the other enantiomer, which was also present in the drug in equal amount may be the cause of birth defects.

Thus, the activity of drugs containing stereocenters can vary between enantiomers sometimes with serious or even tragic consequences.

A large number of other examples exist, in which only one enantiomer is active. For more detail see Chem. Eng. News, 1998, **76**, 83–104.

Though a racemic mixture can be resolved into the enantiomeric forms, this procedure is cumbersone and time consuming. In view of the above, it is necessary to carry out stereoselective synthesis. It is well known that all naturally occurring amino acids are L-amino acids.

7.2 ENANTIOSELECTIVE CONVERSIONS OF FUNCTIONAL GROUPS

For enantioselective synthesis, it is important to convert a functional group enantioselectively into other functional group. Following are given some important enantioselective conversions.

7.2.1 Enantioselective Reduction of Alkynes

Depending on the catalyst and other conditions, it is possible to reduce alkynes enantioselectively to give Z (or cis) or E (or trans) alkenes.

Addition of hydrogen to an alkyne (internal alkyne) in presence of nickel boride (also called P-2 catalyst and can be prepared by the reduction of nickel acetate with sodium borohydride) results in syn addition of hydrogen to give Z (or cis) alkene.

$$Ni(O\overset{\overset{\displaystyle O}{\|}}{C}CH_3)_2 \xrightarrow[\text{C}_2\text{H}_5\text{OH}]{\text{NaBH}_4} \quad \begin{array}{c} Ni_2B \\ p\text{--}2 \end{array}$$

$$CH_3CH_2C\equiv C\,CH_2CH_3 \xrightarrow[\text{(Syn addn.)}]{\text{H}_2/\text{Ni}_2\text{B(P--2)}}$$

3-Hexyne

$$\begin{array}{cc} CH_3CH_2 & CH_2CH_3 \\ \diagdown & \diagup \\ C=C \\ \diagup & \diagdown \\ H & H \end{array}$$

(Z)-3-Hexene
(cis-3-hexene)
97%

Z (or cis) alkenes can also be obtained by the reduction of alkynes with **Lindlar's catalyst** (metallic palladium deposited on calcium carbonate, conditioned with lead acetate and quinoline).

$$R-C\equiv C-R \xrightarrow[\substack{\text{Quinoline}\\ \text{(syn addn.)}}]{\substack{\text{H}_2,\text{Pd/CaCO}_3\\ \text{(Lindlar's catalyst)}}} \quad \begin{array}{cc} R & R \\ \diagdown & \diagup \\ C=C \\ \diagup & \diagdown \\ H & H \end{array}$$

Alkyne   (Z) or (cis) alkene

E (or trans) alkenes are best obtained by the reduction of alkyne with sodium or lithium metal in ammonia or ethyl amine at low temperatures; this results in anti addition of hydrogen atoms to the triple bond.

$$CH_3(CH_2)_2-C\equiv C-(CH_2)_2CH_3 \xrightarrow[\text{2) NH}_4\text{Cl}]{\text{1) Li, C}_2\text{H}_5\text{NH}_2,\, -78°}$$

4-Octyne

$$\longrightarrow \quad \begin{array}{cc} CH_3(CH_2)_2 & H \\ \diagdown & \diagup \\ C=C \\ \diagup & \diagdown \\ H & (CH_2)_2CH_3 \end{array}$$

(E)-4-Octene
(Trans-4-octane)
(52%)

7.2.2 Enzymatic Hydroxylation of Aromatic Rings

A number of aromatic compounds have been oxidised with the enzyme *Pseudomonas putida* to give cis-diols.

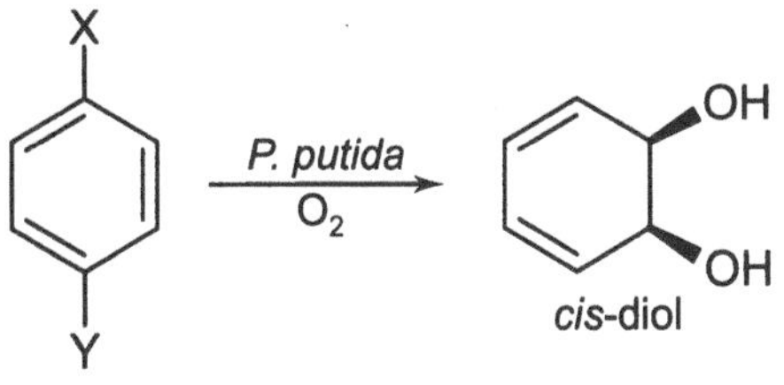

x = y = H

x = Cl, Br, I, F; Y = CH₃

x = H; Y = CH₃

This reaction involves the conversion of a π bond in a benzene ring to a cis-diol.

7.2.3 Enantioselective Reduction of Carbonyl Groups to Give (R) or (S) Alcohols

It has been shown that small ketones like 2-butanone are reduced by the enzyme thermoanaerobium brockii to give R alcohol (2-butanol), but larger ketons like 2-hexanone are reduced to S alcohol. Thus, in these reductions, the selectivity depends on the size and nature of the groups around the carbonyl.

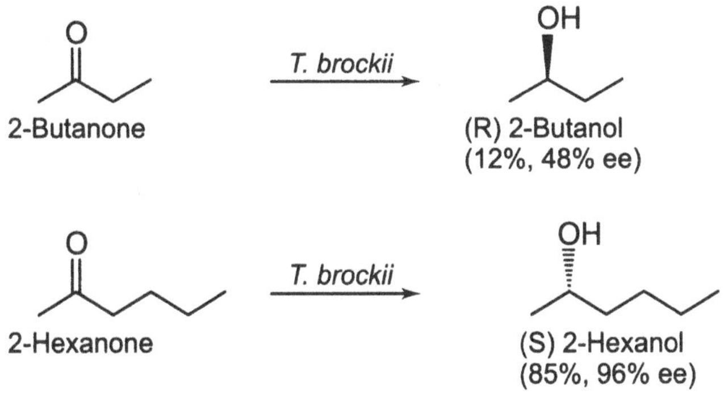

2-Butanone → (R) 2-Butanol (12%, 48% ee)

2-Hexanone → (S) 2-Hexanol (85%, 96% ee)

Bakers yeast has also been used to reduce carbonyl group in β-ketoesters. Thus, reduction of ethyl acetoacetate (β-ketoesters) with Bakers yeast gave the S-alcohol. However, β-Ketovalerate gave the R-alcohol. Thus, the selectivity of reduction changed from S selectivity with small chain esters to R selectivity with long chain esters.

Ethyl acetoacetate → Bakers yeast → S-alcohol 67%

β-Ketovalerate → Bakers yeast → R-alcohol

This is a convenient method for the synthesis of optically active secondary alcohols see also Resolution of carboxylic acids (section 7.2.15).

7.2.4 Conversion of Optically Active Alkyl Halides into Chiral Synthons

Optically active alkyl halides (containing chiral carbon) can be converted into chiral synthons via S_N2 reaction. An example is given below.

In this case, there is inversion of configuration. The alkyl halides can be 1° or 2°. The S_N2 reaction is of great importance for the enantioselective synthesis. For example, (S)-2-methylbutane nitrile can be prepared from (R)-2-bromobutane.

(R)-2-Bromobutane   (S)-2-Methylbutane nitrile

Thus, by reacting an alkyl halide (2° alkyl halide), it is possible to obtain the corresponding ether, thiol, thioether, nitrile, alkyne, esters and azide by

reacting with $\overline{OH}$, $R\overline{O}$, $S\overline{H}$, $R\overline{S}$, $C\overline{N}$, $R—C\equiv\overline{C}$, $R'\overset{\overset{\displaystyle O}{\|}}{C}\overline{O}$ and N_3^-, respectively;

all these are chiral synthons for subsequent enantioselective synthesis.

7.2.5 Resolution of Racemic Ethyl (±)-2-Fluoro Hexanoate into the Enantiomers

The above conversion can be affected by using the enzyme lipase to give pure enantiomers.

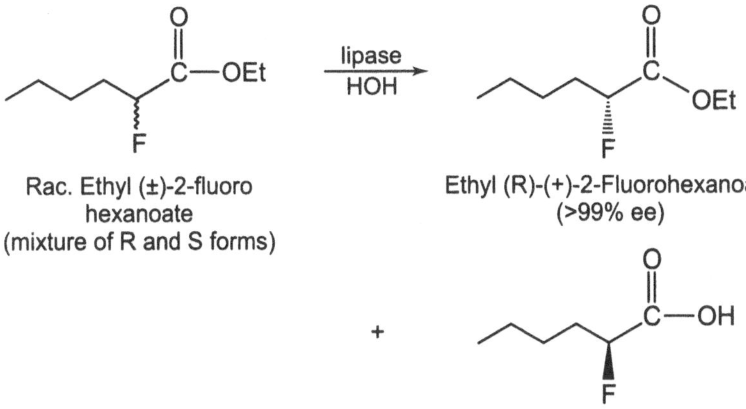

Rac. Ethyl (±)-2-fluoro
hexanoate
(mixture of R and S forms)

$\xrightarrow[\text{HOH}]{\text{lipase}}$

Ethyl (R)-(+)-2-Fluorohexanoate
(>99% ee)

+

(S)-(−)-2-Fluorohexanoic acid
(>69% ee)

7.2.6 Optically Active Secondary Alcohols from Alkenes *via* Hydroboration

Alkenes on hydroboration with an optically active alkyl boranes [like diisopinocampheylborane[6] (Ipc$_2$ BH) and mono-isopinocampheyl borane[7] (Ipc BH$_2$), which are readily obtained by the reaction of borane with (+)- or (-)-α-pinene under appropriate conditions followed by oxidation with alkaline H$_2$O$_2$ give alcohols.

Ipc$_2$BH

Ipc BH$_2$

Thus, Z-alkenes [*e.g.*, (Z)-2-butene] on reaction with (−)–diisopino campheylborane (from (+)-α-pinene), followed by oxidation with alkaline H$_2$O$_2$ gave (R)-2-butanol.

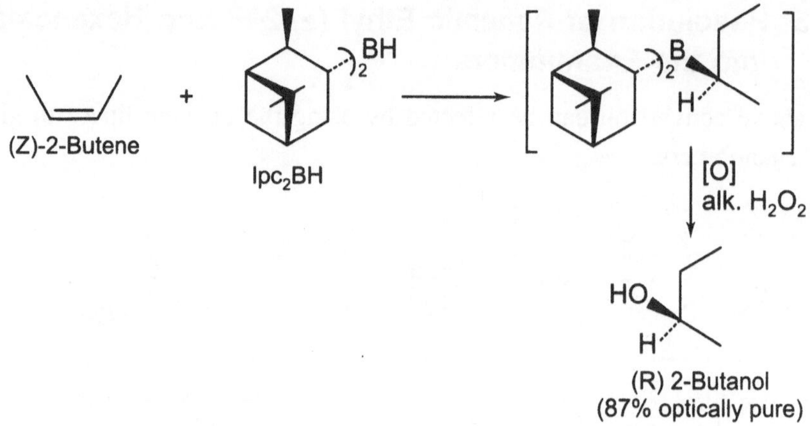

(Z)-2-Butene + Ipc$_2$BH

(R) 2-Butanol
(87% optically pure)

However, E-alkenes [*e.g.*, (E)-2-butene] on reaction with mono isopinocamphenylborane followed by oxidation gave (S)-2-butanol.

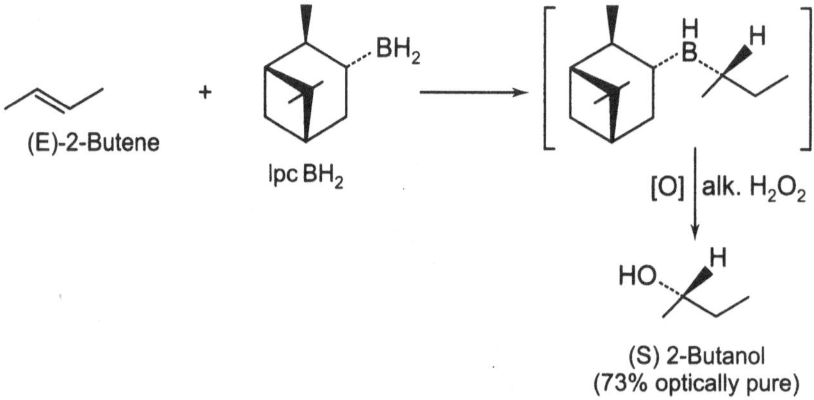

(E)-2-Butene + Ipc BH$_2$

(S) 2-Butanol
(73% optically pure)

With (E)-di-t-butyl ethylene, the corresponding alcohol was obtained in 92 per cent optical purity. For more details about hydroboration, see V.K. Ahluwalia and Rakesh Kumar Parashar, Organic Reaction Mechanism, Narosa Publishing House, 3rd Edn. 2007, page 563–572 and the references cited therein.

Optically active secondary alcohols could also be obtained[8] by the reduction of a number of ketones using trialkyl boron hydride (prepared from (+)-α-pinene and 9-BBN followed by t-butyl lithium).

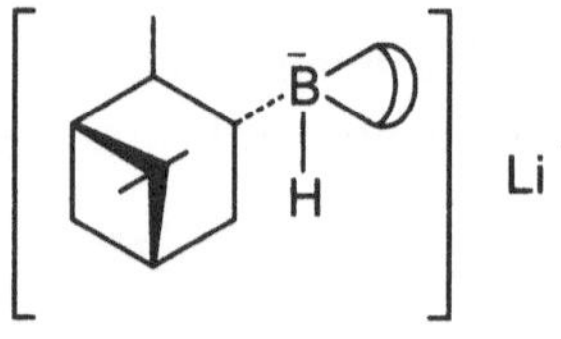

Trialkylboronhydride

7.2.7 Enantioselective Aldol Condensation

It is well known that the aldol condensation involving an achiral aldehyde and an achiral ketone give a mixture of syn and anti aldol products.

However, the aldol reaction involving an achiral aldehyde and an achiral enolate give the aldol products[9,10]. The chiral centre in aldehyde having S configuration is retained in the aldol product. The aldol products are $\underline{S}$RS diasteromers and $\underline{S}$RR diasteromer.

The aldol condensation of achiral aldehyde with an achiral enolate give the aldol products. If the enolate has the chiral centre having R configuration, the aldol products will be RS$\underline{R}$ diastereomer and SR$\underline{R}$ diastereomer.

In case both the aldehyde and the enolate are achiral, configuration of the stereo centre of the aldehyde and its enolate in retained. The aldol products obtained will be $\underline{S}$, R, S, $\underline{R}$ and $\underline{S}$, S, R, $\underline{R}$ diastereomers.

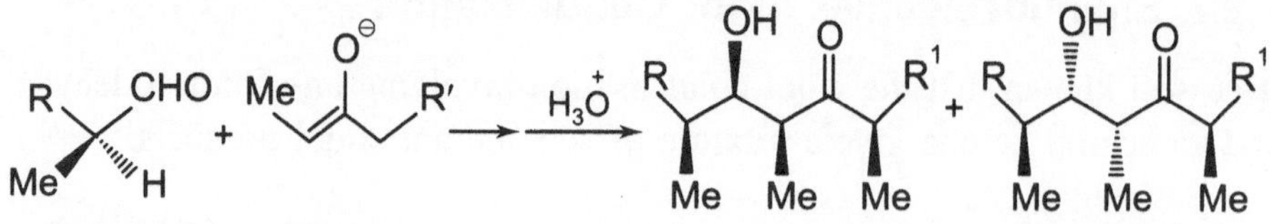

R   CHO   Me   R' $H_3\overset{+}{O}$ S, R, S, R-syn, syn, syn   S, S, S, R-
Me   H   anti, syn, anti

   S   R-
Aldehyde   Enolate

7.2.8 Asymmetric Enamine Synthesis

Michael addition of chiral enamine (prepared from a chiral amine with acrylonitrile or methyl acrylate gave chiral cyclohexanone derivative[12].

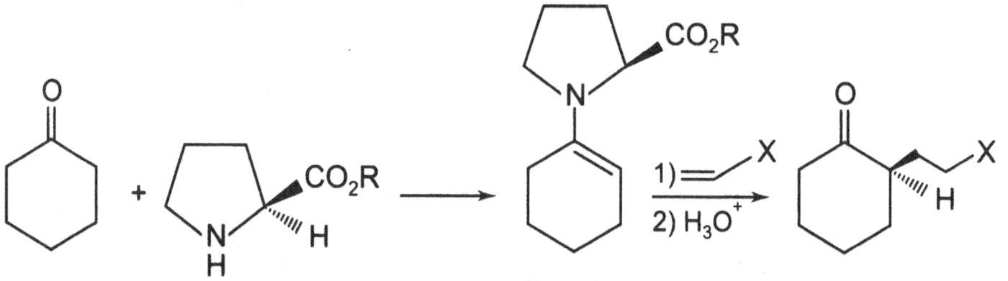

Enamine   X = CN or CO_2Et

Another asymmetric enamine synthesis[13] is given below.

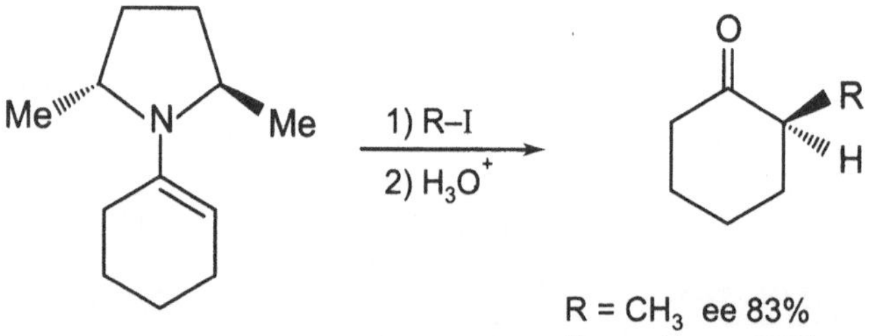

R = CH_3 ee 83%
R = n–Pr ee 93%
R = $CH_2 = CHCH_2Br$ ee 87%

7.2.9 Asymmetric Induction in Friedel-Crafts Reaction

The Friedel-Crafts reaction is known to proceed via a cation and so the reaction with chiral alkyl halides is expected to give racemic products. It has, however been found[14] that if the reaction temperature is kept below 0° and the reaction time is minimum, there is some asymmetric induction. Thus, the reaction of benzene with (S)-2-chlorobutane at 0°, 4 min in presence of $FeCl_3$ give up to 24% ee(R).

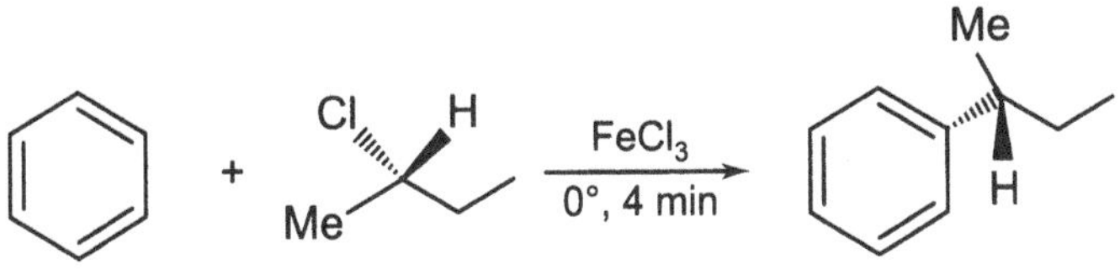

Benzene   (S)-2-Chloro   (R)
butane   24% ee

7.2.10 Asymmetric Baeyer-Villiger Oxidation

It is well known[15] that cyclohexanone on treatment with cyclohexanone oxygenase gives the lactone (ε-Caprolactone).

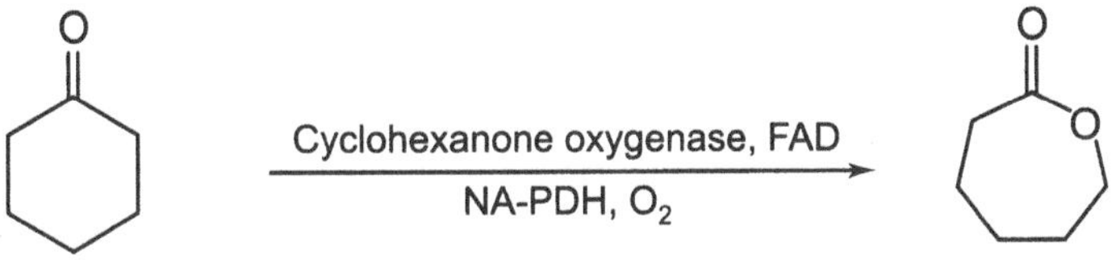

Cyclohexanone   ε-caprolactone

However, 4-methyl cyclohexanone on treatment with cyclohexanone oxygenase obtained from *Acinetobacter*[16] gave the lactone in 80 per cent yield with > 98 per cent ee[17].

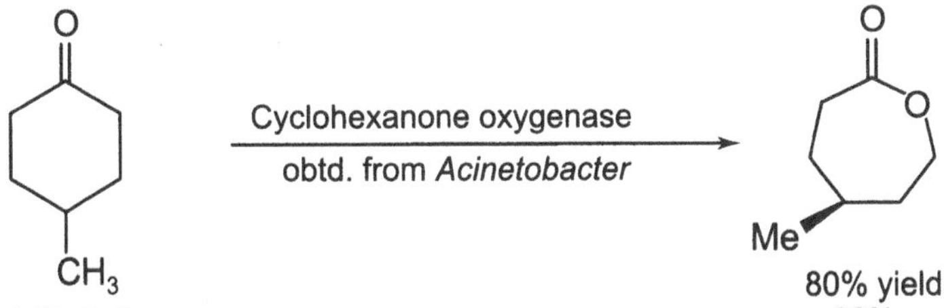

4-Methyl
cyclohexanone

80% yield
> 98% ee
Lactone

7.2.11 Asymmetric Induction in Grignard Reaction

The reaction of a unsymmetrical ketone with Grignard reagent generates a chiral centre.

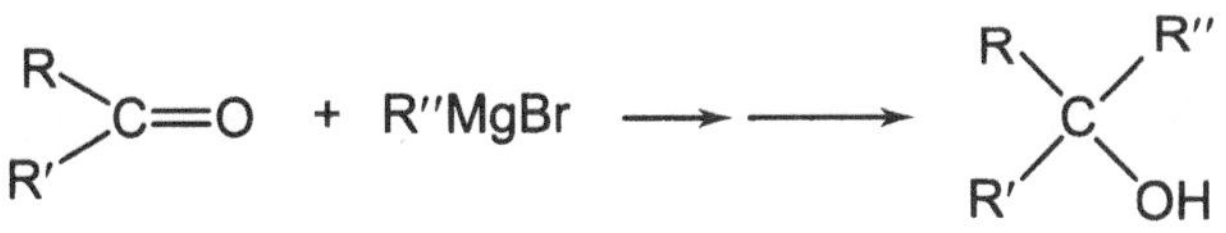

In case a chiral centre is present in either the grignard reagent (but not at the carbon atom bearing Mg) or the carbonyl substrate, diastereomers result. Following four possibilities are:

(*i*) A chiral grignard reagent + carbonyl compound which does not possess a substituent attached to a chiral centre ⟶ one new chiral centre is generated.

(*ii*) Chiral grignard reagent + carbonyl compound which does not possess a substituent attached to a chiral centre (as in (*i*) above) ⟶ diastereomers are generated.

(*iii*) Achiral grignard reagent + carbonyl compound which has a substituent attached to a chiral centre ⟶ diastereomers (with possible diastereoselectively) are generated.

(*iv*) Chiral grignard reagent + carbonyl compound which has a substituent attached to a chiral centres ⟶ good diastereoselectivity results.

7.2.12 Enantioselectivity in Organocuprate Reactions

The organo copper reagents are referred to as **Gilman reactants**[18, 19]. These are prepared as follows:

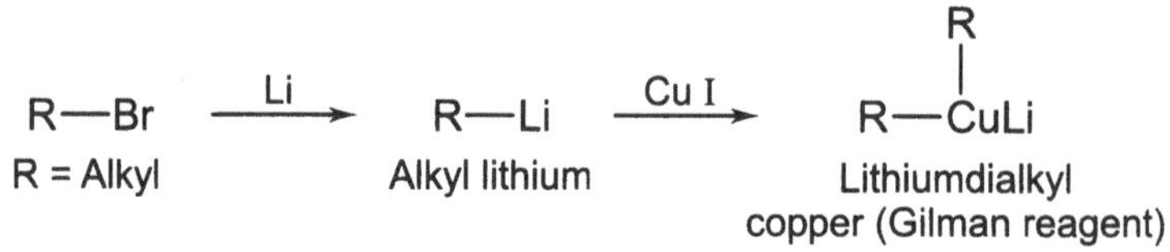

$$R-Br \xrightarrow{\text{Li}} R-Li \xrightarrow{\text{Cu I}} R-\overset{\overset{\textstyle R}{|}}{C}uLi$$

R = Alkyl   Alkyl lithium   Lithiumdialkyl copper (Gilman reagent)

The Gilman reagent react with other alkyl halides to generate a alkane.

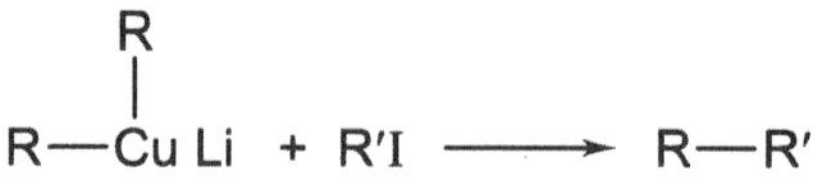

$$R-\overset{\overset{\textstyle R}{|}}{C}u\,Li + R'I \longrightarrow R-R'$$

The above procedure is called the **Corey-Posner, Whitesides-House synthesis.**

Use of R′ I as tertiary halide generates a new stereocentre.

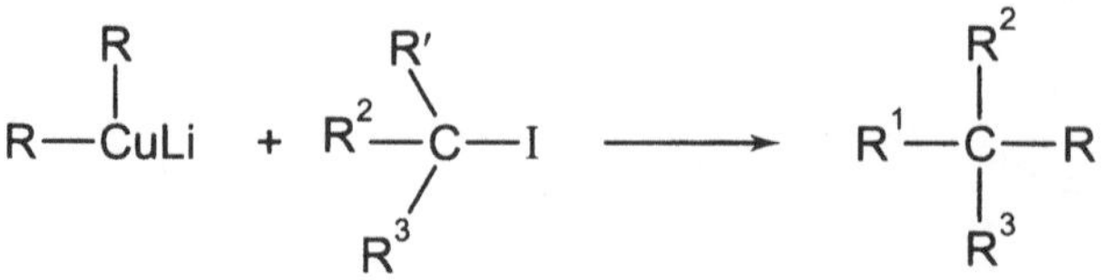

$$R-\overset{\overset{\textstyle R}{|}}{C}uLi + R^2-\overset{\overset{\textstyle R'}{\diagdown}}{\underset{\underset{\textstyle R^3}{\diagup}}{C}}-I \longrightarrow R^1-\overset{\overset{\textstyle R^2}{|}}{\underset{\underset{\textstyle R^3}{|}}{C}}-R$$

7.2.13 Asymmetric Diels-Alder Reaction

A Diels-Alder reaction is a cycloaddition reaction between a conjugated diene and a dienophile.

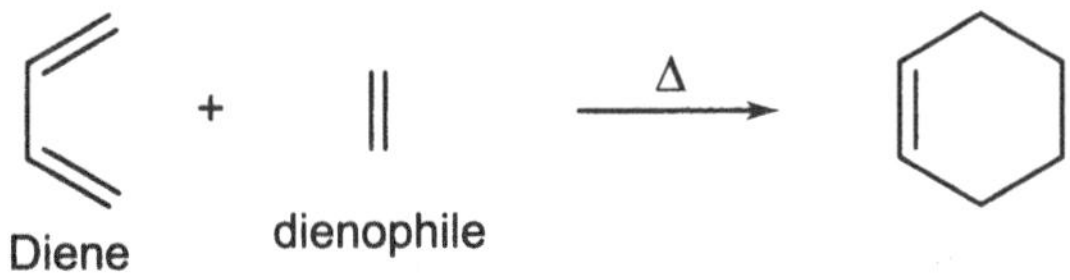

Diene   dienophile

It has been possible to introduce enantioselectivity in Diels-Alder reaction. One method involves the use of chiral auxilaries (A chiral auxiliary is a group present is one enantiomeric form only, that is appended by a functional group to the Diene or dienophile in order to provide a chiral influence on the course of the reaction. After the reaction is over, and the influence of the chiral auxiliary is no longer needed, it is removed by an appropriate reaction. A chiral lewis acid catalyst is used for the reaction. Besides causing extraordinary effective enantioselective product formation, it was recovered and used in future reaction. One example is given below:

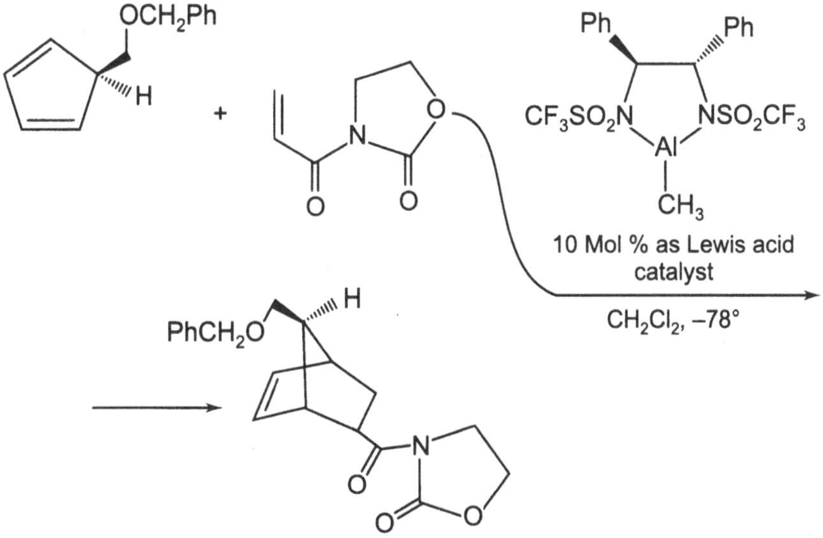

A most common auxiliaries is menthyl, where an acrylic acid derivative is attached to menthol to form a menthyl ester. It has been shown[22] that asymmetric induction is possible with the menthyl ester on reaction with cyclopentadiene to give[23] the adducts.

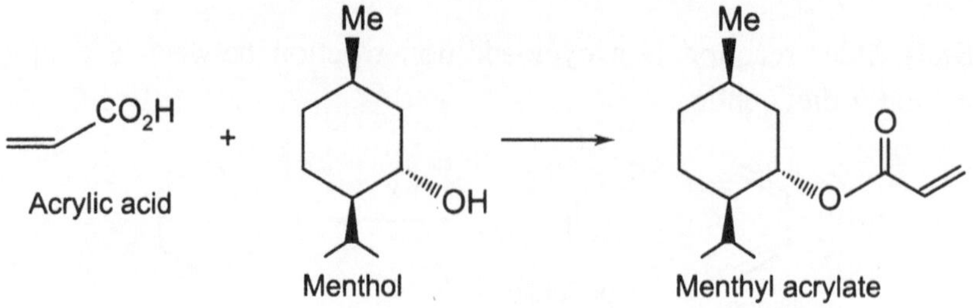

Acrylic acid + Menthol → Menthyl acrylate

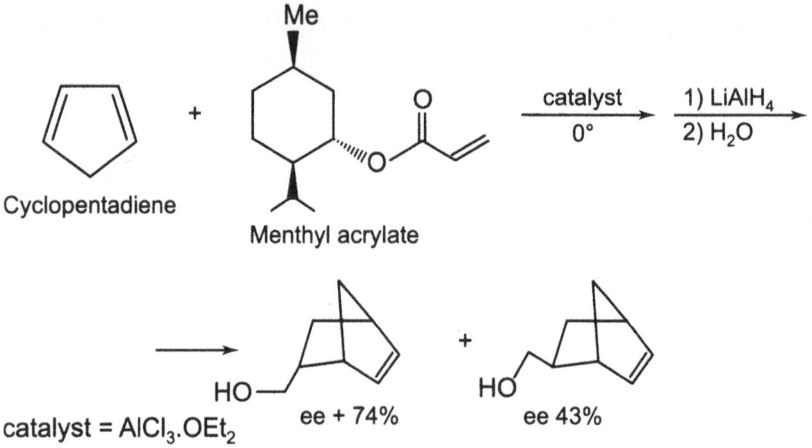

Cyclopentadiene + Menthyl acrylate $\xrightarrow[0°]{\text{catalyst}}$ $\xrightarrow{\text{1) LiAlH}_4}{\text{2) H}_2\text{O}}$

ee + 74%   +   ee 43%

catalyst = AlCl$_3$.OEt$_2$

7.2.14 Hydroxylation of Alkenes

Hydroxylation of alkenes can give either syn diols or anti-hydroxylation depending on the reagent used. For details, see Interconversion of Functional group (section 5.2.2; sub part V). Asymmetric dihydroxylation of alkenes can also be affected. For details, see Interconversion of Functional groups (section 5.2.3(i)).

7.2.15 Resolution of Carboxylic Acid

The carboxylic acid obtained by the various procedures are mixtures of R and S. These can be resolved into the two forms by using an optically active amine. The two salts (R,R) and (S,R) if we use an (R)-amine are separated by crystallisation. The separated pure salts on treatment with acid give the (R) and (S) acid in pure state.

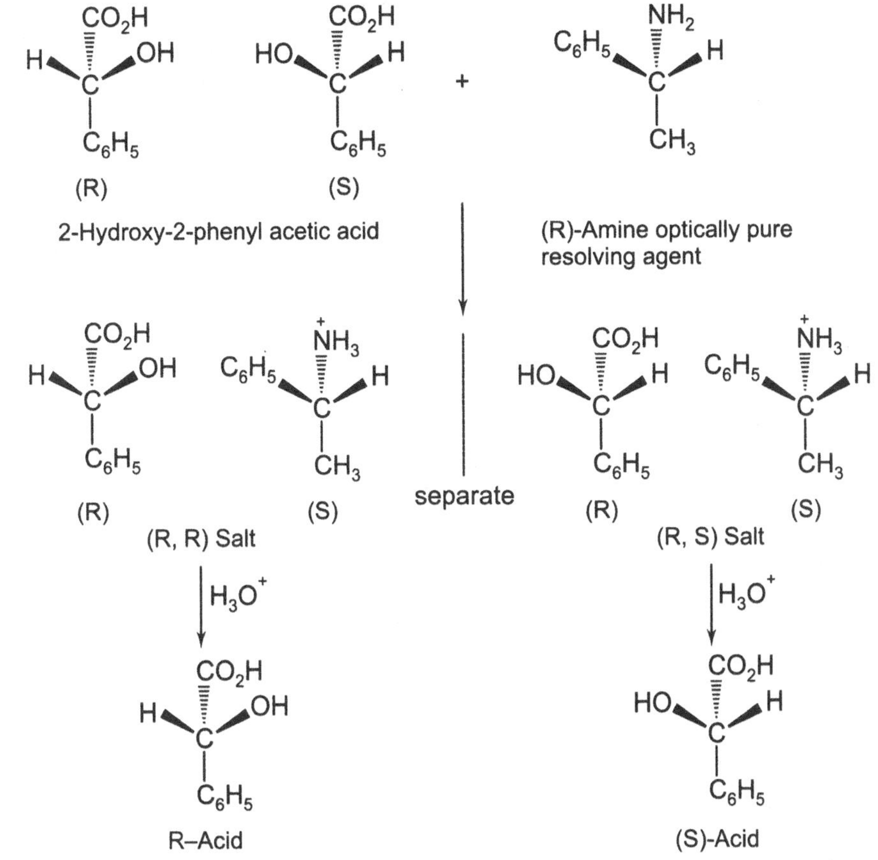
(R)
(S)
2-Hydroxy-2-phenyl acetic acid
(R)-Amine optically pure resolving agent
separate
(R, R) Salt
(R, S) Salt
R–Acid
(S)-Acid

Naturally occurring optically active amines [which are readily available as single enantiomer] like (–)–quinine, (–)–S –strychnine and (–)–brucin are often employed as resolving agents for the resolution of racemic acids.

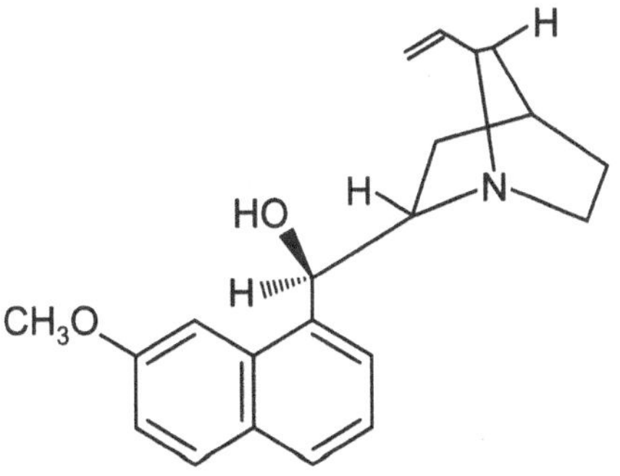

(–)-Quinine

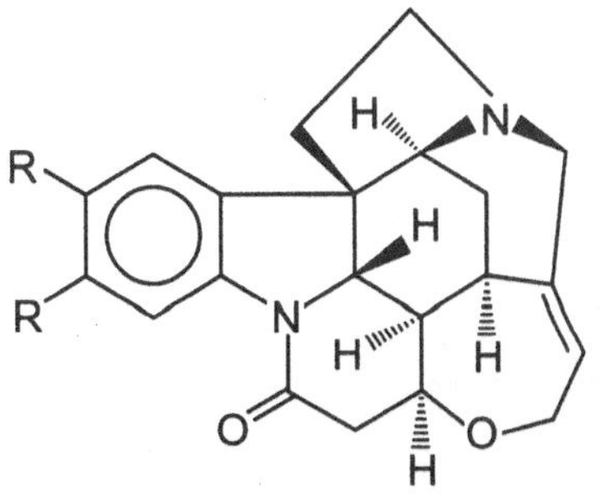

(–)-Strychinine, R = H
(–) Brucine, R = OCH_3

7.3 ENANTIOSELECTIVE SYNTHESIS

7.3.1 Synthesis of Conduritol F and (+)-Pinitol

The cis diol obtained from benzene could be converted into conduritol F in four steps in an overall yield of 25 per cent[24].

The cis-diol from benzene was converted into (+)-pinitol an antidiabetic agent[25].

(+) Pinitol was also prepared from the cis diol obtained from bromobenzene.

7.3.2 Synthesis of 2, 3-Isopropylidene-L-Ribose-γ-Lactone and Acetate of (−)–Specionin

The cis-diol obtained from chloro benzene was converted[26] into 2, 3-isopropylidene-L-ribose-γ-lactone in 4 steps (over all yield 22%).

The same diol has also been used for the synthesis of acetate of (−)−specionin via a synthetic intermediate[27].

Acetate of (−)-specionin

7.3.3 Synthesis of (+)-Lycoricidine

The cis diol obtained[28] from bromobenzene is converted into (+) −Lycoricidine.

Bromobenzene   cis diol   (+)-Lycoricidine

7.3.4 Synthesis of (S) (+) Sulcatol

The (S) hydroxy ester obtained by the reduction of ethylacetoacetate has been used for the synthesis of (S) (+) sulcatol[29].

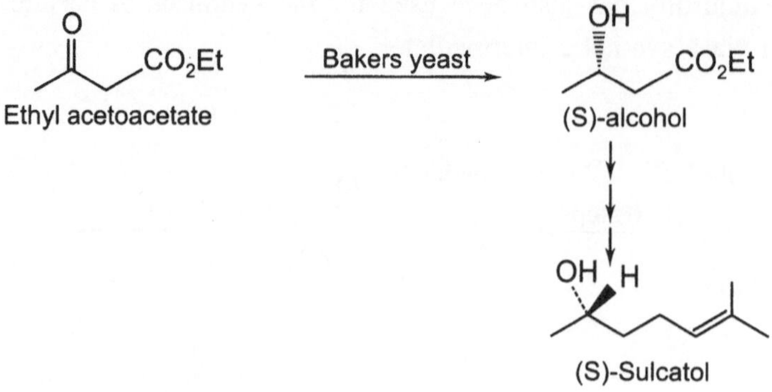

Ethyl acetoacetate → Bakers yeast → (S)-alcohol → (S)-Sulcatol

7.3.5 Synthesis of Thienamycin

Rac. N-Protected aminoglutarate on hydrolysis by procane liver esterase (Plc) followed by deprotection gave the (S)-amino acid; this was converted via the monocyclic β-lactam into thienamycim[30].

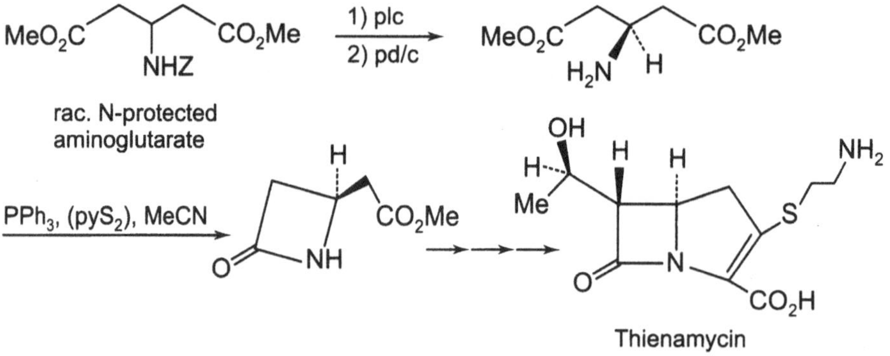

rac. N-protected aminoglutarate

Thienamycin

7.3.6 Synthesis of L-Carnitine

L-Carnitine, an amino acid and a constituent of striated muscle and liver was synthesized[31] by the hydrolysis of the corresponding diester with Corynebacterium equi by following a number of steps.

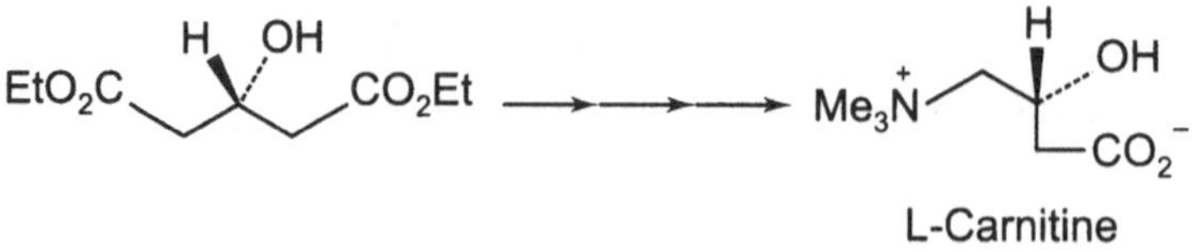

L-Carnitine

7.3.7 Synthesis 6-Aminopenicillic Acid (6APA)

It was obtained[32] on a large scale by the cleavage of 6-acyl group from penicillin G by using an isolated immobilized amidase.

Penicillin G   amidase   6-APA

7.3.8 Synthesis of S-citronellol

Reduction of 7-methyl-3-oxo-oct-6-enoate by Bakers yeast, gave[33] the corresponding (R)-hydroxy acid, which is a synthon for the synthesis of S-citronellol.

7-Methyl-3-oxo-oct-6-enoate   Bakers yeast   (R)-alcohol (97% ee)

(S)-citronellol

7.3.9 Synthesis of Aspartame

It was industrially prepared[34] as shown ahead.

7.3.10 Synthesis of D- or L-Amino Acids

This synthesis has been achieved by the dynamic resolution of amino acids using the enzyme hydantoinase followed by treatment with a second hydrolytic enzyme (carbamolyase) to release the free amino acid of desired configuration. This process has been used on industrial scale[35].

7.3.11 Synthesis of Prednisolone

Prednisolone, a cortisone analog, used as a drug against reheumatoid arthritis is obtained on a commercial scale from Reichsten's compound (a steroid precursor)[36]. This compound is passed though a series of two different columns, each containing a specific enzyme attached to a polymer support.

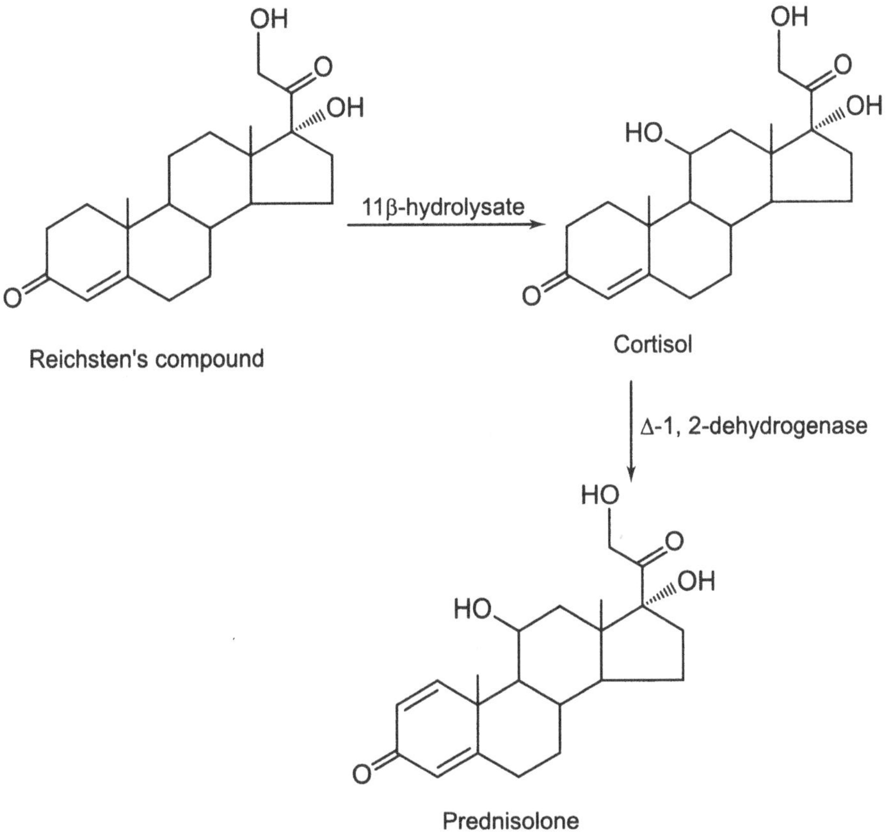

Reichsten's compound

11β-hydrolysate

Cortisol

Δ-1, 2-dehydrogenase

Prednisolone

7.3.12 Asymmetric Synthesis of 2-Alkylcarboxylic Acids

The reaction of 2-alkyl-4, 4-dimethyl-2-oxazolines [which are conveniently prepared[37] by heating 2-amino-2-methyl-1-propanol with carboxylic acid] on treatment with butyl lithium give the lithio derivative. Subsequent reaction of the lithio derivative with alkyl halide give the corresponding 2-alkyl oxazoline. Treatment of 2-alkyl oxazoline with butyl lithium followed by reaction with methyl iodide give the corresponding methyl substituted product, which may be hydrolysed to 2-alkyl carboxylic acid.

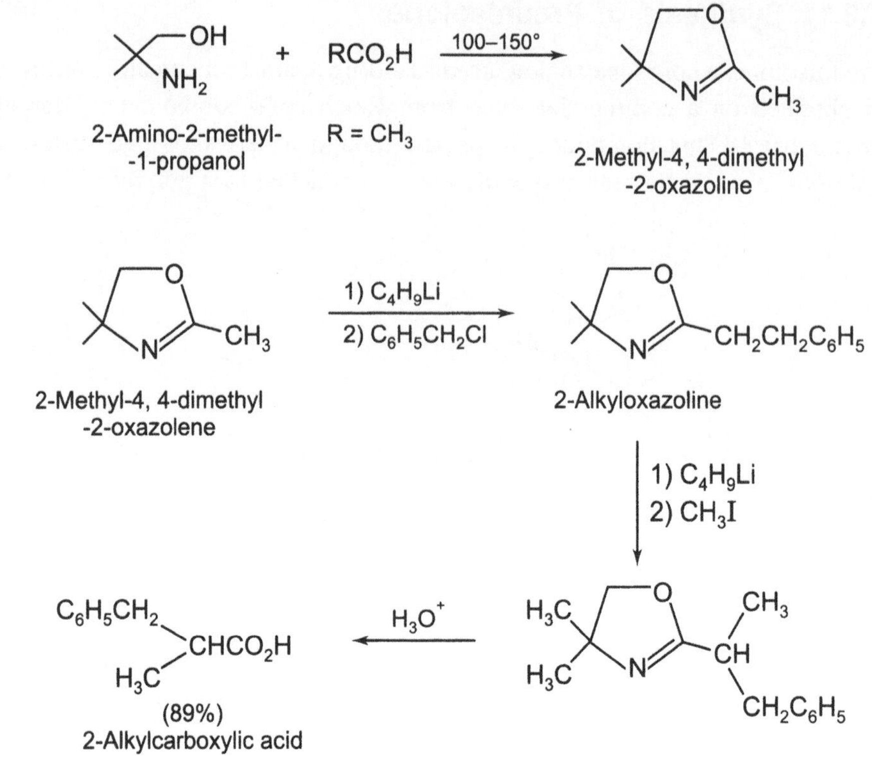
100–150°
RCO2H
R = CH3
2-Amino-2-methyl-
-1-propanol
2-Methyl-4, 4-dimethyl
-2-oxazoline
1) C4H9Li
2) C6H5CH2Cl
2-Methyl-4, 4-dimethyl
-2-oxazolene
2-Alkyloxazoline
1) C4H9Li
2) CH3I
H3O+
(89%)
2-Alkylcarboxylic acid

Use of chiral exazoline in the above procedure gives asymmetric α-alkyl carboxylic acids[38].

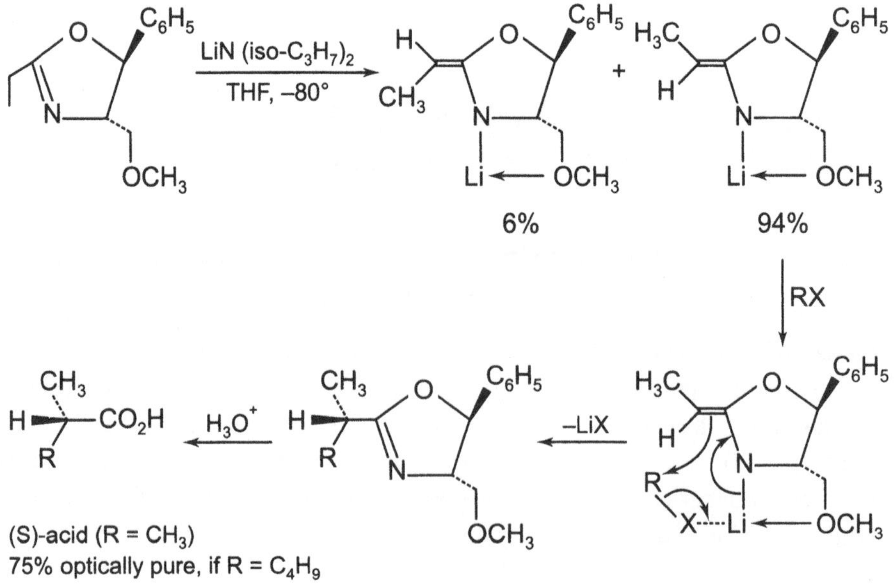
LiN (iso-C3H7)2
THF, –80°
6%
94%
RX
–LiX
H3O+
(S)-acid (R = CH3)
75% optically pure, if R = C4H9

7.3.13 Asymmetric Synthesis of 3-Alkylcarboxylic Acids

The reaction of vinyl-oxazolines with organlithium compounds followed by hydrolysis, give 3-alkyl carboxylic acid[39]. A typical example is given below.

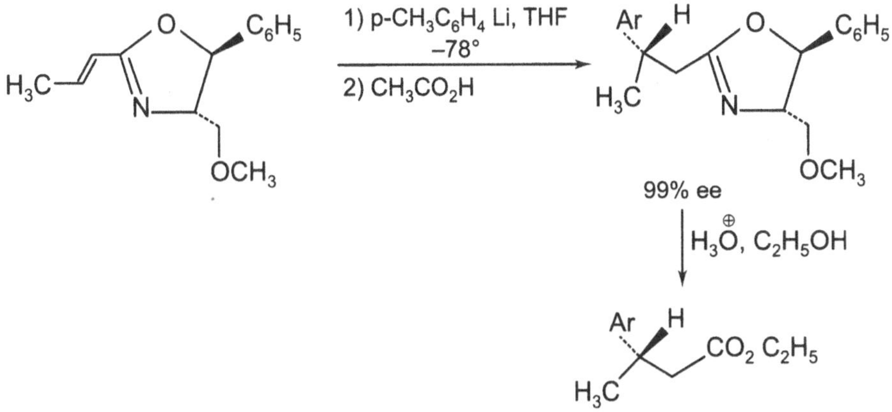

99% ee

3-Alkyl carboxylic acid
ethyl ester

In the above conversion, the oxazolene ring is, infact, a masked carboxyl group. Thus, conversion of an oxazoline is a useful method of protecting carboxyl groups against attack by Li AlH$_4$ or Grignard reagents.

The above 3-alkylcarboxylic acid ethyl ester is a convenient synthon for the synthesis of (+)-ar-turmerone, the chief constituent of the essential oil from rhizomes of *Curcuma longa* Linn.

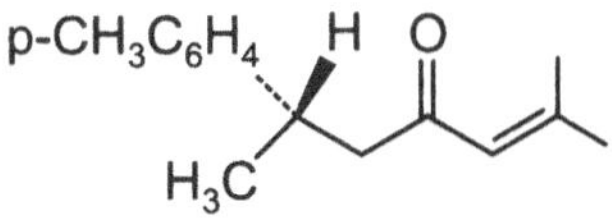

(+)-ar-Turmerone
(93% ee)

7.3.14 Asymmetric Synthesis of 2-Alkyl-3-Hydroxy Ketones

These are conveniently obtained by the reaction of dialkylboran trifluoromethane sulphonate (triflate) with aldehydes to give syn diols with high selectivity. Thus, the reaction of 3-pentanone with dibutylboryl triflate yields[40] the corresponding Z and E-enolates in the ratio 99:1; subsequent reaction with benzaldehyde yields the syn and anti alcohols in a ratio of 7 : 3.

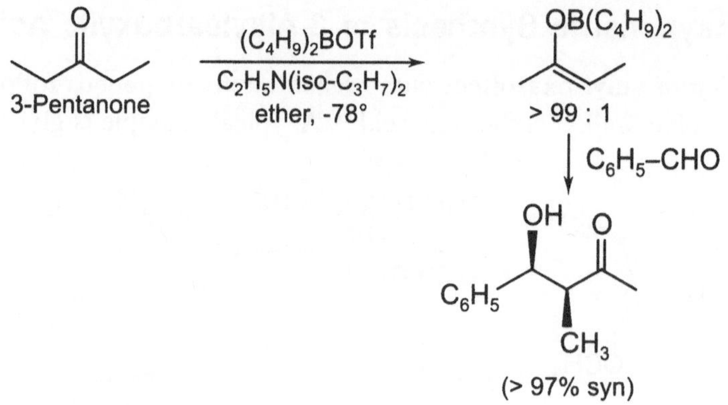

(> 97% syn)

7.3.15 Synthesis of α-Alkyl Amines

The α-alkylation's of amines is an important synthetic transformation. However, the main problem is that the amino group is not sufficiently activating to allow conversion to α-methylamine. The most promising route utilizes the conversion of a secondary amine into α-formamidines. The formed α-form-amides can be coverted[41] into α-alkylamines as shown below.

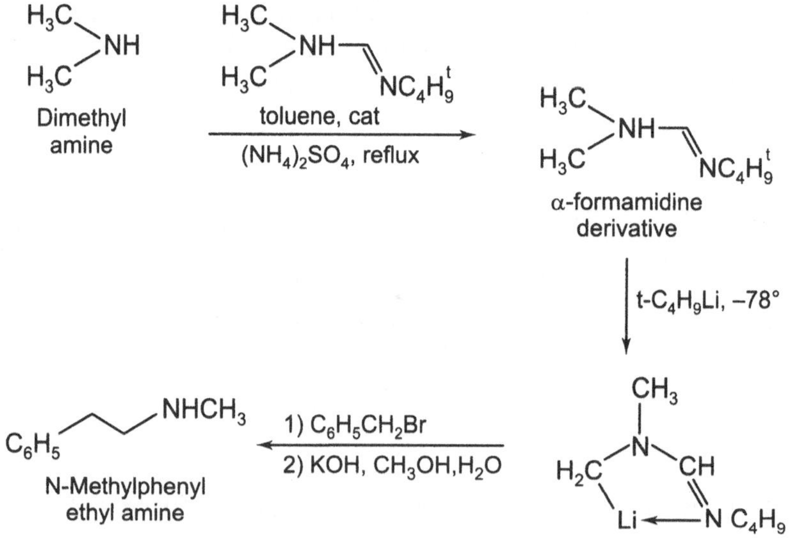

7.3.16 Enantioselective Epoxidations

Asymmetric epoxidations of allylic alcohols with tertiary butyl hydroperoxide in the presence of either (+)-or (-)-diethyl tartrate and titanium tetraiso propoxide yields the corresponding asymmetric epoxides in high optical yields. This procedure is most stereo selective[42] than any other procedure.

7.3.17 Synthesis of Enantiomeric Pure Amino Acids

All naturally occurring amino acids (except glycine) are L-amino acids. However, all procedures for the synthesis of amino acids give racemic mixtures. For getting enantiomeric pure acids either the racemic acid could be resolved or pure enantiomers synthesised.

7.3.17.1 Enantiomeric Resolution

Following methods are used.

(*i*) **Amine salt formation:** The racemic mixture of amino and is mixed with an enantiomer (R or S) of a naturally occurring amine like strychnine or brucine. A diastereomeric pair of salt is obtained. This is separated by difference in their solubility and crystallization. The separated diastereomers (can be identified spectroscopically) on acidification precipitate the respective amino acids.

(*ii*) **Ester formation:** The recemic mixture of amino acids may be converted into diastereomeric esters, which are separated by crystallization and then hydrolysed.

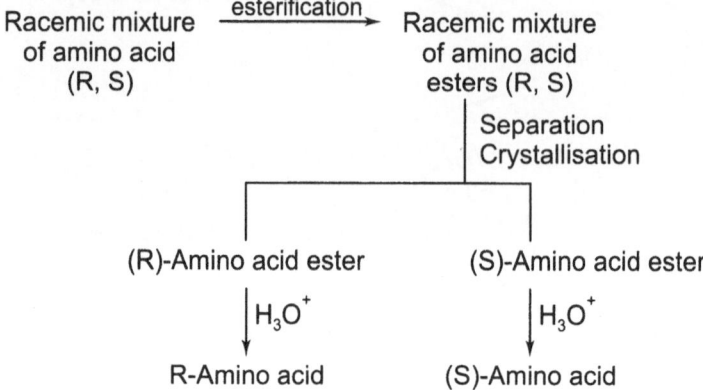

(*iii*) **Resolution by enzymes:** The racemic amino acid is converted into N-acylamino acid and subsequently, treated with the enzyme called deacylase. As the active site of the enzyme is chiral, it hydrolyses only N-aclylaminoacids of the L-configuration and the N-acylaminoacid of D-configuration remains unaffected and so the separation is conveniently affected.

$$\underset{\substack{\text{Rac. Amino acid}\\\text{(DL-Alanine)}}}{\overset{\displaystyle \text{DL–R CHCO}_2^-}{\underset{\displaystyle \overset{+}{N}H_3}{|}}} \xrightarrow{(CH_3CO)_2O} \underset{\substack{\text{N-Acylderivative}}}{\overset{\displaystyle \text{DL–R CHCO}_2H}{\underset{\displaystyle CH_3\,CONH}{|}}}$$

deacylase enzyme

$$\underset{\substack{\text{L-Amino acid}\\\text{(L-Alanine)}}}{H_3\overset{+}{N}-\underset{\displaystyle R}{\overset{\displaystyle CO_2^-}{\underset{|}{\overset{|}{C}}}}-H} \quad + \quad \underset{\substack{\text{D-N-Acylamino acid}\\\text{(D-N-acetylalanine)}}}{H-\underset{\displaystyle R}{\overset{\displaystyle CO_2^-}{\underset{|}{\overset{|}{|}}}}-NHCOCH_3}$$

The liberated amino acids is then precipitated from ethanol while the N-acetyle derivative remains in solution.

7.3.17.2 Stereoselective Synthesis of L-Amino Acids

(*i*) **Reduction of (Z)-3-substituted 2-acylamino propenoic acid:** This reduction is possible using chiral hydrogenation catalyst derived from transition metal. A typical catalyst based on a rhodium complex with (R)-1, 2-bis (diphenyl phosphino) propane, a compound that is called (R)- prophos". When a rhodium complex of norbornadiene (NBD) is treated with (R)-prophos, the (R)-prophos replaces one of the molecules of norbornadiene surrounding the rhodium atom to give a rhodium complex, which is chiral.

$$H_3C$$
$$H\blacktriangleright C-CH_2$$
$$(C_6H_5)_2P \qquad P(C_6H_5)_2 \longrightarrow$$
(R)-Prophos

$$[Rh(NBD)_2] \ ClO_4 \ + \ (R)-prophos \longrightarrow$$

$$\longrightarrow \ [Rh\ ((R){-}prophos)\ (NBD)]\ ClO_4 \ + \ NBD$$
Rhodium complex (chiral)

The chiral hydrogenation catalyst is obtained by treating the chiral rhodium complex with hydrogen in ethanol as solvent. The solution thus obtained contained the active chiral hydrogenation catalyst, which has the composition $[Rh\ ((R)\text{-}prophos)(H)_2\ (EtOH)_2]^+$.

Using this active chiral complex, L-amino acids are obtained as follows:

$$H \qquad CO_2H$$
$$C=C$$
$$R \qquad NHCOCH_3$$

1) $[Rh\ ((R){-}prophos)\ (H_2)(solvent)_2]^+, H_2$
2) OH^-, H_2O, heat and then H_3O^+

(Z)-3-substituted
2-acylaminopropenoic acid
2-Acyl aminopropenoic acid, R = H

$$H$$
$$\longrightarrow RCH_2\blacktriangleright C-CO_2^-$$
$$NH_3$$
$$+$$
L-Amino acids
R = H, L-Alamine

(*ii*) **Corey's method:** Using this method, optically pure α-amino acids are prepared from the corresponding α-amino ands using a chiral reagent. Various steps involved are given as follows:

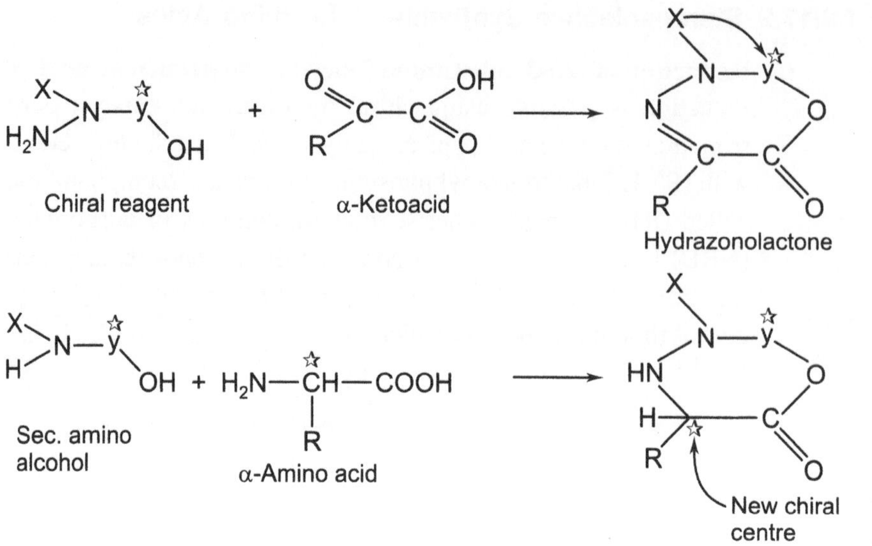
Chiral reagent
α-Ketoacid
Hydrazonolactone
Sec. amino alcohol
α-Amino acid
New chiral centre

Using Corey's method, D-alanine in synthesized as follows:

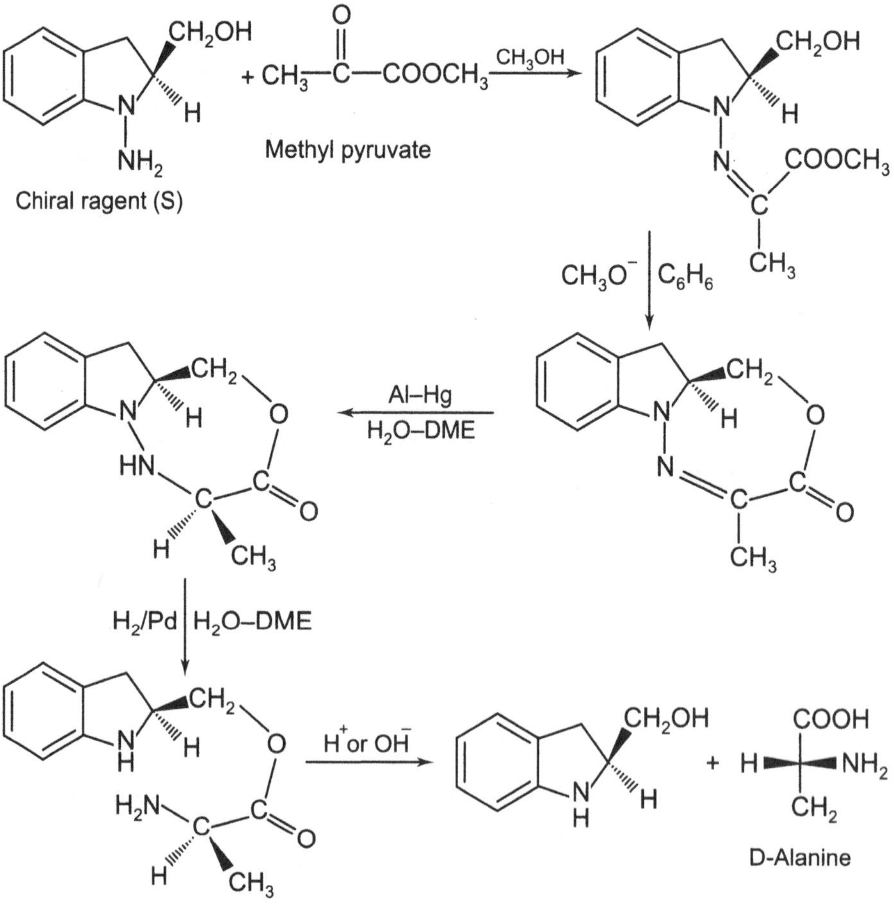
Chiral ragent (S)
Methyl pyruvate
Al–Hg
H2O–DME
H2/Pd
D-Alanine

(*iii*) **Using pseudoephedrine as a chiral auxiliary:** In this procedure, pseudoephedrine is reacted wtih glycine methyl ester to give pseudoephedrine glycinamide, which on alkylation with variety of electrophiles proceed with excellent diastereoselectivity with good yields.

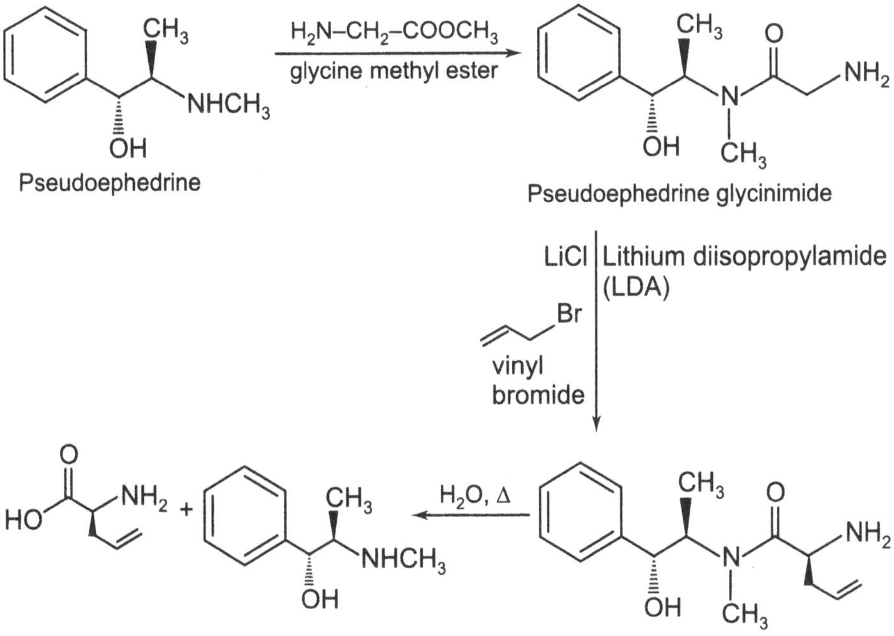
CH3
NHCH3
OH
Pseudoephedrine
H2N–CH2–COOCH3
glycine methyl ester
CH3
O
N
NH2
OH
CH3
Pseudoephedrine glycinimide
LiCl
Lithium diisopropylamide
(LDA)
Br
vinyl
bromide
H2O, Δ
HO
NH2
+

(*iv*) **Schollapf's procedure:** In this procedure an amino acid, Valine is used as a chiral auxiliary, which condenses with glycine to form bis-lactim ether. This on enolisation and electrophilic reaction with a carbonyl compound (aldehyde or ketone) gives a intermediate with high distereoface selectivity.

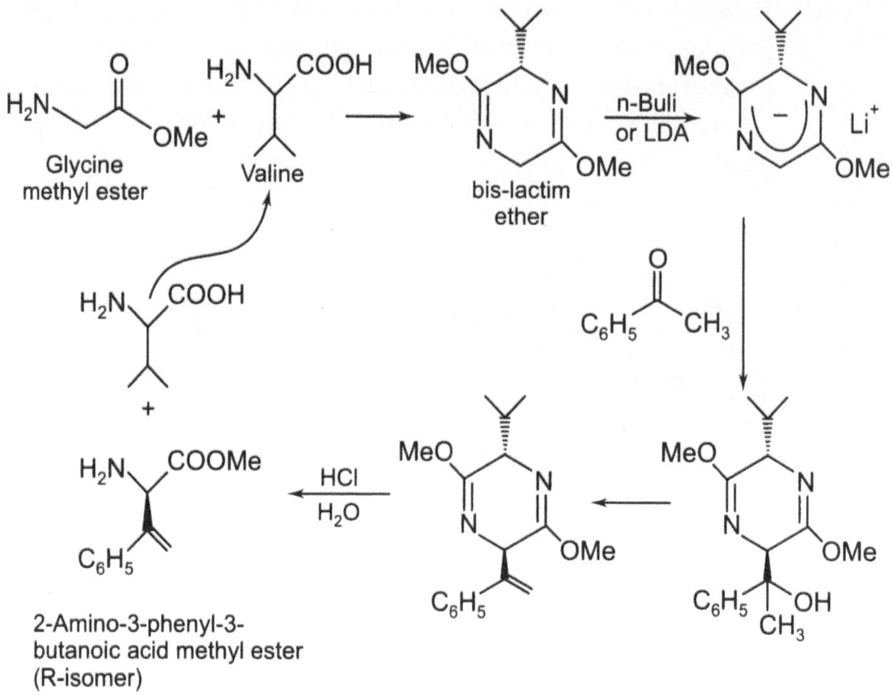

2-Amino-3-phenyl-3-
butanoic acid methyl ester
(R-isomer)

7.3.17.3 Enantioselective Synthesis using Chiral Borane

The reaction of chiral borane, monoisocamphenyl (IpcBH$_2$) with alkenes has been used for enantioselective synthes of ketones in 60–90 per cent ee. For details, see section 7.2.6.

REFERENCES

1. G.H. Ballard, A. Courtis, I.M. Shivlery and S.C. Taylor, J.Chem. Soc. Chem. Commun, 1983, 954.

2. D.T.Gibson, J.R. Koch, C.L. Schuld and R.E. Kallio, Biochemistry, 1968, **7**, 3795.

3. D.T. Gibson, M. Hensley, H. Yoshioka and T.J. Mabry, Biochemistry, 1970, **9**, 1626.

4. E. Kienan, E.K. Hafeli, K.K. Seth and R. Lamed, J. Am. Chem.. Soc., 1986, **108**, 162.

5. R. Zhou, A.S. Gopalan, F. Van Middlesworth, W.-R. Shieh and C.J. Sih, J.Am. Chem. Soc., 1983, **105**, 5925.

6. H.C. Brown, M.C. Desai and P.K. Jadhav, J. Org. Chem., 1982, **47**, 5056.

7. H.C. Brown, A.K. Mandal, N.M. Yoon, B. Schwier and P.K. Jadhav, J.Org. Chem., 1982, **47**, 5069, 5074.

8. H.C. Brown and G.G. Pai, J. Org. Chem., 1982, **47**, 1606.

9. C.Heathcock, C.T. White, J.J. Morrison and D. Van Derveer, J. Org. Chem., 1981, **46**, 1296.

10. P. Fellmann and J.E. Dubois, Tetrahedron, 1978, **34**, 1349; J.E. Dubois and P. Fellmann, Tetraheadron lett. 1975, 1225.

11. D.J. Cram and F.A. Abd Elhafez, J. Am. Chem. Soc; 1952, **74**, 5828; D.J. Cram and K.R. Kopecky, ibid, 1959, **81**, 2748.

12. S. Yamada, K. Hiroi and K. Achiwa, Tetrahedron lett. 1969, 4233; S. Yamada and G. Otani, ibid, 1969, 4237.

13. J.K. Whitesell and S.W. Felman, J. Org. Chem., 1977, **42**, 1663.

14. S. Suga, M. Segi, K. Kitano. S. Masuda and T. Nakajima, Bull. Chem. Soc. Jpn., 1981, **54**, 3611.

15. C.C. Ryerson, D.P. Ballou and C. Walsh, Biochemistry, 1982, **21**, 2644.

16. M. J. Taschner and L. Peddada, J. Chem. Soc. Chem. Common, 1992, 1384.

17. M. J. Taschner and D. J. Black, J. An. Chem. Soc, 1988, **110**, 6892.

18. H. Gilman and J. M. Straley, Rect. Trave. Chem., Pays. Bas., 1936, 55, 821.

19. H. Gilman, R. G. Jones and L. A. Woods, J. Org. Chem., 1952, **17**, 1630.

20. For details see G. H. Posner, Organic Reactions, Willy, New York, 1975, vol. 22, pp 253–400.

21. O. Diels and K. Alder, Ann. 1928, **460**, 98; 1929, **62**, 470; Ber., 1929, **62**, 2081, 2087; J. A. Norton, Chem. Rev. 1942, **31**, 319; D. A. Oppolzer, Angew. Chem. Int. Ed., 1977. **16**, 10.

22. J. D. Morrison and H. S. Mosher, Asymmetric Organic Reaction; ACS. Washington, 1476, p. 256.

23. J. Sauer and J. Kredel, Tetrabedron Lett., 1966, 6359; I. Hamer, J. Org. Chem., 1966. **31**, 2418; O. Ĉevinka and O. Kr̂iẑ, Colect, Czech. Chem. Commun., 1968, **33**, 2342.

24. S.V. Ley and A. J. Redgrave, Synlett, 1990, 393.

25. S. V. Ley, F. Sterenfeld and S. Taylor, Tetrahedron Lett, 1987, **28**, 225

26. T. Hudlicky and J.D. Price, Synlett, 1990, 159; T. Hudlicky, H luna, J. D. Price and F. Rulin, Tetrahedron Lett, 1989, **30**, 4053.

27. M. Natchus, J. Org. Chem, 1992, **57**, 4740.

28. T. Hudlicky and H.F. Oliva, J. Am Chem. Soc, 1992, **114**, 9694 and the reterences cited therin.

29. K. Mori, Tetrahedron, 1981, **37**, 1341.

30. S. Kobayashi, T. Iimori, T. I. Zawa and M. Ohnu, J. Am. Chem. Soc. 1981, **103**, 2406; R. Rossi, A. Carpita and M. Chini, Tetraheclron, 1985, **41**, 627.

31. Carter and Bhattacharya, Biochem Biophys., 1952, 38, 405.

32. D.L. Regan, P. Dunnil and M.D. Lilley, Biotech. Bioeng., 1974, **16**, 333.

33. M. Hirma, M. Shimitzu and M. Iwashila, J. Chem. Soc., Chem. Commun, 1983, 599.

34. K. Oyama, Chirality in Industry, eds. A.N. Collins, G.N. Sheldrake and J. Crosby, Wiley Chichester, 1992, page 237.

35. A.S. Bommarius, K. Drauz, U. Groegar and C. Wandrey in chirality in Industry, eds. A.N. Collins, G.N. Sheldrake and J. Crosby, Willy Chichester, 1972, page 371.

36. K Mosbach and P.O. Larson, Biotechno. Bioeng, 1970, **13**, 19.

37. A.I. Meyers and E.D. Mihlich, Angew. Chem. Internal Edn., 1976, **15**, 270.

38. A.I. Meyers, Acc. Chem. Res., 1978, **11**, 375.

39. A.I. Meyers and R.K. Smith, Tetrahedron, 1979, 2749.

40. S. Masamune, Organic Synthesis Today and Tommarrow, eds. B.M. Trost and C.R. Hutchinson, Oxford, Pergamon, 1981, page 197; D.A. Evans, J. Bartroli and T.L. Shih, J. Am. Chem. Soc., 1981, **103**, 2127; S. Masamune and Y. Choy, Aldrichimica Acta, 1982, 15, 47.

41. P.D. Edwards, W.F. Rieker and T.R. Baiey, J. Am Chem . Soc., 1984, 106. 3270.

42. K. Kasturi and K.B. Sharpless, J.Am. Chem. Soc., 1980, 102, 5974., B.E. Rossiter, T. Kasturi and K.B. Sharpless, J. Am Chem. Soc. 1981, 103, 464; K.B. Sharpless, S.S. Woodward and M.G. Finn, Pure and Appl. Chem., 1983, **55**, 1823.

8

Retrosynthesis

8.1 INTRODUCTION

Only in some cases, it is possible to visualize, the steps by which a particular compound can be synthesized. For example, a hydrocarbon containing an odd number of carbon atoms can be conveniently synthesized from and alkyl halide and reacting it with lithium diethyl cuprate (Gillman reagent) the procedure is known as Corey-Posner, Whitesides-House Synthesis Various steps involved are given below (Fig. 8.1).

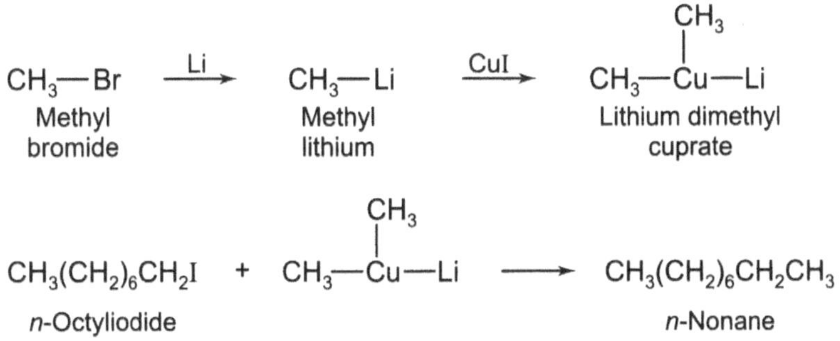

$$CH_3—Br \xrightarrow{\text{Li}} CH_3—Li \xrightarrow{\text{CuI}} CH_3—\overset{\overset{\displaystyle CH_3}{|}}{Cu}—Li$$

Methyl bromide — Methyl lithium — Lithium dimethyl cuprate

$$CH_3(CH_2)_6CH_2I \;+\; CH_3—\overset{\overset{\displaystyle CH_3}{|}}{Cu}—Li \;\longrightarrow\; CH_3(CH_2)_6CH_2CH_3$$

n-Octyliodide — n-Nonane

Fig. 8.1

The above procedure is very convenient and is useful for the synthesis of numerous hydrocarbons.

However, in most of the cases it is not possible to visualize the various steps for the synthesis of a particular compound. In such cases, since we know the objective, we proceed backwards. As a first step, the immediate precursors that could be made to react to give the target molecule. It may, however be possible to have more than one pathways by which the target molecule can be synthesized. As a second step, the immediate precursor (1st precursor) becomes the target molecule and the next set of precursors are identified to get the 1st precursor. This process is repeated a number of times until final campounds that are simple and are readily available are reached. The final compound thus becomes the starting compound (Fig. 8.2)

Retrosynthesis

Target molecule $\Longrightarrow$ 1st precursor $\Longrightarrow$ 2nd precursor $\Longrightarrow\Longrightarrow\Longrightarrow$ starting compound

Synthesis

Fig. 8.2

The procedure described above (Fig. 8.2) is called **retrosynthetic analysis**. The arrows, (open arrows) used is called **retrosynthetic arrow**. The retrosynthetic analysis is commonly called a **retrosynthesis**.

The term retrosynthetic analysis was coined by E.J. Corey, who was the first person to state its principles. On the basis of his work, E.J. Corey was awarded Nobel Prize (1990). For more details, see E.J. Corey and X.-M Cheng; The Logic of chemical Synthesis, Wiley, New York, 1989. In fact, it was Corey and his workers who made the designing of complex organic synthesis systematic to be aided by computer.

8.2 DESIGNING A RETROSYNTHESIS

As already stated, the first step for designing a retrosynthesis is retrosynthetic analysis (Fig. 8.2) in which plausible precursors (that can be converted into the target molecule) are identified. These precursors then become the target molecules, and next set of precursors are identified from which the 1st precursor can be obtained. This process gives a number of starting materials from which the 1st precursor and the target molecule can be obtained. The whole process is represented in Fig. 8.3 and is referred to a synthetic tree.

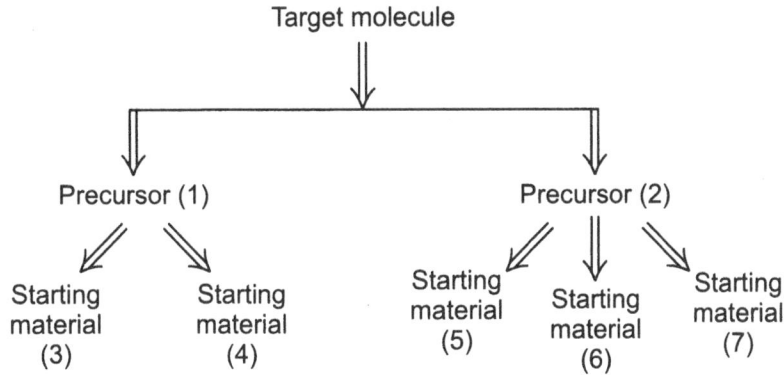

Fig. 8.3

Various procedures for converting the starting materials (3, 4, 5, 6 and 7) into the precursor (1) (from 3 and 4) and precursor (2) (from 5, 6 and 7) are considered. This is followed by procedures for converting the precursors (1 and 2) into the target molecules.

On the basis of the above, final decision is made for the exact pathway for the synthesis of the target molecule.

8.3 BASIC REQUIREMENTS FOR RETROSYNTHESIS

Following are given the basic requirements for retrosynthesis

8.3.1 Functional Group Transformation

In case the target molecule has a functional group, the precursor also must contain a functional group, which can be converted into the functional group of the target molecule. For example, if the target molecule has a hydroxyl group, the precursor molecule must contain a carbonyl group. Another example is the conversion of an alkyne (functional group —C $\equiv$ C—) into the corresponding alkane. As an illustration, hydroxy group can be converted into keto group by oxidation or reaction with a suitable grignard reagent. In view of the above, knowledge of functional group transformation is necessary (For detail, see chapter 5).

8.3.2 Reactions Involving Carbon-Carbon Bond Formation

Most of the synthetic pathways involve the formation of carbon-carbon bond. So, a knowledge of the various reactions which can be used for carbon-carbon bond formation is necessary (for details, see chapters 2, 3 and 4)

8.3.3 Disconnection of Bonds

For identifying the precursors, it is necessary to understand, which bond is most likely to break? It has found that disconnections are most suitable at or adjacent to a double or triple bond, a ring junction, a branch point next to a heteroatom or a functional group. As an illustration, in case of a simple alcohol, the possible disconnection sites are as given below.

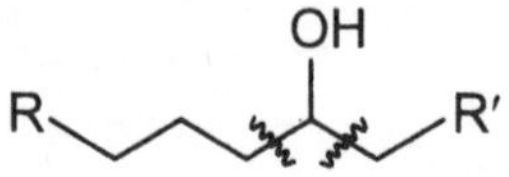

Thus, in retrosynthetic analysis, the alcohol can be formed in two ways from two fragments as shown below (Fig. 8.4).

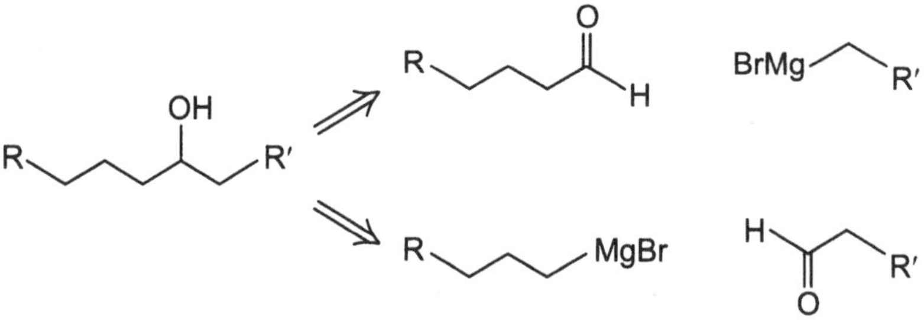

Fig. 8.4

The two possible routes for the synthesis of the target molecule are given below (Figs. 8.5 and 8.6).

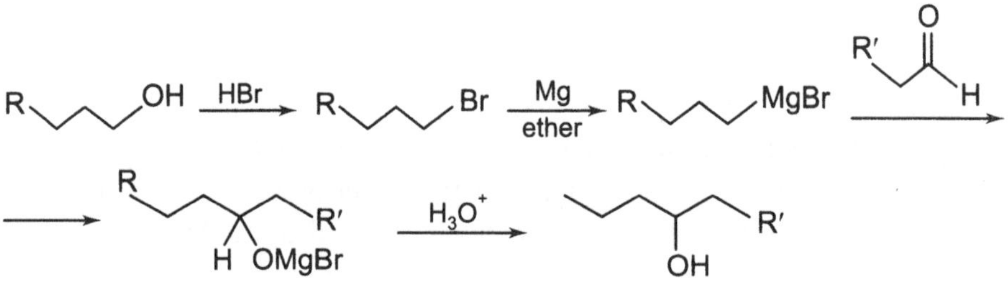

Fig. 8.5

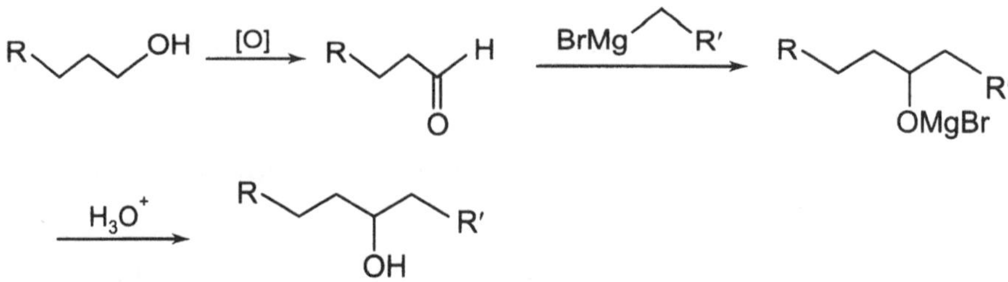

Fig. 8.6

The final route chosen for the synthesis of the target molecule depends on the availability or convenience of the preparation of the two fragments.

8.3.4 Reaction Conditions

For carrying out any synthesis whether of precursors or target molecule, it is most advisable to follow the basic principles of green chemistry as proposed by Paul T. Anastas[1]. Some of the important principles are given below.

(*a*) **Atom economy[2]:** The basic concept is that there should be maximum incorporation of the reactants (starting materials and reagents) into the final product. This concept is very important in dealing with waste minimization. It is, of course, well known that rearrangement and addition reactions have a higher atom economy than in which small portion of one of the reactants is transferred to the product. Examples of latter type of reactions include substitution reactions and elimination reactions.

(*b*) **Selection of an appropriate solvent[3, 4]:** The solevent used for any reaction should not cause any environmental pollution and health hazard. The solvents (known as green solvents) which do not cause environmental pollution include water, super critical carbon dioxide, ionic liquids and polyethylene glycol.

(*c*) **Energy requirement for synthesis:** In any chemical synthesis, the energy requirement should be kept to a minimum. Energy to a reaction can be supplied[3, 5, 6] by photochemical means, microwaves or sonication.

(*d*) **Use of catalyst:** It is well known that use of a catalyst facilitates transformations without the catalyst being consumed in the reaction. The common catalyst used are phase transfer catalysts[6], crown ethers[6] and biocatalysts (enzymes)[4] besides the usual catalyst like Pd-BaSO$_4$, Ni-H$_2$, *etc.* Some green catalysts are also known[8]. Even polymer supported catalysts can also be used[8]. There include[8] polystyrene aluminium chloride, polymeric super acid catalysts, polystyrene metalloporphyrins, polysupported photosensitizers and polymer supported phase transfer catalysts.

(*e*) **Reagents:** Besides the ordinary reagents used for chemical synthesis, there are groups of reagents (which though are ordinary reagents) are bound to polymer support. Such reagents are called polymer supported reagents[9]. The main advantage of these reagents is that any excess of the reagent can be recovered by filtration and can be used again. In such cases, the isolation of the product is very easy. Some of such reagents include, polymer supported peracids, chromic acid and polymeric thioanisolyl resin, Poly-N-bromosuccinimide (PNBS), polymeric organotin dihydride reagent (as a reducing agent), polystyrene carbodiimide, polystyrene anhydride, polystyrene Wittig reagent, polymeric phenylthiomethyl lithium reagent and polymer supported peptide coupling agent *etc*.

(*f*) **Protecting groups:** At times, in a chemical synthesis, it is necessary to protect one of the functional group in case there are more than one functional group in a subtract molecule. A number of protecting groups are available (Table 8.1).

The selected protecting group should be easy to introduce under mild condition and should not react under the reaction condition. They can be removed under the conditions that do not affect the stereo chemistry of the product.

(For a general discussion on protecting group see chpater 6) Since a protection and deprotection involve two extra steps, and the overall atom economy is adversely affected, these should by avoided if possible.

Besides the reaction conditions cited above the selectivity of each reaction that is chosen has to be considered. The reaction could be chemo selective transformation, regioselective product, stereospecific product or enantioselective product.

The simplest problem to resolve is the chemoselectivity, which refers to the differentiation between the reactivity pattern of functional group. For example, primary alcohols can be oxidised to aldehydes, but tertiary alcohols are inert to oxidations. Thus, the oxidation of a primary alcohol in presence of a tertiary alcohol is a chemoselective transformation. Similarly, a triple bond is reduced to double bond by Pd-BaSO$_4$ (The double bond, if already present in the substrate remains unaffected).

The orientation in addition reactions refers to regionselectivity. As an example, the addition of HBr to alkene gives a Markovnikov addition product. However, the above addition in presence of peroxide gives anti-Markovnikov addition product. Both are **regioselective reactions** (Fig. 8.7).

Table 8.1: Some commonly used protecting groups in chemical synthesis

Functional Groups	Reagents	Protected Form	Conditions for Removal
R—OH	Cl—SiMe$_3$, Et$_3$N	R—O—SiMe$_3$	F$^-$
	Br—CH$_2$C$_6$H$_5$, base	R—O—CH$_2$—C$_6$H$_5$	H$_2$, Pd catalyst (hydrogenolysis)
	H$^+$	R—O— (tetrahydropyranyl)	H$_3$O$^+$
	Cl—CH$_2$—O—CH$_3$, base	R—O—CH$_2$—O—CH$_3$	H$_3$O$^+$
R—C(=O)—R'	HOCH$_2$CH$_2$OH , H$^+$	(1,3-dioxolane)	H$_3$O$^+$
	HSCH$_2$CH$_2$SH , ZnCl$_2$	(1,3-dithiolane)	HgCl$_2$, CH$_3$CN; H$_2$O,CaCO$_3$

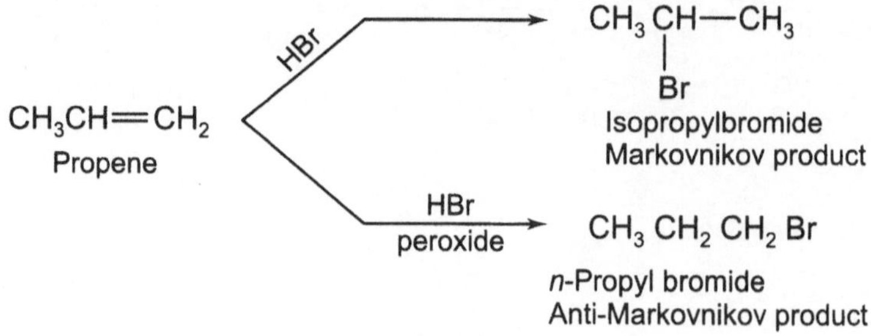

Fig. 8.7

Hydroboration is also a regioselective process. The addition of BH_3 to alkene follows Markovnikov addition rule. Since boron is more electropositive than hydrogen, therefore, attachment of boron takes place readily to the less substituted carbon of the double bond (Fig. 8.8).

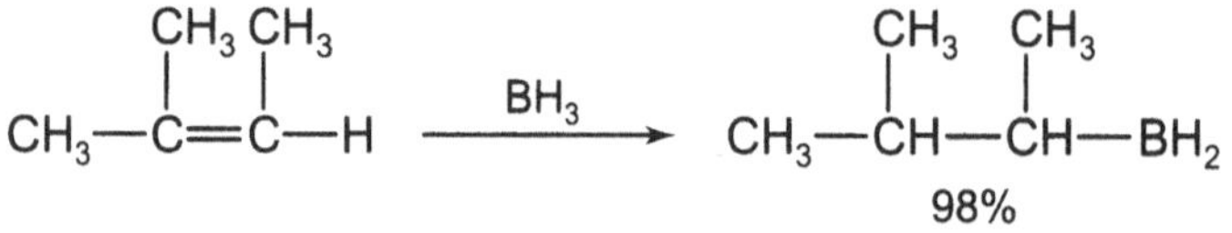

Fig. 8.8

Thus, addition of borane to alkene followed by oxidation gives the alcohol (anti Markovnikov product). However, hydration of alkenes in presence of sulphuric acid yield alcohol in agreement with Markovnikov rule (Fig. 8.9).

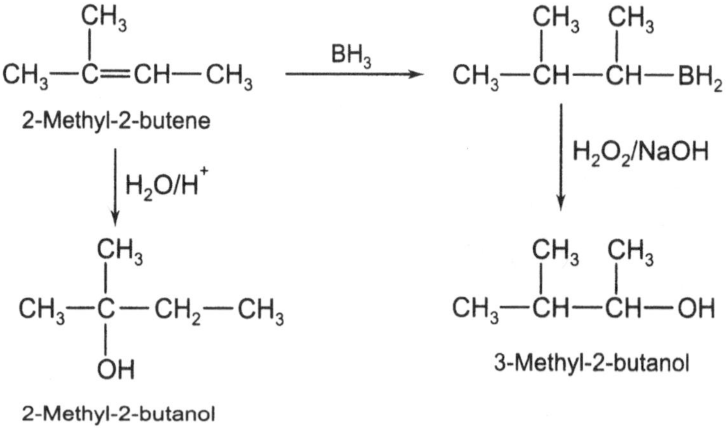

Fig. 8.9

In case of optically active compound, the formation of product depends on the stereoselectivity of reaction. A number of reactions are **stereospecific**. A typical example is an S_N2 reaction which goes by inversion of configuration of a stereogenic centre (Fig. 8.10).

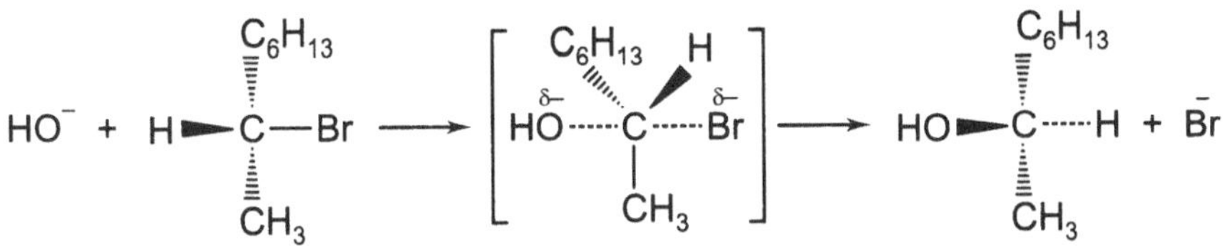

(–)-2-Bromooctane
$[\alpha]_D = -39.6°$

(+)-2-Octanol
$[\alpha]_D = +10.3°$

Fig. 8.10 S_N2 Mechanium (Complete inversion of configuration)

Creation of a new chiral centre in the molecule is determined by enantio selectivity. An example is the reaction of cyclohexene with chiral organoborane reagent to give a single enantiomer.

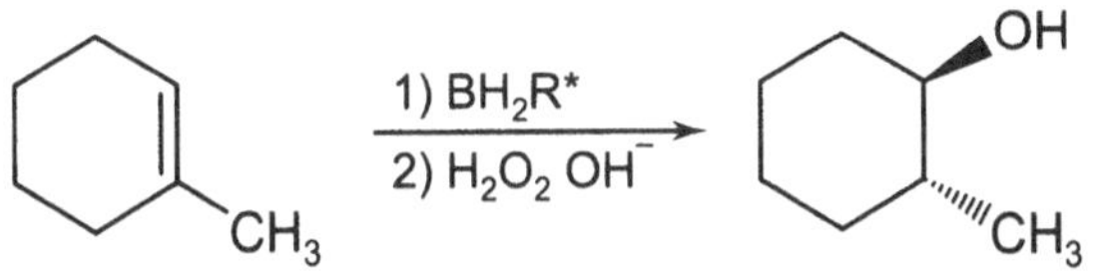

1-Methyl cyclohexene

A number of chiral molecules exist in nature. These are valuable synthons for making other chiral molecule, which is enantiometrically pure. An example is given below.

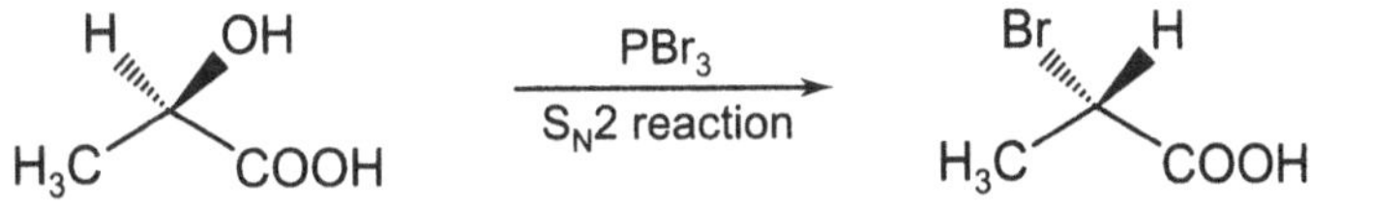

(R)-2-Hydroxypropionic acid (D-Lactic acid)

(S)-2-Bromopropanoic acid

It is possible that in a chemical reaction, the existing stereocentre in the substrate molecule is not affected. This is another procedure to make optically active compound. A example is given below.

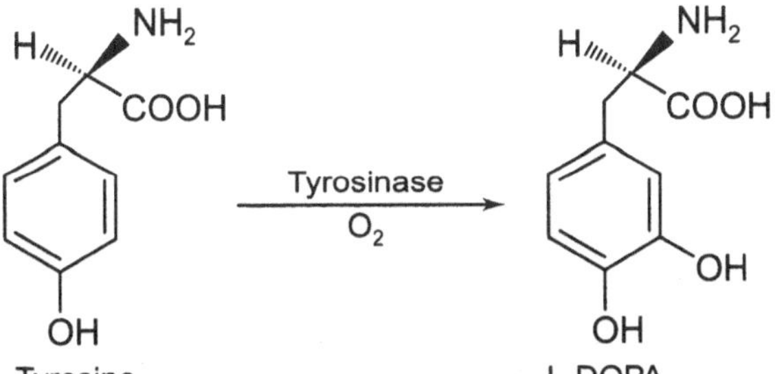

L-Tyrosine

L-DOPA

8.4 SELECTION OF A CONVENIENT ROUTE FOR SYNTHESIS

On the basis of retrosynthesis, we know the possible routes which can be used for synthesis of a target molecule. The actual route selected depends on the following:

(*i*) Availability of starting materials.

(*ii*) Number of steps involved.

(*iii*) Reaction conditions.

(*iv*) Ease of purification of the intermediates and final product.

8.5 SOME EXAMPLES OF RETROSYNTHESIS

2-Methyl Heptane

The alkane, 2-Methyl heptane does not have a functional group. However, it can be obtained from an appropriate alkyne by reduction with Pt/H$_2$. Thus, the first retrosynthetic step is the conversion of alkane (I) into an alkyne (II, III or IV). The precursor alkyne (II, III or IV) can in turn be obtained from an alkynide anion and an alkyl halide). Thus, the retrosynthetic analysis of 2-methylpentane (I) is represented as shown below (Fig. 8.11).

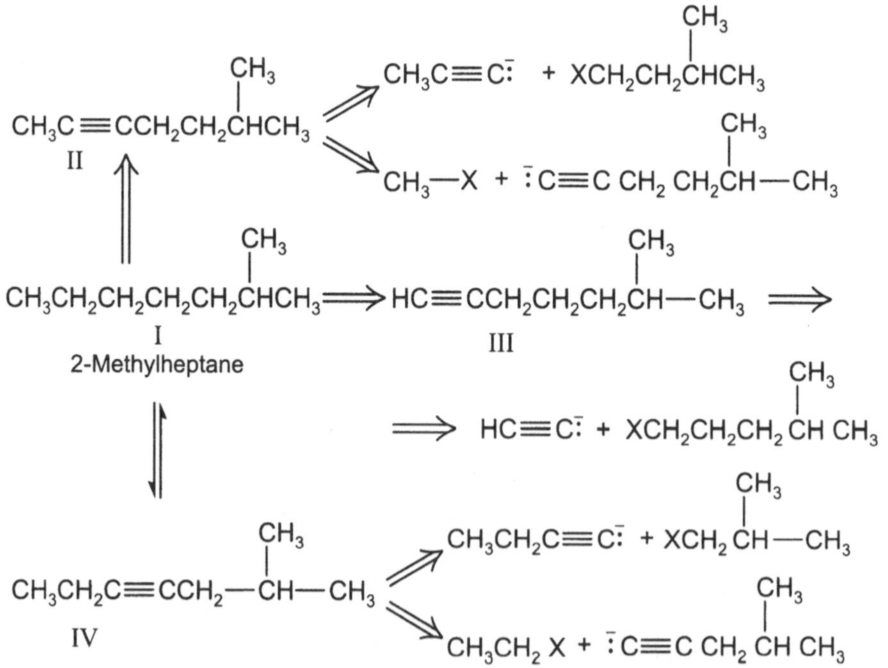

Fig. 8.11 Retrosynthesis of 2-methylheptane

3-Methyl-2-Butanol

The retrosynthesis of 3-methyl-2-butanol is represented below:

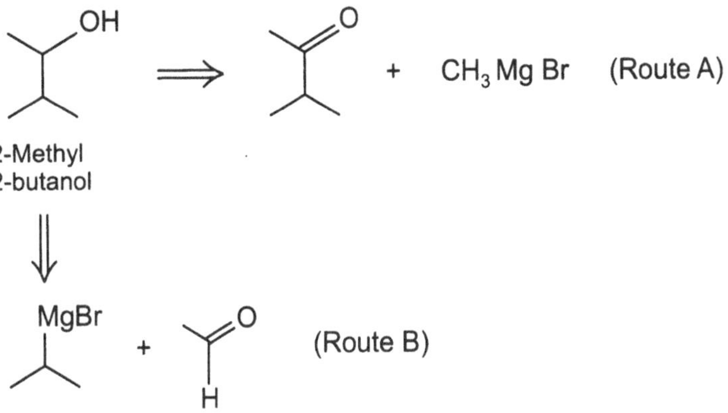

2-Methyl
2-butanol

+   CH$_3$ Mg Br   (Route A)

MgBr   +   (Route B)

The route A is represented as shown below.

2-Methylpropanal

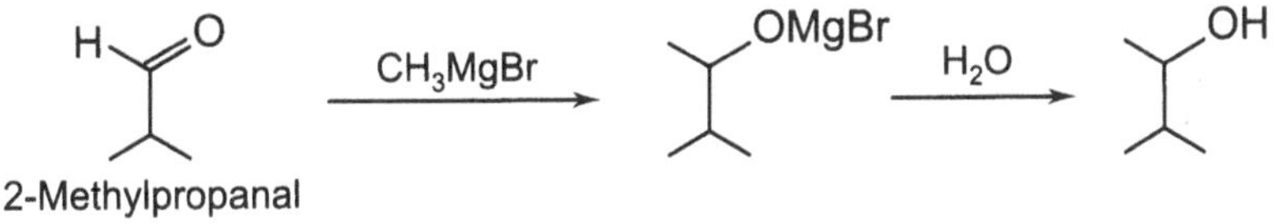

The starting 2-methyl propanal is obtained as follows.

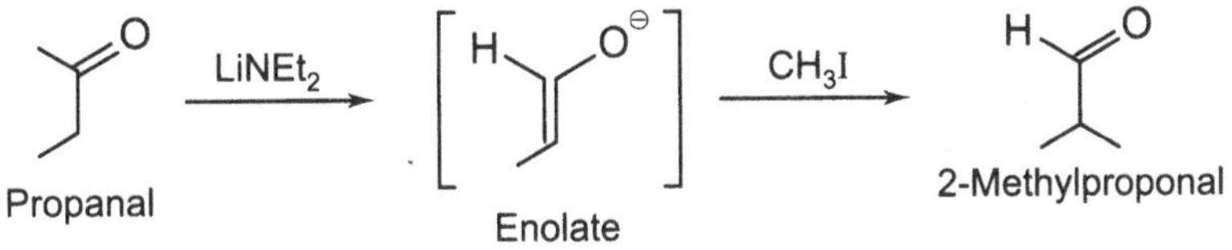

Propanal   Enolate   2-Methylproponal

The route B is represented as shown below.

Acetaldehyde   3-Methyl-2-butanol

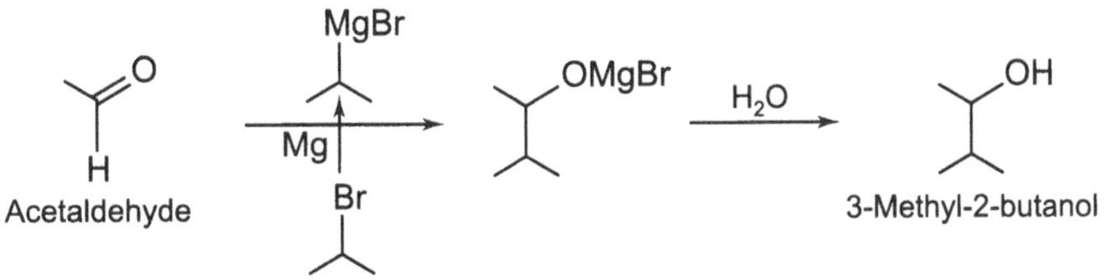

On the basis of the above retrosynthesis routes (A and B), it is found that route B is more convenient than route A.

1-Cyclohexyl-1-Propyne

Retrosynthetic analysis

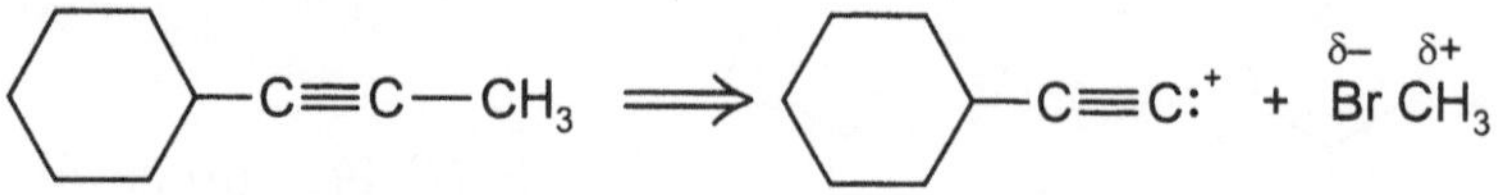

Synthetic route

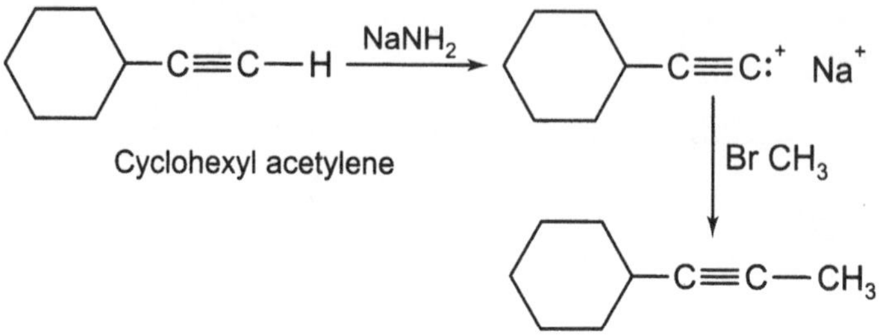

Cyclohexyl acetylene

1-Cyclohexylpropyne

2-Bromo-2-Methyl Pentane

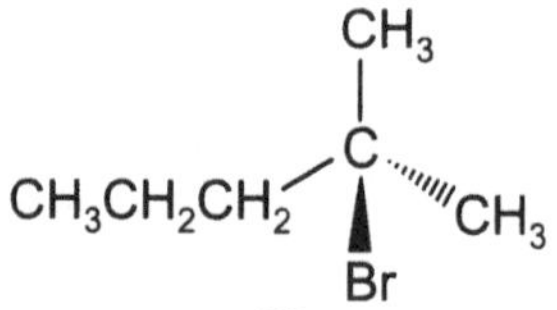

(I)

The 3° alkyl bromide can be obtained from two precursors (II) or (III). The two precursors (II and III) can be obtained as shown below.

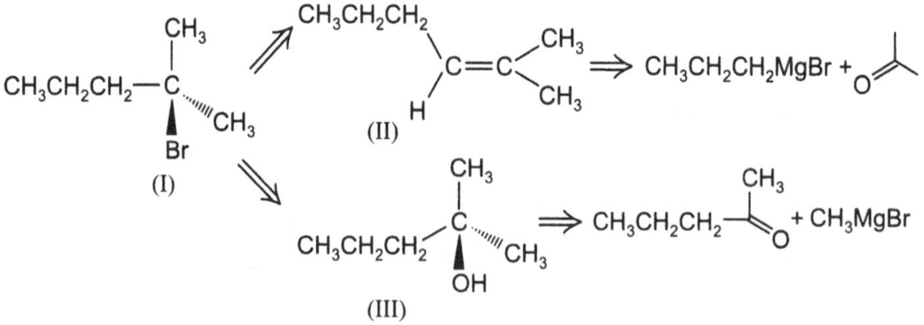

(I) (II) (III)

The route to be followed for the synthesis of the target molecule (I) is given below.

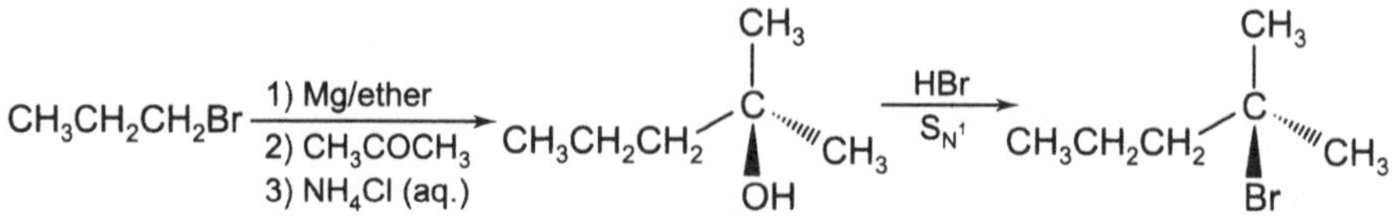

I

Epoxide

Retrosynthetic analysis

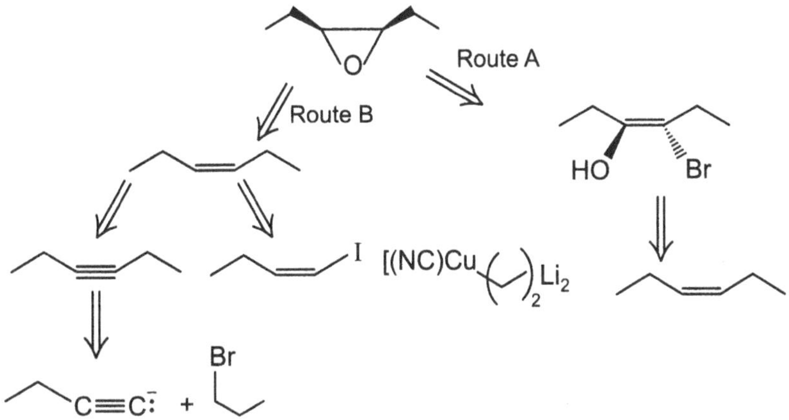

The synthesis following route B is represented below.

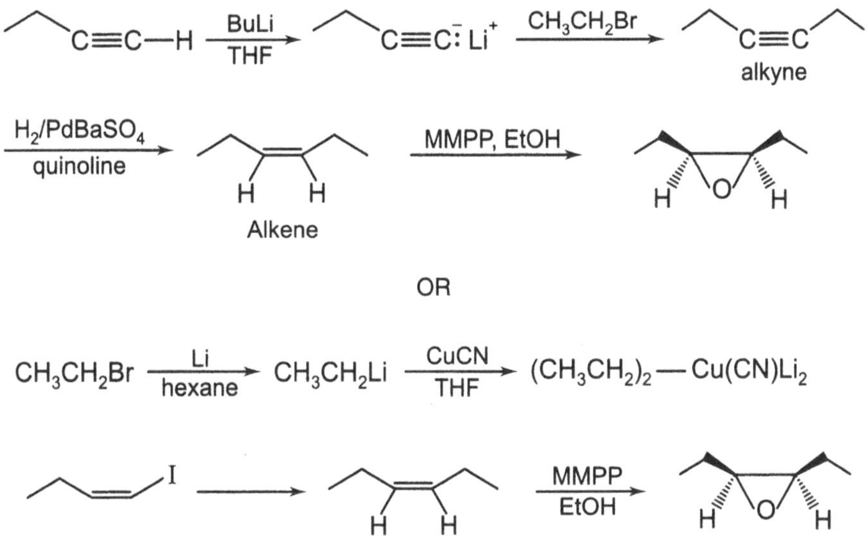

OR

MMPP-Magnesium monoperoxyphthalate

The synthesis using A is given below using the same alkene as given in the above synthesis.

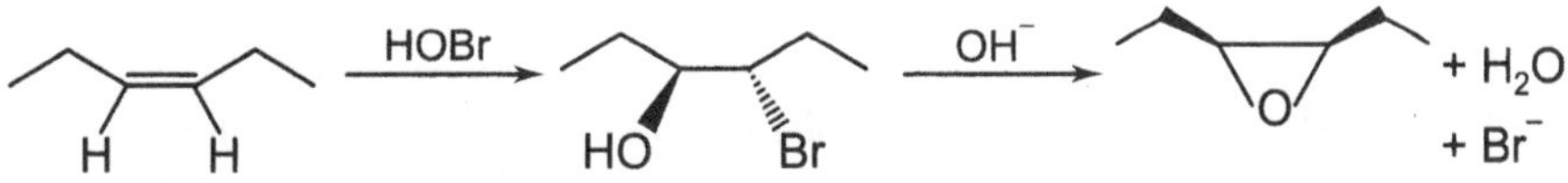

On the basis of the above synthetic pathways, the convenient pathway has to be selected.

Using retrosynthetic approach, a number of complex molecule have been synthesized. These include:

Gibberellic acid[11]

Prostaglandin E_2[12]

Longifolene[13]

Vernolepin[14,15,16]

Saxitoxin[17]

REFERENCES

1. Paul T. Anastas and John C. Warner, Green Chemistry, Theory and Practice, Oxford University Press, New York 1998.
2. Barry M. Trost, Science, 1991, **254**, 1471–1477.
3. V.K. Ahluwalia and M. Kidwai, New Trends in Green Chemistry, Anamaya Publishers, New Delhi 2004 and the references cited therein.
4. V.K. Ahluwalia and R.S. Varma, Green solvents, Narosa Publishing House, New Delhi, 2009.
5. V.K. Ahluwalia and R.S. Varma, Alternate Energy Processes in Chemical Synthesis, Microwave, Ultrasonic and Photo Activation, Narosa Publishing House, New Delhi, 2008.
6. V.K. Ahluwalia and R. Aggarwal, Organic Synthesis. Special Techniques, Narosa Publishing House, New Delhi 2001 and the references cited therein.
7. V.K. Ahluwalia, Enzymes in organic synthesis, Narosa Publishing, House, 2010, and the references cited therein.
8. Reference. 6, page 27 to 37. and the references cited therein.
9. Reference 6, page 22-26 and the references cited therein.
10. For more details about hydroboration. See V.K. Ahluwalia and R.K. Parashar, Organic Reaction Mechanism, Narosa Publishing House, New Delhi, 2002, page 563-575, and the references cited therein.
11. E.J. Corey, R.L. Danheiser, S. Chandrasekaran, P. Siret, C.E. Keck and J.-L. Eras, J. Am. Chem. Soc, 1978, **100**, 8031.
    E.J. Corey, C.R. L. Danheiser, S. Chandrasekaran, G.E. Keck, B. Gopalan, S.D. Larsen, P. Siret and J. –L. Gras, J. Am. Chem. Soc., 1978, **100**, 8034.
12. R.E. Donaldson and P.L. Fuchs, J. Am. Chem.. Soc., 1981, **103**, 2108.
13. E.J. Corey, M. Ohno, R.B. Mitra and P.A. Vatakencherry, J. Am. Chem. Soc., 1964, **86**, 478; E. J. Corey; R.B. Mitra and H. Uda, J. Am. Chem. Soc; 1964, **86**, 485.
14. S. Danishefsky, P. Schuda and K. Kato, J. Org. Chem; 1976, **41**, 1081.
15. P.A. Grieco, M. Nishizawa, T. Oguri, S.D. Burke and N. J. Marinovic, J. Am. Chem. Soc. 1977, **99**, 5773; P.A. Grieco, P.A. Noguez and Y. Masak, J. Org. Chem. 1977, **42**, 495.
16. H. Iio, M. Isobe, T. Kawai and T. Goto, Tetrahedron, 1979, **34**, 941; M. Isobe, H. Iio, T. Kawai and T. Goto, Tetrahedron Lett, 1977, 703.
17. Y. Kishi, Helerocycles, 1980, **14**, 1477; S. Hannick and Y. Kishi J. Org. Chem., 1983, **48**, 3833; M. Taguchi, H. Yazawa, J.F. Arnett and Y. Kishi, Tetrahedron Lett., 1977, 627; H. Tanino, T. Nakala, T. Kaneko and Y. Kishi, J. Am. Chem. Soc. 1977, **99**, 2818.

□□□

Index